Prostate Cancer: Shifting from Morphology to Biology

Stefania Staibano
Editor

Prostate Cancer: Shifting from Morphology to Biology

Springer

Editor
Stefania Staibano
Biomedical Advanced Sciences
Pathology Section
University Federico II of Naples
Naples, Italy

Please note that additional material for this book can be downloaded from http://extras.springer.com.

ISBN 978-94-007-7148-2 ISBN 978-94-007-7149-9 (eBook)
DOI 10.1007/978-94-007-7149-9
Springer Dordrecht Heidelberg New York London

Library of Congress Control Number: 2013946016

Printed on acid-free paper

Springer is part of Springer Science+Business Media (www.springer.com)

This book is dedicated to my husband Amando, my children Amanda and Giulio and my mother, Angela: they endlessly turn my life into a wonderful happy journey. The book is dedicated also to the memory of Nino, my father, who died too early, but goes along with me every day, warning my soul forever.

Foreword

For many years in the past, few patients with prostate carcinoma were treated by surgery.

Two were the main reasons. First, diagnosis often occurred at such an advanced stage of the illness as to make radical treatment impossible. Second, the few patients who underwent surgery were routinely left incontinent.

An alternative to less-than-adequate surgical treatment was hormonal therapy which was the fruit of Huggins' 1941 study. It offered a satisfactory cure, indeed it is still efficacious at present. Hormonal therapy had the advantage of prolonged patient survival while affording fairly good quality of life.

For a long time the clinical approach to prostate carcinoma has been based on these fundamentals.

A preference for surgical treatment of prostate cancer began with the identification of the prostate-specific antigen, used in a mostly efficacious but not completely conclusive way, in efforts to isolate the illness in a precocious state. The decisive surgical innovation, however, came about in the 1980s with the introduction of techniques that made it possible to safeguard urinary continence in the first place and secondly erectile function. Subsequently surgical therapy of prostate carcinoma has exploded. Its merits are better cures for many patients than in the past and innumerable resultant studies on all aspects of this disease.

The present volume is up-to-date and comprehensive as a treatise, but agile and fluent as a manual. It considers prostate carcinoma in all its modern pathological aspects, from classical morphology to immunohistochemistry and molecular biology.

A clinician will gain from this text a comprehensible synthesis of the immense and complex literature on the biology of the prostate cancer.

Additionally much attention is placed on arguments that have important clinical implications as, for example, those on new specific markers of aggressive disease (circulating cancer cells), metastatic dissemination, or irreversible castration resistant disease.

The volume has also the merit of transmitting the enthusiasm of the young research scientists, who co-authored the book, in regard to the preliminary results of on-going research. Their hopes for future positive results with new medical treatments, though not yet in hand, would obviate surgery for prostate cancer. Such discovery would close the circle on how to treat the disease. It may prove that prostate cancer surgery will lead eventually to more efficacious medical treatment—the method described at the beginning of this foreword.

Full Professor of Urology
Clinica Urologica II, University of Florence
Department of Urology, University of Florence
Viale Pieraccini, 18, IT-50139 Florence, Italy
April 24, 2013

Giulio Nicita

Preface

"Prostate Cancer: Towards the Molecular Pathology" is an ambitious title for an extraordinarily complex subject. This title symbolizes the purpose to offer a virtual journey on the bumpy roads that cross the morphological and molecular events underlying the metamorphosis of prostate tissue from normal to cancerous.

This book does not pretend to be a comprehensive opera, but aims to provide a description of the protean morphological, genetic and epigenetic events related to this process, supplying a *fil rouge* which may help the reader to orientate into this rapidly evolving field.

It is meant to propose a user-friendly approach to the integrated management of patients with prostate cancer, whether the reader comes from a background of urology, histopathology, molecular pathology, clinical and experimental oncology, or radiotherapy.

The topics, included in the different chapters of the book, will demonstrate that common morphological features may underlie different pathogenetic events, and that the resulting picture derives, ultimately, from the complex and variable interplay between a plethora of intersecting molecular pathways.

I am grateful to the friends and colleagues that authored these chapters.

They all shared my aim to offer their expertise to provide a synthetic, yet representative, up-to-date text on the most debated questions and promising trends concerning the morpho-molecular correlation in prostate cancer.

I wish to thank, also, the younger researchers who actively participated into this project with outgoing passion. We all hope that this book will provide the reader, whether clinician or laboratory worker, with the essential elements for a current approach to prostate cancer.

Naples, Italy
March 28, 2013

Stefania Staibano

Acknowledgements

The book has undergone linguistic editing by Dr. Amanda Tedeschi.

Our sincere gratitude goes to Ilse Hensen, of the Springer editorial staff, for her precious collaboration during all the phases that have lead to publication.

Molecular pathways schemes have been conceived and executed by Francesco Merolla, M.D., Ph.D., with the collaboration of Maria Siano, M.D.

Contents

Contributors

Paolo Antonio Ascierto Unit of Melanoma, Cancer Immunotherapy and Innovative Therapy, Department of Melanoma, National Institute of Tumors Fondazione "G. Pascale", Naples, Italy

Rita Bisogni Department of Biochemistry and Medical Biotechnology, Federico II University, Naples, Italy

Michele Caraglia Department of Biochemistry and Biophysics, Second University of Naples, Naples, Italy

Angela Celetti Department of Cellular and Molecular Biology and Pathology, CNR, Institute of Experimental Endocrinology and Oncology 'G. Salvatore' EOS-CNR, Naples, Italy

Paolo Chieffi Department of Experimental Medicin, Second University of Naples, Naples, Italy

Gaetano De Rosa Department of Advanced Biomedical Sciences, Pathology Section, Faculty of Medicine and Surgery, University of Naples "Federico II", Naples, Italy

Gaetano Facchini Department of Urogynecologic Oncology, National Institute of Tumors Fondazione "G. Pascale", Naples, Italy

Renato Franco Department of Laboratory Medicine, National Institute of Tumors Fondazione "G. Pascale", Naples, Italy

Antonio Maria Grimaldi Unit of Melanoma, Cancer Immunotherapy and Innovative Therapy, Department of Melanoma, National Institute of Tumors Fondazione "G. Pascale", Naples, Italy

Gennaro Ilardi Department of Advanced Biomedical Sciences, Pathology Section, Faculty of Medicine and Surgery, University of Naples "Federico II", Naples, Italy

Massimo Mascolo Department of Advanced Biomedical Sciences, Pathology Section, Faculty of Medicine and Surgery, University of Naples "Federico II", Naples, Italy

Francesco Merolla Department of Advanced Biomedical Sciences, Pathology Section, Faculty of Medicine and Surgery, University of Naples "Federico II", Naples, Italy

Sisto Perdonà Department of Urogynecologic Oncology, National Institute of Tumors Fondazione "G. Pascale", Naples, Italy

Maria Fiammetta Romano Department of Biochemistry and Medical Biotechnology, Federico II University, Naples, Italy

Simona Romano Department of Biochemistry and Medical Biotechnology, Federico II University, Naples, Italy

Daniela Russo Department of Advanced Biomedical Sciences, Pathology Section, Faculty of Medicine and Surgery, University of Naples "Federico II", Naples, Italy

Maria Siano Department of Advanced Biomedical Sciences, Pathology Section, Faculty of Medicine and Surgery, University of Naples "Federico II", Naples, Italy

Ester Simeone Unit of Melanoma, Cancer Immunotherapy and Innovative Therapy, Department of Melanoma, National Institute of Tumors Fondazione "G. Pascale", Naples, Italy

Stefania Staibano Department of Advanced Biomedical Sciences, Pathology Section, Faculty of Medicine and Surgery, University of Naples "Federico II", Naples, Italy

Silvia Varricchio Department of Advanced Biomedical Sciences, Pathology Section, Faculty of Medicine and Surgery, University of Naples "Federico II", Naples, Italy

MariaLuisa Vecchione Department of Advanced Biomedical Sciences, Pathology Section, Faculty of Medicine and Surgery, University of Naples "Federico II", Naples, Italy

Introduction

One more book on prostate cancer. Do we really need it, in a scenario crowded by so many outstanding specific textbooks?

At first glance, "*no, thanks*" is the only reasonable answer.

However, every month the flow of Web references on prostate cancer biology continues to steeply rise, leading to an endless overload of information.

So, a new book could attempt not only to fix the most significant news on this tumor, but to guide the reader towards a new approach that originates from a strict integration between surgical and molecular pathology, genetics and epigenetics, equally acting in determining the overall biology of prostate cancer.

This wide-angle vision of the problem is essential, at least for two order of pragmatic and therapeutic problems.

The first one derives from the remark that about 90 % of prostate cancers are detected by means of screening, and most of patients actually die from other causes (that is to say "with", and not "for" the tumor), with up to two million survivors in the United States alone. The expected costs of care for the next 5 years outnumber $2 billions, so the search of a definitive consensus about the optimal screening and treatment approaches is necessary to avoid unneeded therapies and expenses.

The latter resides in the observation that subsets of prostate cancer kill about 1 in 36 men in US, representing the sixth leading cause of death from all cancer in men worldwide, and that we are still far to cure them. A lot of work is still needed to correctly identify the patients to be treated with alternative individual therapies, and to find the most effective combinations of drugs. These new promising therapeutic agents could get us close to the goal to induce their permanent remission, strikingly expanding the life expectancy of patients.

The approach chosen in this book to address the updated information on prostate cancer morphology and biology was exquisitely pragmatic.

The 18 chapters begin with a concise overview of the background underlying the treated specific topics, then they proceed with the main text body. Most of the concepts related to molecular alterations have been graphically resumed and simplified in diagrams and schemes, which can be found at the end of each chapter. An on-line version of these graphics, supplying their step-by-step construction, is available on web (http://extras.springer.com). Morphology, as a rule, has been illustrated with histologic images.

The references included in each chapter have been selected according to the state of knowledge at the time of writing.

This book has represented an extraordinary cooperative working experience for the authors, editor and publisher. We hope that it may have the power to spread, to the reader, the enthusiastic sensation that we can really win the final battle against prostate cancer.

Part I
Clues to Morphological Diagnosis and Prognosis Evaluation

Chapter 1
Update on Diagnostic Criteria, on Biopsy and Surgical Specimen: Preinvasive Lesions, from Epithelial Cell Hyperplasia to Carcinoma In Situ and Invasive Carcinoma – First-Line Immuno-Phenotyping of Prostate Diseases

Massimo Mascolo, Daniela Russo, and Gaetano De Rosa

Abstract Prostate cancer (PCa) is the first most common malignant neoplasm among men in Western countries and the fifth cause of cancer worldwide. Up to date, there is no single tumor biomarker that accurately predicts patient's clinical outcome. On the basis of these considerations, an optimal characterization of patients with PCa represents an increasingly exciting challenge for surgical pathologists. The ability to discriminate between low and high aggressive PCa represents an highly debated issue, which is critical for the therapeutic choices. This chapter focuses on the recent findings about the prostate morphological and immunohistochemical findings, with a brief reference to recent development of molecular markers with diagnostic and/or prognostic relevance, in the hope of guiding the approach to the pathological interpretation of prostate bioptic and surgical tissue specimens.

Abbreviations

PCa	Prostate cancer
PSA	Prostatic specific antigen
DRE	Digital Rectal Examination
TRUS	Transrectal ultrasonography
PCA3	Prostate Cancer gene 3
HG-PIN	High grade prostatic intraepithelial neoplasia
ASAP	Atypical small acinar proliferation
TURP	Transurethral resection of the prostate
BPH	Benign prostatic hyperplasia

M. Mascolo (✉) • D. Russo • G. De Rosa
Department of Advanced Biomedical Sciences, Pathology Section, Faculty of Medicine and Surgery, University of Naples "Federico II", via S. Pansini, n.5, Naples, Italy
e-mail: mmascol@gmail.com; danielarusso83@yahoo.it; gaderosa@unina.it

S. Staibano (ed.), *Prostate Cancer: Shifting from Morphology to Biology*,
DOI 10.1007/978-94-007-7149-9_1, © Springer Science+Business Media Dordrecht 2013

TUIP	Transurethral incision of the prostate
EAU	European Association of Urology
AMACR	α-methylacyl-CoA racemase, P504S
PSAP	Prostate-specific acid phosphatase
PSMA	Prostate-specific membrane antigen
P501S	Prostein
PAP	Human prostatic acid phosphatase
AP	Acid phosphatase
GSTP1	The Glutathione S-transferase P1
CGA, GRN-A	The Chromogranin A
PSCA	The Prostate Stem Cell Antigen
EPCA	The Early Prostate Cancer Antigen
Cav-1	The Caveolin-1
PCA3/DD3	The Prostate cancer antigen 3
TMPRSS2-ERG	The Transmembrane protease serine 2-Ets Related Gene
GOLPH2	The Golgi phosphoprotein 2
DAB2IP	The DAB2 interacting protein
FASN	The Fatty acid synthase
PIN	Prostatic intraepithelial neoplasia
AJCC/UICC	The American Join Committee on Cancer/Union Internationale Contre le Cancer
ISUP	The International Society of Urological Pathology

Prostate cancer (PCa) is a major public health problem worldwide, being one of the most common malignancy accounting about 900,000 estimated new cancers (Ferlay et al. 2010; Center et al. 2012), and its incidence rate varies according to geographic locations and race (Ferlay et al. 2010; Siegel et al. 2012). PCa, in fact, represents the first most common malignant neoplasm (other than skin cancer) among men in Western countries and the fifth worldwide (Ferlay et al. 2010; Parkin et al. 2010; Siegel et al. 2012). Over the past two decades, the broad use of prostatic specific antigen (PSA) in the individual and mass screening for PCa, combined to the standardization of prostate needle biopsy methodology, has resulted in a significant increase of early diagnosis and in a "stage migration", leading to a progressive reduction of "incurable" mortal carcinomas (Noldus et al. 2000; Jani et al. 2001; Ung et al. 2002; Derweesh et al. 2004; Moore et al. 2009). Nevertheless, PCa with an estimated 258,000 deaths per year in the world constitutes the sixth leading cause of death from cancer, representing the second one in developed countries (Ferlay et al. 2010; Center et al. 2012; Siegel et al. 2012). The expected decrease of PCa mortality is hypothesized to be the result of a higher therapeutic success rate due to an increase of early stage PCa diagnosis and, especially, of the major ability to identify patients with a higher risk of progression from the great majority of "insignificant" neoplasms (Andriole et al. 2009; Schroder et al. 2009). An early diagnosis of more aggressive PCa before that any symptoms of tumor progression are seen, allows, in fact, to adopt the most appropriate therapeutic management. Despite extensive technical advances, to date, the efficacy of PSA or, generally speaking, of a single tumor biomarker, to accurately predict patient's clinical outcome is still limited (Bjartell et al. 2011; Pomerantz et al. 2011). Basing on this consideration, the best possible characterization of each patient with PCa represents an increasingly exciting challenge for surgical pathologist. Its pivotal and not always comfortable role is not only limited both in the cancer diagnosis, especially when the cancer focus is small, but also and in the identification of patients with the greatest risk of systemic progression (Montironi et al. 2011a). The ability to discriminate low from high aggressive PCa represents today's most debated issue, which influences the therapeutic approach of single patient. Furthermore, recent data confirmed that PCa with a combined Gleason score of 6 or less almost never metastasize to lymph nodes; conversely PCa with at least one pattern of 4 or 5 can metastasize (Ross et al. 2012). In fact, patients with high risk PCa have to be treated with conventional therapeutic options, such as radical prostatectomy or radiotherapy, and the others with a "watchful waiting strategy", which consists in the active surveillance of the patient by a close follow-up using serum PSA and repeated biopsies (Bill-Axelson et al. 2011; Montironi et al. 2011a), adopting the standard curative treatment only when in these patients an upgrading at repeat biopsy is observed (Montironi et al. 2011a). Prostate needle biopsy represents the primary approach in the establishment of a definitive diagnosis of PCa. In developed countries, pathologists have to confront in daily practise with a rising number of prostate biopsies. The objective of this chapter is to provide a contemporary update of the current evidences regarding the prostate morphological and immunohistochemical findings, with a brief reference to recent development of molecular markers, in the hope of guiding pathologists toward accurate interpretation of biopsies and prostatectomy. We discuss also the surprising interactions between clinical and pathological characteristics, with particular attention to their role in defining prognosis and therapeutic management of PCa patients.

1.1 Physical Examination of Patients

PCa represents a heterogeneous disease, with different clinico-pathological presentations and prognostic features. To provide the best management for patient with PCa, obtain a complete clinical history is crucial, and it has to be supported by imaging studies and integrated with histological, immunophenotypical and molecular data. To date, no definitive cause for PCa development has been identified, but three well established risk factors appear to have a role in its development: age, ethnic group and heredity. The risk to develop PCa becomes, in fact, higher in men over age 50, with nearly 65 % of cases occurring in men age 65 and older, that live in developed countries, with a positive family history for cancer (Quinn and Babb 2002; Siegel et al. 2012). African-Americans have the highest chance of developing PCa and twice the risk of dying for it, while Asian men have the lowest risk (Gronberg 2003; Williams and Powell 2009; Kheirandish and Chinegwundoh 2011). In addition, several exogenous factors including sexual behaviour, use of alcohol, a high-fat diet and occupational exposure are considered important in determining the risk of developing clinical PCa (Kolonel et al. 2004). The majority of physicians try to obtain evidence of PCa through Digital Rectal Examination (DRE), serum concentration of PSA, PSA velocity overtime, free/total PSA ratio, and transrectal ultrasonography (TRUS) (Stamey et al. 1987; Partin et al. 1993; Basler and Thompson 1998; Jacobsen et al. 1998; Mahon 2005). In addition, the finding of PCA3 (Prostate Cancer gene 3), highly over-expressed in patients affected by PCa, has been recently introduced into clinical routine as a specific urine marker for PCa (Auprich et al. 2011). A needle baseline prostate biopsy is performed if the results of DRE and/or PSA and/or TRUS are not within normal range. Its main objective is to diagnose PCa; the histotype, the extent and the grade of carcinoma in needle are, in fact, important findings in better establishing the prognosis and the contemporary management of the patient, helping to choose through a wide spectrum of potential therapeutic strategies available: active surveillance, radiotherapy, adjuvant hormonal therapy and different types of surgery.

1.2 Surgical Techniques: Needle Prostate Biopsy, Transurethral Incision and Resection of the Prostate, and Open or Radical Prostatectomy

A needle core prostate biopsy can be done, with similar adequate results, throughout transperineal or tranrectal routes (Hara et al. 2008; Takenaka et al. 2008). The decision to proceed to a needle biopsy is generally established on the basis of the serum PSA kinetics and/or suspicious DRE and/or TRUS findings. A lesion guided biopsy is performed on palpable nodules or TRUS identified lesion; conversely the great majority of "invisible lesions" are generally detected by systematic TRUS needle biopsy (El-Hakim and Moussa 2010). The first systematic bioptic approach used was the sextant scheme (3 cores from each lobe: 1 from the base, the mid and the apex) (Hodge et al. 1989). Then, to reduce the number of not diagnosed cancer, a laterally directed biopsy method was proposed (Stamey 1995). The currently applied strategy, also called "extended prostate biopsy", is usually done with 18-gauge biopsy needles and provides a minimum of 10 to a maximum of 18 cores (in glands $\geq$50 cm^3) (Eichler et al. 2006; Scattoni et al. 2008). Prostate tissues from at least five regions, including the lateral peripheral zone and midline, are obtained. The choice of the method to use depends mainly on the clinical (prostate size and patient age) and laboratory data (PSA range) of the patient. Although multivariate analysis revealed no significant difference between 18 and 22 cores compared to

10–12 cores approach, the highest relative positive rate was reported with the 18–22 biopsy scheme (Eichler et al. 2006). A repeat biopsy within 3–6 months is indicated when a rising and/or persistent PSA and/or suspect DRE, high grade prostatic intraepithelial neoplasia (HG-PIN) and atypical small acinar proliferation (ASAP) are found (Epstein and Herawi 2006). Patients with sustained suspicion of PCa, despite the second negative repeated biopsy, should be subjected to saturation biopsy, an aggressive biopsy procedure with up to 45 cores obtained (Rabets et al. 2004; Walz et al. 2006). To date, prostate needle procedure provides a proper evaluation of the posterior gland, the site where the majority of PCa develops. Nevertheless, taking in consideration the microscopic anatomy of prostate (three anatomically distinct glandular zones, peripheral, transition and central area, and the anterior fibro-muscular stroma), this methodology needs to further improvements to allow a more precise detection of the transition zone and the anterior localized PCa (Fine and Reuter 2012). Finally, an extended repeat or a saturation biopsy is strongly encouraged in young patients or healthy men with a strong suspicion of cancer (Chun et al. 2010).

Transurethral resection of the prostate (TURP) is the most common surgical method to treat conditions like obstructive prostatic hypertrophy, due to benign prostatic hyperplasia (BPH), and is considered the primary choice in the treatment for prostate sized 30–80 ml. This procedure removes the portion of the gland immediately surrounding the urethra, including transition and periurethral zones, the bladder neck and the anterior fibro-muscular stroma, through a urethral endoscopic approach. Tissues from seminal vesicles, central and peripheral zones, are generally untouched. The optimal number of chips required for an adequate histologic examination is almost a full cassette for each 5 g of tissue (Henson et al. 1994; Heidenreich et al. 2011).

Transurethral incision of the prostate (TUIP) is an endoscopic procedure reserved to the treatment of smaller prostate ($\leq$30 g). It is associated with few complications, such as fever, and, if applied to appropriate patients, it provides the same clinical benefit as TURP, even if a higher long-term failure rate for this approach is described (Orandi 1985).

The open prostatectomy represents the standard procedure for resecting larger prostate (>80–100 ml). This surgical approach involves the removal of the inner portion of the prostate through an incision under the lower abdominal area.

Finally, while TURP, TUIP and open prostatectomy, constitute the conventional surgical options for BHP according to European Association of Urology (EAU) guidelines, the radical prostatectomy is the most common surgical procedure for the treatment of advanced PCa. It consists in the removal of the whole prostate gland and, according to the clinic and imaging findings, of the seminal vesicles, deferent ducts and loco-regional lymph nodes. The radical prostatectomy can be performed through a retropubic or a perineal approach. The first allows the evaluation of lymph node status also via frozen section before the complete removal of prostate; the second does not permit this assessment.

1.3 Histology

All specimens (needle core biopsies, TURP, enucleation, radical prostatectomy; Fig. 1.1) must be fixed in 4 % formaldehyde solution and sent to the pathology laboratory. Needle biopsy samples consist of thin cores taken from different prostate sites. It is necessary to count all the cores, measure each core in length and, then, carefully transfer each of them into separate tissue cassettes (up to 3 cores for each cassette), paying particular attention not to crush the tissue (Iczkowski et al. 2002a). Multiple tissue fragments constitutes TURP specimen; and measure their aggregate dimension, register their weight and put the chips in separate cassettes (up to about 5 g for each cassette) is needed. The enucleation specimen is generally constituted by either partially or totally not orientable nodules. Specimen size and weight must be noted. Radical prostatectomy specimen must be oriented, measured in three dimensions and weighted. It is important to evaluate the additional organs attached (seminal vesicles, vas deferens, bladder neck), count and measure the lymph nodes. To achieve an optimal fixation, samples have to be completely submerged in a volume of formalin at least ten times their volume for about 24 h and those regarding the radical prostatectomy, a proper fixation tank should contain at least 500 ml of formalin.

1.3.1 Sampling

A fundamental task for the appropriate management of a patient with PCA is the adequate sampling. Although PCa sampling protocols vary from laboratories, depending on pathology's own experience and laboratory management, some sampling steps must be performed systematically. After inking the specimen, an adequate sampling of a prostate removed for BHP requires serial sections of whole gland at intervals of 3–4 mm in thickness. Unlike simple gland removal, radical prostatectomy usually includes the seminal vesicles and vas deferens. These

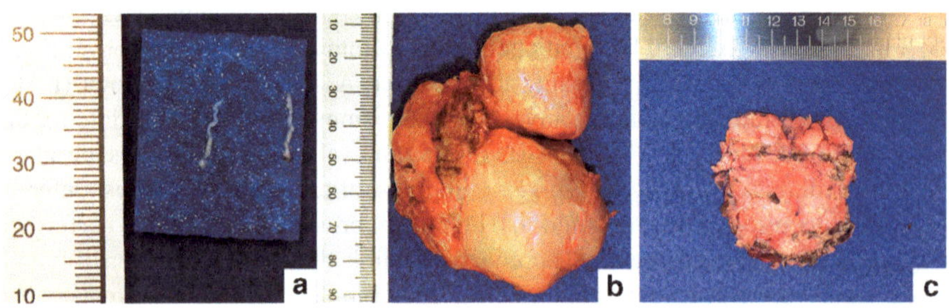

Fig. 1.1 Prostate surgical samples: (**a**) prostate needle cores; (**b**) open prostatectomy; (**c**) radical prostatectomy

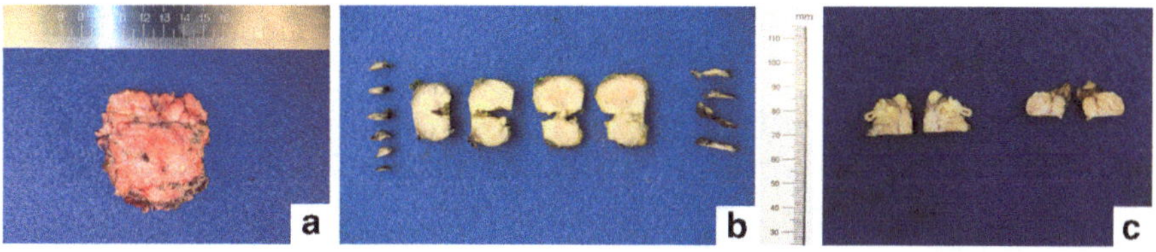

Fig. 1.2 Phases of radical prostatectomy sampling (a) sample of radical prostatectomy; (b) total prostate sampling; (c) seminal vesicles previously removed

structures can be of variable size depending on the surgical technique used. After fixation, the vesicles and the vas deferens must be removed and, the weight and the three prostate dimensions, noted (Fig. 1.2). Then, it is necessary to ink the specimen with at least two colours, so to distinguish the right from the left lobe. This step is critical for the correct histologic evaluation of surgical margin tumor involvement and it is performed by 99.6 % of European pathologists. In fact, the presence of tumor on surgical margin increases the probability of cancer recurrence after radical prostatectomy. At the ISUP consensus meeting, a full agreement on the modified cone method as the best apex cutting protocol was reached. A section is removed from the top, then, cut and embedded sagittally. After the sampling of the apex has been carried out, the prostate is sectioned every 3–4 mm. If the tumor is identified, a description with its location (e.g., posterior, postero-lateral, lateral, anterior and apex, mid, base), size (greatest diameter) and consistency (e.g., firm, fleshy) is necessary. It is also necessary to note the number, appearance and diameter of lymph nodes, if received. The last section of the base should be treated like the cone apex. The prostate sampling should be total but, if the partial embedding method is applied, it must be documented in the pathology report.

1.3.2 Processing

From each formalin-fixed, paraffin embedded tissue block, serial sections of 4 μm thickness are cut. The first two sections are stained with haematoxylin and eosin (h&e; Fig. 1.3) for histological diagnosis. In unresolved cases, the remaining slides may be used for immunohistochemistry.

1.3.3 Histology of PCa

PCa diagnosis depends on its presence in biopsy cores, TURP or prostate specimens. Detection of all phases in prostate carcinogenesis, from normal to hyperplastic to pre-neoplastic and, ultimately to cancer, may be identified in histologic slides (Che et al. 2003; Chappell and McLoughlin 2005a, b). The needle biopsy

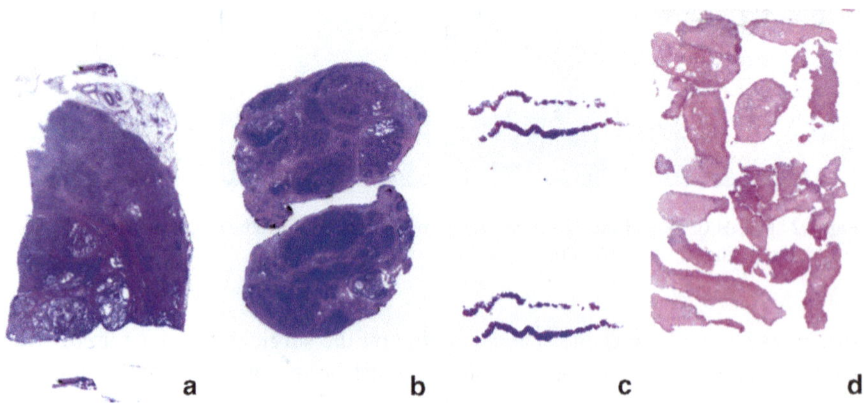

Fig. 1.3 Haematoxylin and eosin slides of: (**a**) radical prostatectomy specimen; (**b**) open prostatectomy; (**c**) needle core biopsy; (**d**) transurethral biopsy of prostate

Table 1.1 Pathognomonic criteria for cancer

Pathognomonic criteria for cancer	Major criteria for cancer	Minor criteria for cancer	Criteria against cancer
Perineural infiltration	Architecture of the glands	Intraluminar contents	Corpora amylacea
Glomerulations	Absence of basal cells	Mitotic figures	Atrophic cytoplasm
Mucinous fibroplasias	Nuclear atypia	Crystalloids	Pale to clear cytoplasm
Mucin extravasation	–	Nuclear hyperchromasia	–
–	–	Amphophilic cytoplasm	–
–	–	Adjacent HG-PIN	–

constitutes the most common tool for establishing a definitive diagnosis of PCa (Humphrey 2007). The importance of prostate biopsy interpretation, especially for early lesions, has been long recognized. A correct biopsy analysis requires a methodical (from low, to medium, to high magnification) and thorough approach evaluating several architectural and cytological features of glands. Both major and minor criteria for the PCa diagnosis have been defined (Epstein 2004) (Table 1.1). The glands' architecture (infiltrative growth pattern, single or cords of atypical cells, solid nests with or without necrosis), presence or absence of basal cells and nuclear atypia (nuclear enlargement, hyperchromasia and prominent nucleoli) are defined as major histological criteria (Fig. 1.4) (Totten et al. 1953; Helpap 1988; Algaba et al. 1996; Varma et al. 2002; Aydin et al. 2005). Intraluminar mucin secretions and crystalloids, hyperchromatic nuclei, mitosis, amphophilic cytoplasm and adjacent HG-PIN are, instead, considered as minor histological criteria (Fig. 1.4) (Ro et al. 1986; Epstein and Fynheer 1992; Epstein 1995b; Vesalainen et al. 1995; Henneberry et al. 1997). The presence of corpora amylacea, an atrophic cytoplasm, or pale to clear cytoplasm, are to be considered features

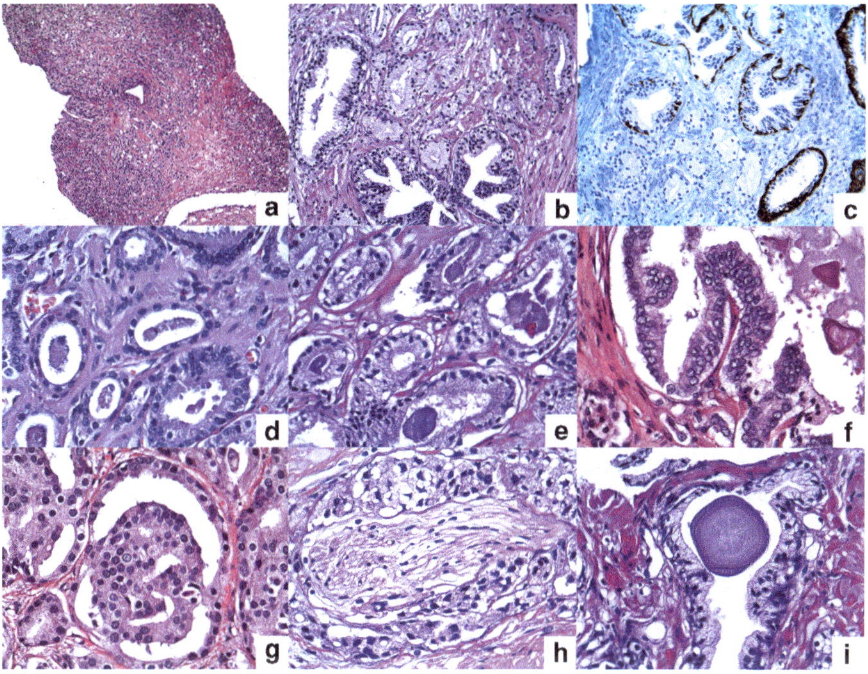

Fig. 1.4 Architectural and cytological features of PCa: (**a**) Architectural pattern of a Gleason 6 PCa; (**b**) Haematoxylin eosin staining; (**c**) Cytokeratin 34BE12 staining: note the absence of basal cells; (**d**) Cytological features of PCa: note the difference of nuclear size and tumor cells with nucleolus; (**e**) Intraluminar *pink* amorphous secretions; (**f**) Adjacent HG-PIN; (**g**) Glomerulation; (**h**) Circumferentially perineural infiltration; (**i**) Corpora amylacea

against cancer. Whenever the histopathological diagnosis may be in doubt, serial deeper cutting and immunohistochemistry are mandatory (Algaba et al. 1996; Novis et al. 1999; Iczkowski 2006). Finally, suspicious lesions may be resolved by intradepartmental and/or external consultation (Novis et al. 1999). In fact, in needle prostate biopsy specimens, the pathologist encounters a wide spectrum of lesions, including benign or hyperplastic glands, associated or not to inflammatory infiltrates (acute or chronic or granulomatous prostatitis), atrophy, atypical adenomatous hyperplasia, LG- and HG-PIN, adenocarcinoma, but especially the small atypical focus represent a diagnostic challenge. In this case, the diagnosis is a difficult task to accomplish: sometimes, cancer cannot be definitely diagnosed, due to the lack of clear diagnostic morphological features of malignancy. Four findings are considered pathognomonic of malignancy: perineural infiltration, glomerulations, mucinous fibroplasias and mucin extravasation (Fig. 1.4). Perineural invasion is seen in approximately 11–37 % of prostatic needle biopsies showing extensive adenocarcinoma, but only in 0–3 % of cases with limited cancer (Humphrey 2007). The glands should almost completely surround the nerve (Baisden et al. 1999). Such findings must be distinguished from perineural indentation or intraneural involvement by benign prostate glands (Ali and Epstein 2005) and have been associated with extraprostatic extension. Glomerulations consists of glands containing balls or tufts of cancer cells, resembling renal glomeruli, seen in approximately 3–15 % of prostatic needle biopsies (Pacelli et al. 1998; Baisden et al. 1999; Varma et al. 2002). Mucinous fibroplasias, also named collagenous micronodules, is a rare diagnostic finding in PCa (about 1–2 % of prostatic needle biopsies), which consists of nodular aggregates of paucicellular hyalinised stroma, within or outside cancer glands, often representing the organization of intraluminar mucin (McNeal et al. 1991; Thorson et al. 1998; Baisden et al. 1999).

1.4 Immunophenotyping of PCa

Immunohistochemistry has assumed an increasingly prominent role in diagnostic prostate pathology. It is a very valuable and frequently used tool in the differential diagnosis between PCa and its benign mimickers. Furthermore, this method permits to discriminate poorly differentiated PCa from other malignant tumors such as colonic and urothelial carcinomas, that secondarily involve or metastasize to the prostate gland (Bates and Baithun 2002; Srigley 2004; Herawi et al. 2005; Chuang et al. 2007; Osunkoya et al. 2007). The most commonly immunohistochemical antibodies used to differentiate PCa from its benign mimickers are high molecular weight cytokeratin, such as anti-cytokeratin 34BE12 or cytokeratins 5/6, p63 or in alternative its homologue p53, and α-methylacyl-CoA racemase, P504S (AMACR) (Fig. 1.5) (Kahane et al. 1995; Wojno and Epstein 1995; Zhou et al. 2004). The first two proteins are helpful for establish whether basal cells are present or not, that is an important key difference between benign and malignant prostate glands. In fact, the vast majority of invasive cancer does not present basal cells. However, negative staining can be seen in 5–23 % of benign prostate glands. In addition a fragmented basal cell layer can be detected in partial atrophy, adenosis and HG-PIN. In these circumstances an immunostaining with AMACR must be added. Racemase is a positive marker for PCa that complements negative basal cell staining (Beach et al. 2002; Jiang et al. 2002; Luo et al. 2002; Rubin et al. 2002; Magi-Galluzzi et al. 2003a). Nevertheless, AMACR may be also positive in HG-PIN and, occasionally, in some PCa mimickers, such as atrophy and adenosis (Bjartell et al. 2011). A staining cocktail obtained using these antibodies can improve the diagnostic accuracy (Bjartell et al. 2011). Other markers helpful in discriminating between benign and malignant acini are PSA, prostatic-acid phosphatise (PAP) and S-100. To determinate the endocrine nature of the cancer we can use chromogranin and neuron specific-enolase. It should be emphasized that immunohistochemical staining are not necessary in each case of PCa. The most common antibodies performed to distinguish PCa from other malignant tumors involving prostate are cytokeratins 7 and 20, cytokeratin 34BE12, p63, prostate-specific antigen (PSA), prostate-specific acid phosphatase (PSAP), prostate-specific membrane antigen

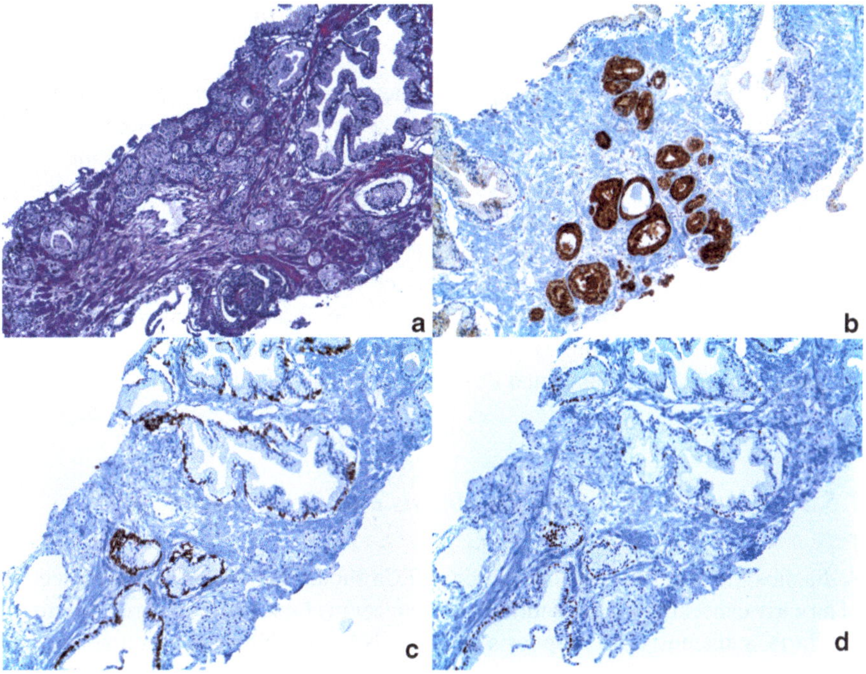

Fig. 1.5 Gleason 6 (3 + 3) PCa: (**a**) Haematoxylin eosin staining; (**b**) AMACR staining (positive granular cytoplasmic staining in the neoplastic glands); (**c**) Cytokeratin 34BE12 staining (absence of basal cells in neoplastic glands); (**d**) P63 staining (absence of basal cells in neoplastic glands)

(PSMA), prostein (P501S), TTF-1 and CDX-2 (Kunju et al. 2006; Leite et al. 2008a). The wide availability of reliable antibodies helps to provide a reliable diagnosis in most of the challenging cases.

1.5 Molecular Features

A tumor biomarker can be used to predict the prognosis and/or the clinical efficacy of the antitumor drugs in patients with cancer (Madu and Lu 2010). Prognostic biomarker gives informations on the tumor aggressiveness, helping to identify poor from good prognosis cancer; predictive biomarkers express patient's response to a particular anticancer drug (sensitivity or resistance); pharmacodynamic biomarker provides evidences of immediate pharmacological response to a drug, possibly determining the proper dosage to use (Madu and Lu 2010). An ideal prostate biomarker should be tissue specific, easy to measure in an accessible biological fluid or tissue, useful in the early detection and monitoring of PCa and helpful to distinguish PCa from normal or hyperplastic tissue (Madu and Lu 2010).

1.5.1 Molecular Biomarkers in PCa Screening

Over the past 25 years, an increasing early diagnosis and a successfully reduction of PCa death rate has been reached thanks to the wide diffusion of PSA screening program. Despite its limitations, serum PSA (a serine protease with a molecular mass of approximately 30 kDa that belongs to the human kallikrein proteases family) remains, to date, the most useful marker for detecting, staging, and monitoring PCa, especially in its early stage (Stamey et al. 1987; Partin et al. 1993). The main advantages of the PSA testing are the superior sensitivity, a high degree of patient acceptance and its cheapness; conversely, the main disadvantage is represented by its variable ability to distinguish between benign conditions (BPH and prostatitis) showing, in some cases, abnormal PSA values, and neoplastic diseases.

1.5.2 Biomarkers for the Diagnosis and Prognosis of PCa

A diagnostic marker for early detection of PCa should allow the accurate detection of an early cancer or of lesions turning into cancer in about 100 % of cases. Biomarkers for PCa diagnosis and prognosis include DNA and RNA-based biomarkers, and protein markers. The first described serum marker for PCa was the human prostatic acid phosphatase (PAP), named also serum acid phosphatase (AP), even if it has resulted not sensitive enough for PCa screening (Gutman and Gutman 1938; Bishop et al. 1985; Veeramani et al. 2005). Currently PSA test has, in fact, replaced the use of acid phosphatase determination, due to its more sensitivity and specificity. This test measures levels of PSA in the blood, in both its form: free and complexed with α1-antichymotrypsin or α2-macroglobulin. The ratio of free PSA in total PSA can help differentiate PCa from benign processes when a slightly elevated serum PSA value (from 4 to 10 ng/ml) is observed. Patients with a free PSA less than 10–25 % of the total PSA have, in fact, a higher probability to have PCa. To date, none of the biological markers used is sensitive and specific enough to provide an early diagnosis of PCa and to predict the natural course of the cancer. It is therefore urgently necessary to characterize and identify novel and existing diagnostic and prognostic markers starting from those reported below.

The α-Methylacyl Coenzyme A Racemase (AMACR; *diagnostic marker*) plays a key role in peroxisomal beta-oxidation of branched-chain fatty acids. It works as a growth promoter, androgen independent, in PCa (Kuefer et al. 2002) and, unlike normal or hyperplastic tissues, it has been demonstrated to be overexpressed in the majority of PCa (Kristiansen 2012).

The Glutathione S-transferase P1 (GSTP1; *diagnostic marker*) plays an important role in the detoxification. It is able to detect the presence of PIN and PCa and to differentiate patients with hyperplastic from neoplastic prostate. Since GSTP1 gene resulted hypermethylated in PCa, PCR detection of this gene in urine could be used to better stratify patients undergoing prostate biopsy (Lin et al. 2001).

The Chromogranin A (CGA, GRN-A; *diagnostic and prognostic marker*), an acidic glycoprotein, is a marker of neuroendocrine differentiation, playing an important role in the resistance to androgen-deprivation therapy. It has been proposed that its elevated expression in plasma and tissue could represent an independent factor in hormone refractory PCa and be associated with decreased patient survival (Deftos 1998; Kamiya et al. 2008; Berruti et al. 2010).

The Prostate-specific Membrane Antigen (PSMA; *diagnostic and prognostic marker*) is a type 2 membrane glycoprotein selectively found in prostate tissues and circulating PCa cells. It shows increasing levels of expression from hyperplastic low-grade to high grade PCa. It is also overexpressed in PCa metastases (Bostwick et al. 1998).

The Prostate Stem Cell Antigen (PSCA; *prognostic marker*) is a cell surface glycoprotein predominantly prostate specific. Its high expression is associated positively with adverse PCa features, such as increasing Gleason score, worsening clinical and pathological stage and androgen-independence. PSCA may have the potential to be a prognostic marker and a therapeutic target (Reiter et al. 1998; Zhigang and Wenlv 2006).

The Early Prostate Cancer Antigen (EPCA; *diagnostic and prognostic marker*) is a nuclear matrix protein, whose preoperative level predicts the increased risk for the development of subsequent cancer in patients with HG-PIN and the presence of incidental carcinoma in patients undergoing TURP for BPH (Zhao and Zeng 2010; Zhao et al. 2010). Furthermore, recent studies have demonstrated that significantly high preoperative serum EPCA levels are associated with PCa progression (Paul et al. 2005; Zhao et al. 2011, 2012).

The B7-H3 (*diagnostic and prognostic marker*) first identified in 2001, is a member of the B7-CD28 family, a group of proteins that enhances T cell proliferation and IL-2 production (Zang et al. 2007). The B7-H3 was found overexpressed in PCa compared to BPH, significatively linked to the PCa spread and poor outcome (Yuan et al. 2011). Finally, the lack of change in its expression in patients with androgen deprivation therapy makes the B7-H3 an attractive target for new combined treatments (Roth et al. 2007; Chavin et al. 2009).

The Sarcosine (*diagnostic and prognostic marker*) is an N-methyl derivative of glycine, strongly increased during PCa progression. Recently, serum and urine levels of sarcosine have been proposed as useful indicators for the presence of PCa (Sreekumar et al. 2009).

The Caveolin-1 (Cav-1; *prognostic marker*) is an integral membrane protein of vesicular structures called caveolae, expressed in two isoforms (caveolin-1α and caveolin-1β). Early and recent studies have shown that Cav-1 regulates many signalling proteins involved in cell survival and angiogenic activities (Tahir et al. 2001). Cav-1 is secreted by PCa cells and its increased levels are associated with PCa progression (Freeman et al. 2012).

The Prostate cancer antigen 3 (PCA3/DD3; *diagnostic and prognostic marker*) is a non-coding mRNA segment from chromosome 9q21-22, found overexpressed in the majority of PCa (de Kok et al. 2002; Hessels and Schalken 2009). To date, the evaluation of urine PCA3 level, in combination with PSA kinetics values and DRE, is used to counsel or confirm biopsy indications, especially in the gray area of PSA value ranging from 4 to 10 ng/ml (Rigau et al. 2010; Auprich et al. 2011). A PCA3 score >25 has been proposed as a strong indicator for tumor volume (Ploussard et al. 2011).

The Transmembrane protease serine 2 (TMPRSS2-ERG Gene Fusion Rearrangement; *diagnostic marker*) is an androgen-regulated type II transmembrane-bound serine protease, located on chromosome 21. It was found expressed in normal prostate and overexpressed in neoplastic prostate tissue, particularly in poorly differentiated and untreated PCa cells. Fusions between the TMPRSS2 gene and several oncogenes have been demonstrated in many PCa (Vest et al. 2010). In particular, in early PCa it was demonstrated that the fusion between TMPRSS2 and the transcription factor genes ERG and ETV1 resulted in an overexpression of ETS family members in cancer cells and, consequently, in tumor progression (Kumar-Sinha et al. 2008). Finally, the TMPRSS2/ERG gene fusion protein, combined with PCA3 determination, increases the usefulness of serum PSA in predicting PCA (Tomlins et al. 2011).

The Ki-67 (*prognostic marker*) was found progressively increased from benign prostatic hyperplasia to adenosis, HG-PIN and high-grade cancers (Haussler et al. 1999). Furthermore, patients with high proliferation index showed a significant higher incidence of metastatic PCa in the pelvic lymph nodes (Revelos et al. 2005).

The Golgi phosphoprotein 2 (GOLPH2; *diagnostic marker*) is a novel Golgi membrane protein, that codes for type II Golgi membrane antigen GOLPH2/GP73. In a recent study evaluating the expression of several potential PCa markers, an increase in the levels of GOLPH2 was found (Kristiansen et al. 2008; Laxman et al. 2008).

The DAB2 interacting protein (DAB2IP; *diagnostic marker*) is a RasGTPase-activating protein that acts as a tumor suppressor. The human DAB2IP gene has frequently down-regulated in PCa cell lines (Chen et al. 2003).

1.5.3 Tissue Biomarkers

Basal cell markers represent, undoubtedly, the immunohistochemical cornerstone of PCa diagnostics. Of all the immunohistochemical markers, in fact, high molecular cytokeratins (CK34BE12, CK 5/6) and p63 are the best known and widely used (Brawer et al. 1985). A wide range of other markers (P-cadherin, D2-40, CD109 or BCL-2), that reliably label basal cells in the prostate, can be employed diagnostically (Jarrard et al. 1997; Ramos Soler et al. 2006; Hasegawa et al. 2007; Kuroda et al. 2010) but, to demonstrate the prostatic origin of the tumor, several additional markers have been proposed:

The Prostate-specific antigen (PSA), even if not entirely prostate-specific (also detected in ovarian and breast carcinomas), is most commonly used to immunohistochemically confirm the prostatic origin of metastatic cancers.

The Prostate-specific membrane antigen (PSMA) is a folate dehydrolase antigen that is strongly expressed in most PCa and their metastases and has therefore been suggested as novel diagnostic marker. PSMA is generally up-regulated during PCa progression; it is not prostate-specific and widely found in several cancers, such as gastric and renal tumors and urothelial carcinomas (Wright et al. 1995).

The Prostein is located at the Golgi apparatus and it is androgen-regulated. Its biological function is still unclear. Nonetheless, this protein has been successfully used to discriminate, by immunohistochemistry, PCa from tumors of colon and bladder (Xu et al. 2001; Kalos et al. 2004).

The homeobox gene NKX3.1 (chromosome 8p21.2) is an androgen-regulated and mainly prostate-specific gene, which was observed primarily in secretory prostatic benign and neoplastic epithelia. It is rarely found in benign testis and invasive lobular carcinomas of the breast (Gelmann et al. 2003). Some authors proposed NKX3.1 as new promising prognostic marker due to its loss in the high grade PCa (Bowen et al. 2000; Bethel et al. 2006; Kristiansen 2012).

The Alpha-methylacyl-CoA racemase (AMACR) is found expressed in numerous normal tissues, including liver, renal tubules and gallbladder epithelium, in dysplastic tissues or malignant tumors such as colon and renal carcinoma, with the highest level (>95 % of cases) described for PCa. In this malignancy it represents a useful marker, reaching clinical value (Went et al. 2006; Kristiansen 2012).

The GOLM1 (GOLPH2, GP73) is a Golgi phosphoprotein of 73 kDa that was first found expressed in hepatocellular disease, especially in liver carcinoma, whose functions are still unknown. Several studies concerning PCa profiling have described an overexpression of GOLM1 mRNA in these malignancies (Lapointe et al. 2004; Kristiansen 2012).

The Fatty acid synthase (FASN) protein is encoded by the FASN gene located on chromosome 17q25 and represents a crucial enzyme for the de novo synthesis of fatty acids. FASN protein was found overexpressed in PCa, in which a close concordance between AMACR and FASN expression has been observed (Shurbaji et al. 1996; Tischler et al. 2010; Kristiansen 2012).

The truncated ERG product is found overexpressed in approximately 50 % of all PCa cases harbouring TMPRSS2/ERG gene fusion and can be used as a diagnostic marker since its antibody is commercially available (Miettinen et al. 2011).

1.6 Epithelial Prostatic Diseases

1.6.1 Normal Prostatic Epithelium

Normal prostate is composed of rounded and irregularly branching glands, medium to large in size, embedded in a fibromuscular stroma (Fig. 1.6). The gland consists of two cell layers: an inner layer constituted of differentiated secretory cells and an outer layer, called basement membrane, containing mainly basal cells; several intermediate cells are present between basal and secretory cells, and represent the intermediate steps in basal to secretory cells differentiation. A minor cell population is constituted by neuroendocrine cells.

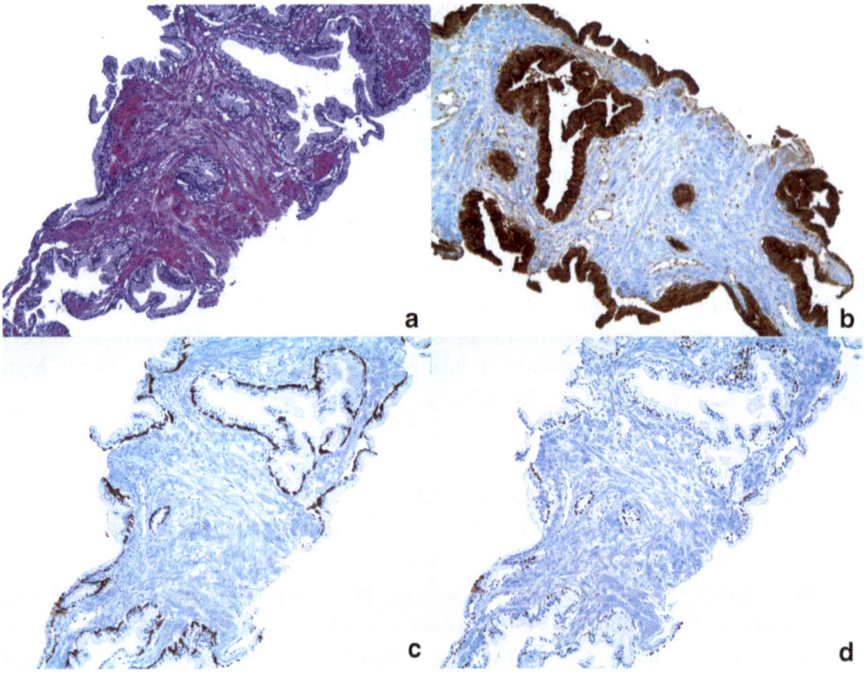

Fig. 1.6 Normal prostate tissue: (**a**) Haematoxylin eosin staining; (**b**) PSA staining (positive in prostate tissue); (**c**) Cytokeratin 34 BE12 staining (highlighting basal cells); (**d**) P63 staining (highlighting basal cells)

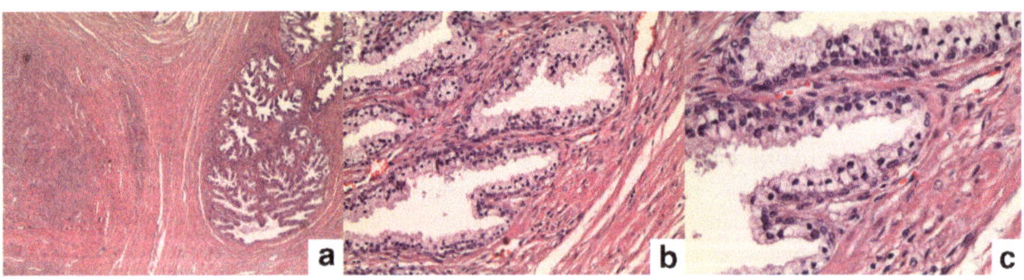

Fig. 1.7 Benign prostatic hyperplasia (**a–c**: from low, to medium, to high magnification)

1.6.2 Benign Prostatic Hyperplasia

Benign prostatic hyperplasia (BPH) is a benign disease, also named benign prostatic hypertrophy, characterized by the proliferation of epithelial and stromal compartments, resulting in cellular accumulation, glands enlargement and stromal nodules (Fig. 1.7). It is generally caused by aging (50 % of men show histopathological BPH by age 60, but the great majority by age 85) and it depends on testosterone and dihydrotestosterone production.

1.6.3 Prostatic Intraepithelial Neoplasia

Prostatic intraepithelial neoplasia (PIN) is defined as a neoplastic proliferation of the epithelial cells confined to pre-existing prostatic ducts, ductules or acini and it has been widely accepted as the main precursor of PCa (Dovey et al. 2005; Ayala and Ro 2007; Chin et al. 2007; Godoy and Taneja 2008; Bostwick and Cheng 2012). First described by McNeal in 1969 (McNeal 1969), it was named PIN by Bostwick and Braver in 1987, and categorized into three grades: PIN 1, PIN 2, and PIN 3, basing on the degree of prostatic cell changes, respectively mild, moderate and severe (Bostwick et al. 2004; Ayala and Ro 2007). To date, compelling evidence widely confirmed that only HG-PIN, corresponding to PIN 2 and 3, may precede PCa. Pathologists, in fact, do not routinely report low-grade PIN (Bostwick and Cheng 2012). The isolated HG-PIN incidence is reported ranging from 4.4 to 22.5 %, with an average of 9 % of prostate biopsies (Bostwick et al. 2004; Bostwick and Cheng 2012). Like PCa, HG-PIN tends to be more common in prostate peripheral zones and to be multifocal (Haggman et al. 1997; Pierorazio et al. 2007). The incidence of PCa detected in repeat biopsies following the previous diagnosis of HG-PIN varies from 22 to 79 % (Borboroglu et al. 2000; Dovey et al. 2005; Gokden et al. 2005; Joniau et al. 2005; Moore et al. 2005; Herawi et al. 2006). Furthermore, the presence of multiple HG-PIN foci increases the probability of concomitant PCa (Bostwick and Brawer 1987) Nonetheless, to date, no single pathologic, molecular or clinical variable helps to identify patients at higher risk for developing PCa among those with HG-PIN (Abdel-Khalek et al. 2004; Roscigno et al. 2004; Singh et al. 2004; Chin et al. 2007; Loeb et al. 2007). Histologically, HG-PIN is characterized by one or more large glands lined by pseudostratified epithelium constituted by proliferating cells with morphologic features similar to those found in PCa. The main difference between HG-PIN and PCa consists in the integrity of basal cell layer: in fact, unlike HG-PIN where the basal layer is present, *in* most cases fragmented, the PCa glands show an absent basal membrane (Haggman et al. 1997). HG-PIN is characterized by increased cell size, typically overlapping, large and hyperchromatic nuclei, changes in chromatin structure, and more prominent nucleoli (Fig. 1.8). Rarely, HG-PIN can also present small, neuroendocrine, mucinous, squamous, apocrine and signet ring cells differentiation, this latter almost always associated with an adjacent invasive signet ring cell carcinoma (Reyes et al. 1997; Montironi et al. 2007).

Several HG-PIN patterns have been described: the most frequent types show a flat, stratified, branching, micropapillary or cribriform architecture, but other unusual histologic patterns, such as foamy gland, signet-ring cells and mucinous have rarely been seen (Bostwick et al. 1993a; Epstein 2009). HG-PIN can be distinguished from PCa by immunohistochemical examination. In fact, unlike PCa, HG-PIN is positive for both cytokeratin 34B12 and p63 antibodies, identifying a basal membrane with interruptions (Bishara et al. 2004; Montironi et al. 2011b). In difficult cases, the alpha-methylacyl-CoA racemase (AMACR) can be added; even if it is frequently expressed in HG-PIN (Kumaresan et al. 2010; Montironi et al.

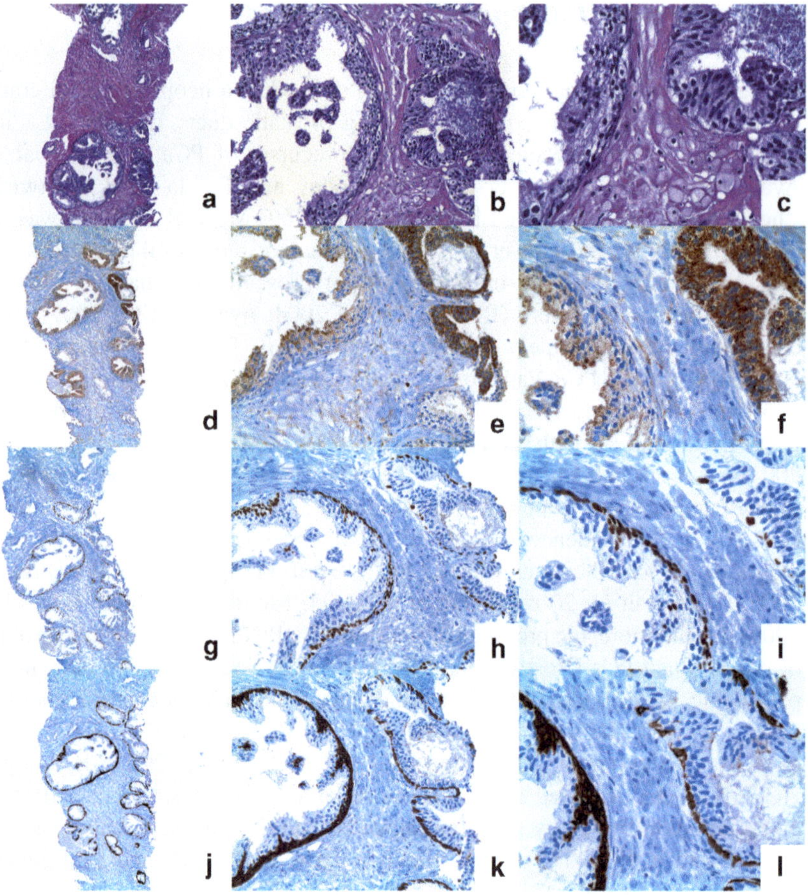

Fig. 1.8 Low-grade versus high-grade prostatic intraepithelial neoplasia (LG-PIN on *left* side vs HG-PIN on *right* side of the picture). (**a–c**) Haematoxylin eosin staining (from low, to medium, to high magnification); (**d–f**) AMACR staining (from low, to medium, to high magnification): AMACR is negative in LG-PIN; conversely HG-PIN shows a positive granular cytoplasmic staining for this; (**g–i**) p63 staining (from low, to medium, to high magnification): p63 staining shows a continuous basal membrane in LG-PIN; conversely, in HG-PIN, p63 staining is fragmented, highlighting a basal membrane with interruptions; (**j–l**) Cytokeratin 34 BE12 staining (from low, to medium, to high magnification): note respectively a continuous basal membrane in LG-PIN and a basal membrane with interruptions in HG-PIN

2011b). The prognostic significance of HG-PIN in clinical decision making remains under discussion. In cases in which small atypical glands are adjacent to HG-PIN, repeated biopsy is recommended within 3–6 months of the initial diagnosis (Naya et al. 2004; Epstein and Herawi 2006; Girasole et al. 2006; Herawi et al. 2006; Meyer et al. 2006; Netto and Epstein 2006; Schoenfield et al. 2007; Godoy and Taneja 2008). The prognostic value of HG-PIN for diagnosing PCa depends on the number of samples obtained during the first and subsequent biopsies and the time interval between the biopsies. Conversely, clinical parameters such as serum PSA level, DRE and imaging studies are not as effective in identifying which patients with HG-PIN are more likely to show carcinoma on repeated biopsy. Therefore, repeated biopsies are generally recommended within a year of the initial diagnosis of HG-PIN, particularly if it was shown in three or more cores (Naya et al. 2004; Epstein and Herawi 2006; Girasole et al. 2006; Herawi et al. 2006; Meyer et al. 2006; Netto and Epstein 2006; Schoenfield et al. 2007; Godoy and Taneja 2008). HG-PIN is considered an essential early step for prostate carcinogenesis, with genetic and phenotypic alterations that are intermediate between the abnormalities of normal tissue and cancer (Bostwick et al. 2004). Among these, the most frequent are the increased expression of AMACR, Rb inactivation, loss of p27KIP1, PTEN, hypermethylation of the promoter region of GSTP1 and the fusion of TMPRSS2-ERG genes (Zynger and Yang 2009). Furthermore, approximately 20 % of HG-PIN shows the TMPRSS2-ERG fusion gene abnormality, which is also observed in approximately 50 % of PCa (Mosquera et al. 2008). Elevated expression of p16, p53, Bcl-2, and MYC can be seen in both HG-PIN and PCa. A decreased expression of NKX3.1 and p27 genes in both HGPIN and PCa has been described (Montironi et al. 2011b). Proliferative index and apoptosis show a trend to increase from normal through HG-PIN to PCa (Ananthanarayanan et al. 2006).

1.6.4 Atypical Small Acinar Proliferation (ASAP)

ASAP is not a histopathological entity, but mainly a diagnostic term, which identifies the presence of a small foci of atypical glands with features highly suggestive but not convincing for a definitive diagnosis of PCa (Fig. 1.9) (Cheville et al. 1997; Iczkowski et al. 1997). Such lesions are identified in about 1.5–9 % of prostate needle biopsy (Cheville et al. 1997; Iczkowski et al. 1997, 2002b; Ouyang et al. 2001; Fadare et al. 2004). In these cases, the pathologist does not make prudently a diagnosis of cancer, but raises the suspicion, so the biopsy has to be remade within 3–6 months. It has been reported that the cancer detection rate varies between 21 and 51 % on the second biopsy in patients with ASAP (Borboroglu et al. 2000; Leite et al. 2008b). This finding significantly predicts the presence of cancer verified, in fact, in the following biopsies (Brausi et al. 2004; Gupta et al. 2004). Histologically ASAP is identified as a small size focus (less than two dozen acini) with distorted acini characterized by the lack of convincing features of malignancy (nuclear and nucleolar enlargement). Immunohistochemically, such lesions are usually negative for high molecular weight cytokeratin and p63 and show a focally, significant immunopositivity, for racemase.

1.6.5 Prostate Carcinoma

An adequate and accurate characterization of individual PCa is critical to define the risk of tumor progression and to determine the best therapeutic approach and the correct management of each single patient (Edge et al. 2010; Cheng et al. 2012).

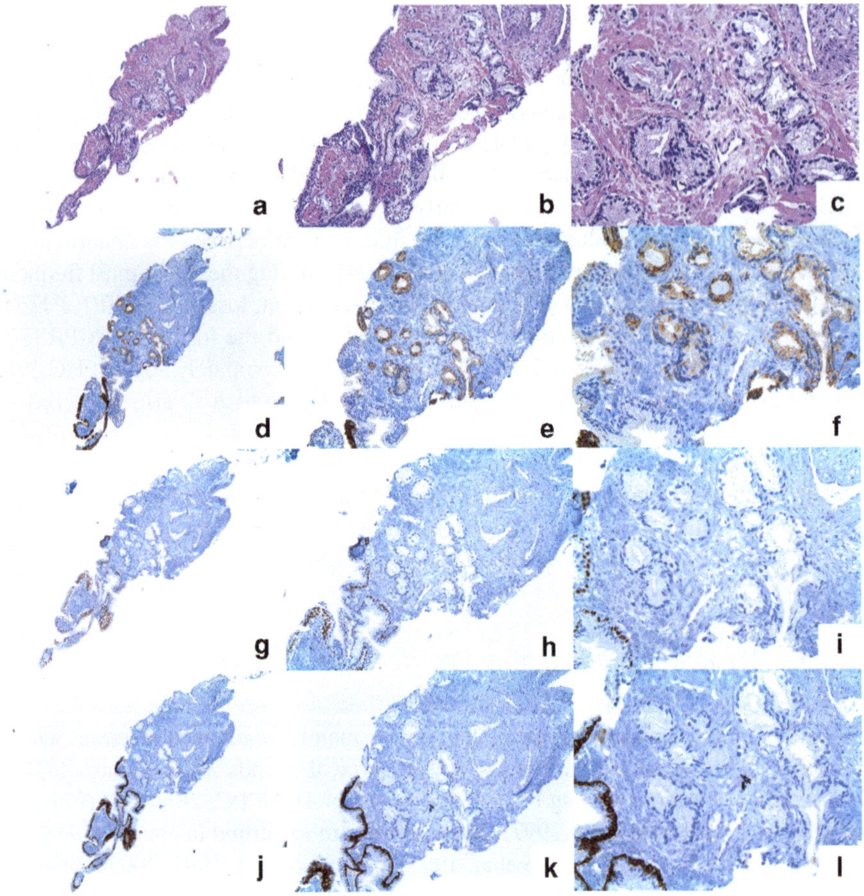

Fig. 1.9 Atypical small acinar proliferation (ASAP). (**a–c**) Haematoxylin eosin staining (from low, to medium, to high magnification); (**d–f**) AMACR staining (from low, to medium, to high magnification): ASAP shows a positive granular cytoplasmic positivity for AMACR; (**g–i**) p63 staining (from low, to medium, to high magnification): ASAP is negative for p63; (**j–l**) Cytokeratin 34 BE12 staining (from low, to medium, to high magnification): note the absence of basal cells in atypical glands

The pathology report should provide detailed informations on tumor's site (mono o bilateral involvement), histotype (e.g. acinar adenocarcinoma, duct carcinoma, signet ring cells carcinoma), histological grade (according to the modified Gleason score, with primary and secondary patterns), extension depending on the type of surgical specimens (needle, TURP, enucleation or radical prostatectomy), surgical margin status (presence of carcinoma at inked margin, presence of extraprostatic extension (extraprostatic adipose tissue), seminal vesicle invasion, number of total and metastatic lymph nodes (Epstein et al. 2005a) and the staging according to the 2010 revision of the American Join Committee on Cancer/Union Internationale

Contre le Cancer (AJCC/UICC) for radical prostatectomy). It should also include the presence of angioinvasion, perineural invasion and tumor necrosis that are generally optional (Montironi et al. 2012b).

1.6.5.1 Prostate Carcinoma Variants

The large majority of PCa is represented by conventional acinar histotype (Eble et al. 2004; Humphrey 2012). Different variant of acinar carcinoma, such as foamy, atrophic, pseudohyperplastic, signet ring, oncocytic, colloid and lymphoepithelioma-like, are anyway identified (Humphrey 2012). The remaining 5–10 % of PCa is constituted by non acinar variants, such as sarcomatoid, ductal urothelial, squamous, small-cell, basal cell and clear cell carcinoma (Humphrey 2012). These latter have different clinical, biological and genetic features compared to acinar carcinoma, including prognosis (Han et al. 2009; Humphrey 2012).

1.6.5.2 Grading

PCa should be graded according to the current Gleason score system, modified by the International Society of Urological Pathology (ISUP), the most widespread method of PCa grading significantly associated with prognosis (Srigley et al. 2005). The Gleason score is the sum of the primary and secondary most common histologic patterns in PCa, rated on a scale of 1–5, with 1 being the well differentiated and 5 the undifferentiated carcinomas (Fig. 1.10). A higher score is more likely to be seen with disease not confined to the prostate and is correlated with poorer response to treatment of localized disease.

In case of radical prostatectomy, if a higher tertiary pattern is seen, it should not contribute to the final score as a secondary pattern, as this would probably result in the over-grading of the tumor (Trock et al. 2009). The tertiary pattern would have to be reported as the third most prevalent pattern of carcinoma.

Contrary to the recommendations for radical prostatectomy, for needle biopsy, the lower-grade secondary pattern comprising <5 % of biopsy material should be ignored (the primary and the secondary patterns are considered of equal grade), any proportion of higher secondary pattern should be reported and, when present, the tertiary higher pattern should contribute to the final score, referred as the secondary pattern (Amin et al. 2005; Delahunt et al. 2012). In case of biopsy, the patterns 1 and 2 should not be reported.

1.6.5.3 Tumor Extent and Prostate Cancer Volume

One of the most important pathologic parameter in choosing therapeutic strategy and determining the biological course of each PCa is provided by its extension in prostate surgical samples, such as TURP, needle cores biopsy or prostatectomy

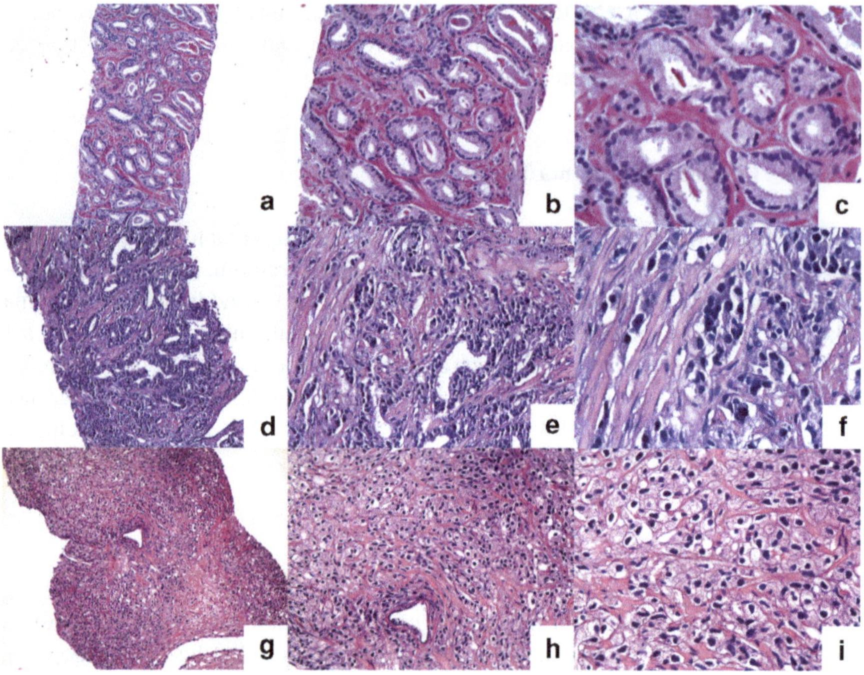

Fig. 1.10 PCa grading (haematoxylin eosin staining). (**a–c**) Gleason 6 (3 + 3) PCa (from low, to medium, to high magnification); (**d–f**) Gleason 7 (4 + 3) PCa (from low, to medium, to high magnification); (**g–i**) Gleason 10 (5 + 5) PCa (from low, to medium, to high magnification)

(Montironi et al. 2012b). In core needle biopsy, the tumor extent is mainly reported as millimetres of cancer per core or as percentage of cancer per core (Montironi et al. 2012b). It is also registered as the number of tumor-infiltrated cores, fraction of positive cores, total percentage of cancer in the entire specimen and total millimetres of cancer among all cores (Montironi et al. 2012b).

PCa can discontinuously involve a single core. The two options generally used are to measure it as if all separate cancer foci were one single continuous foci, or collapse all foci and ignore the amount of intervening benign prostate tissue (Fig. 1.11) (Karram et al. 2011; Montironi et al. 2012b). In radical prostatectomy the tumor extent is reported as percentage of one or both prostatic lobes involved.

The prognostic value of PCa volume in radical prostatectomy is still discussed: several studies showed its independent prognostic value, others denied it (Kikuchi et al. 2004; van Oort et al. 2008; Yadav et al. 2009; Wolters et al. 2010). Nevertheless, a cut-off value of 0.5 ml has proposed as significant to distinguish PCa with a high risk of progression from the great majority of "insignificant" neoplasm (Stamey et al. 2000). It may be useful to provide the size of the largest tumor nodule identified.

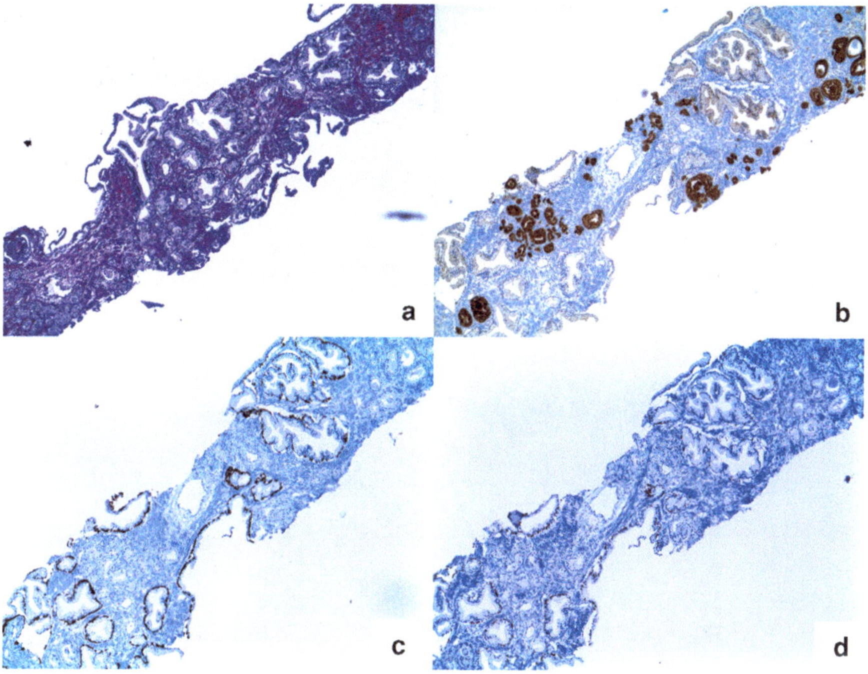

Fig. 1.11 PCa discontinuously involved a single core: (**a**) Haematoxylin eosin staining; (**b**) AMACR staining (positive granular cytoplasmic staining in the neoplastic gland); (**c**) Cytokeratin 34 BE12 staining (absence of basal cells); (**d**) P63 staining (absence of basal cells)

1.6.5.4 Surgical Margin Status and Extraprostatic Extension

The necessity of providing the surgical margin status in radical prostatectomy specimens and its predicting value is widely demonstrated; however, great variability in its report by pathologists of different institutions still exists. The margin is considered negative if the tumor is not present at the inked surface of the specimen; conversely it is evaluated as positive if the neoplastic cells touch the inked tissue (Fig. 1.12) (Epstein et al. 2005b). The presence of tumor at the margin must be not considered as extraprostatic extension (Chuang and Epstein 2008). Surgical margin status is a significant predictor of disease recurrence, independent of the pathological stage (Kausik et al. 2002). In addition, it is recommended to provide the tumor extent at the surgical margin (linear extent in millimetres, or number of blocks with positive margin involvement) because significantly associated with tumor recurrence (Epstein 1990; Epstein et al. 1996; Epstein and Sauvageot 1997).

Extraprostatic extension is defined as the presence of cancer beyond the prostate gland capsule, including anterior muscle, large neurovascular bundles, seminal vesicle stroma and adipose tissue tumor infiltration (Edge et al. 2010). It is regarded as one of the most important prognostic factor for PCa (Cheng et al. 1999). In a

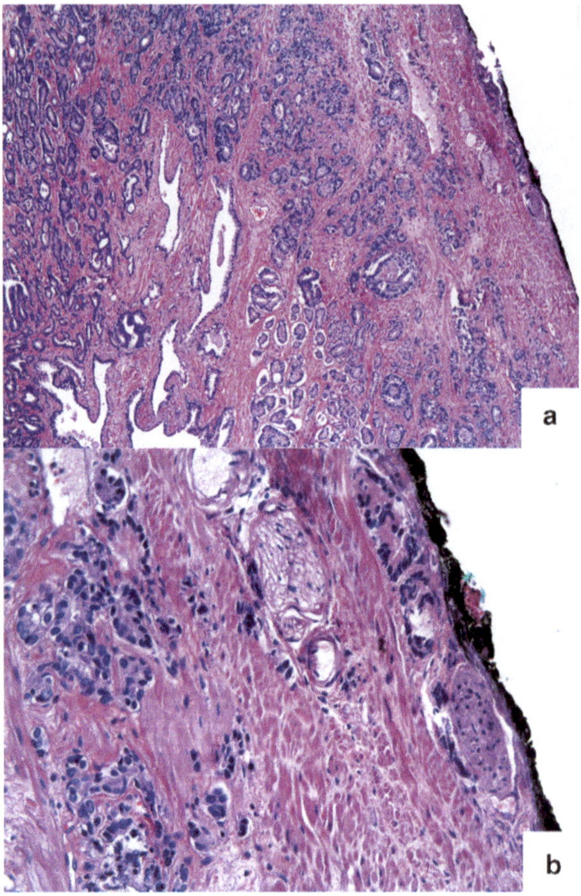

Fig. 1.12 Prostate surgical margin: (**a**, **b**) PCa is present at the inked surface of the specimen (haematoxylin eosin staining; **a**: low magnification; **b**: high magnification)

needle biopsy specimen, the presence of adipose tissue within prostatic parenchyma is extremely rare, but the recognition of tumor involvement of fat has to be interpreted as extraprostatic extension. A significant association between extraprostatic extension on prostate cores and high-stage PCa in subsequently prostatectomy was reported (Miller et al. 2010). Cancer involving the skeletal muscle on needle biopsy specimen is not, instead, diagnostic of extraprostatic extension. Furthermore, the presence of carcinoma in skeletal muscle on needle biopsy is not associated with high grade PCa at subsequent radical prostatectomy (Ye et al. 2010). Seminal vesicles stroma invasion must, conversely, be considered extraprostatic extension (Fig. 1.13). However, it is important to distinguish between seminal vesicles from ejaculatory ducts invasion, because the latter is not considered as extraprostatic disease (Fine and Reuter 2012). A helpful criterion for differentiating seminal vesicle from ejaculatory ducts is the presence of a coat of smooth muscle encircling a central lumen, not recognizable in seminal vesicle.

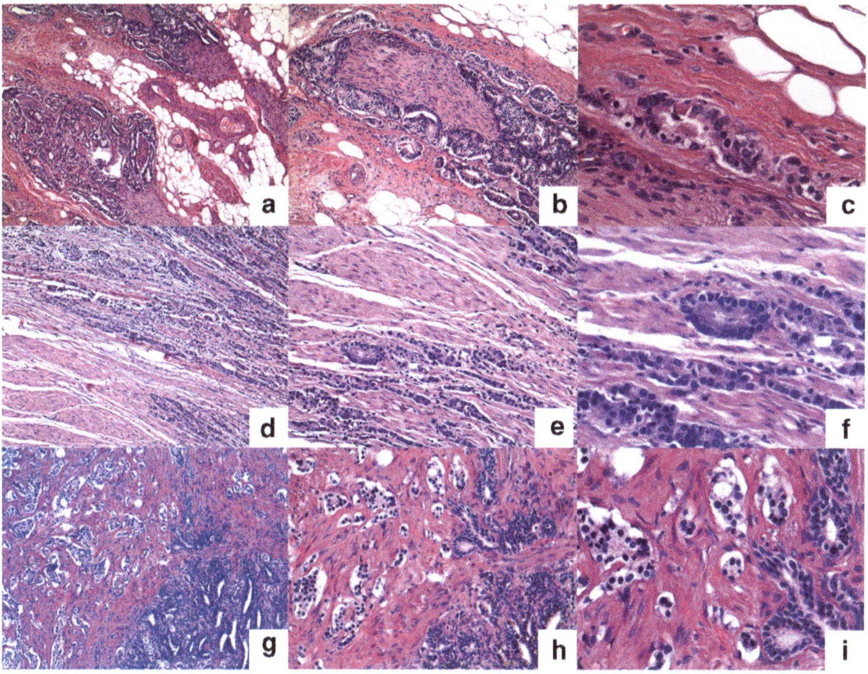

Fig. 1.13 Extraprostatic PCa extension (haematoxylin eosin staining): (**a**–**c**) Adipose tissue PCa infiltration (from low, to medium, to high magnification); (**d**–**f**) Smooth muscle PCa infiltration (from low, to medium, to high magnification); (**g**–**i**) Seminal vesicle stroma PCa infiltration (from low, to medium, to high magnification)

1.6.5.5 The TNM Staging

To provide an accurate staging of PCa the 2010 revision of the American Join Committee on Cancer/Union Internationale Contre le Cancer (AJCC/UICC) is widely preferred. Its application allows the identification of patients at the greater risk of progression and therefore is critical for the therapeutic approach and the correct management of each patient with PCa (Cheng et al. 2012). The current TNM staging system differs from previous 2002 version mainly for the definition of pT3a stage, now including both extraprostatic extension and microscopic bladder neck invasion (Edge et al. 2010; Cheng et al. 2012). The main debated issue of the 2010 modified version of the TNM system remains the stage pT2 and its subclassification in T2a (half of a single lobe involvement), T2b (more than half lobe involvement) and T2c (both lobes involvement, also focally) tumors, due to the tendency of PCa to be multifocally (Cheng et al. 2005, 2012; Andreoiu and Cheng 2010). The pathological staging (pTNM), mainly based on the evidence acquired from surgery and from pathological evaluation, consists in the pathologic assessment of the primary tumor (pT; size and eventual extension of the primary tumor beyond the contour of the prostate gland), of the regional lymph nodes (pN) and of distant metastasis (pM). The evaluation of local node status requires the removal of nodes adequate to validate the node metastasis absence (pN0). It is recommended to report its location and extension, since its correlation to the risk of recurrence (Wheeler et al. 1998; Marks et al. 2007).

1.6.5.6 Other Factors

Although the prognostic value and clinical usefulness of the following factors has not been sufficiently established, it may be useful to note in pathologic report perineural invasion, neuroendocrine differentiation, microvessel density, chromatin texture, proliferation markers and prostate-specific antigen derivatives (Bostwick et al. 2000; Srigley et al. 2005).

1.7 Mimickers

PCa mimickers are lesions of potential diagnostic difficulty that show architectural and cytological features overlapping with PCa histopathologically, including anatomic structures or inflammatory and reactive conditions (Srigley 2004; Hameed and Humphrey 2010; Montironi et al. 2012a). PCa mimickers can be differentiated by their growth patterns as small gland, large and cribriform gland or solid and non glandular mimickers (Srigley 2004; Hameed and Humphrey 2010; Montironi et al. 2012a). In the first subgroup are included the conditions that simulate low-grade PCa (Gleason pattern ≤ 3), further subdivided in lesions of prostatic epithelial origin such as atrophy, atypical adenomatous hyperplasia, sclerosing adenosis and basal cell hyperplasia and of non prostatic epithelial origin, such as ejaculatory ducts epithelium and seminal vesicle, mesonephric remnants, nephrogenic adenoma, mucinous metaplasia, Cowper and colonic glands. The large and cribriform pattern mimickers encompass conditions as reactive epithelial atypia, basal cell or clear cell cribriform hyperplasia, while the solid and non glandular pattern subgroup includes granulomatous prostatitis, prostatic xanthoma, dense inflammation, signet ring change in non epithelial cells and paraganglia (Srigley 2004; Hameed and Humphrey 2010; Montironi et al. 2012a). Many of these lesions are readily identifiable and separated from PCa on routine haematoxylin and eosin-stained sections. In challenging cases immunohistochemistry can be helpful (Paner et al. 2008).

1.7.1 Adenosis/Atypical Adenomatous Hyperplasia

Adenosis, also referred to as atypical adenomatous hyperplasia, is a common finding in routine practice, easily encountered in TURP and prostatectomy samples, ranging an incidence of about 20 % (Fig. 1.14) (Bostwick et al. 1993b; Gaudin and Epstein 1995; Cheng et al. 1998; Lotan and Epstein 2008). It is mainly localized near the apex, the transitional zone and periurethral area. It consists of a well-circumscribed

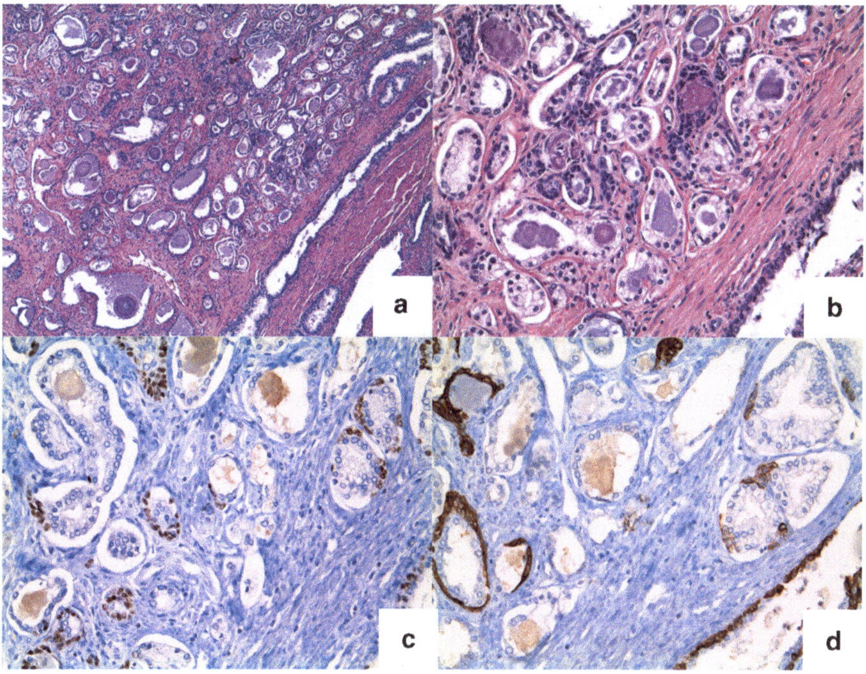

Fig. 1.14 Prostate adenosis: (**a**, **b**) Haematoxylin eosin staining (**a**: low magnification; **b**: high magnification); (**c**, **d**) p63 and CK34BE12 staining (showing a discontinuous but present basal layer)

proliferation of large complex glands, typical of hyperplasia, admixed with round glands, with similar cytoplasmic and nuclear findings (crystalloids and basophilic mucin occasionally present, small nuclei with fine chromatin, very small nucleoli). These findings may be very helpful to distinguish this lesion from low-grade PCa. In challenging cases, serial sections of suspicious foci and immunohistochemistry may be useful; in fact anti-cytokeratin 34BE12 and p63 are used to show the presence of an intact or fragmented basal cell layer. AMACR is often positive, so is not really useful (Qian et al. 1995; Yang et al. 2002; Kunju et al. 2003; Shah et al. 2004). Due to its overlapping features with low-grade PCa, this lesion has been considered a potential precursor for low-grade cancer, but being not shown associated with an increased risk of carcinoma on follow-up (Epstein 1995a), a close follow-up may be suggested. When associated to PCa, this is generally a low-grade carcinoma (Gleason score 2–4) (Helpap et al. 1997).

1.7.2 Atrophy and Post-atrophic Hyperplasia

Atrophy is a common microscopic finding, increasing in incidence with advancing age. It consists of well formed, often little distorted, prostate glands, not particularly crowded, lined by flattened epithelium. Different forms of prostatic atrophy have been described, which may coexist in the same prostate sample. It is most commonly seen in the peripheral zone, but it can be observed also in the central and the transitional zone. It is generally characterized by glands lined by secretory cells showing a marked reduction in cytoplasm and hyperchromatic nuclei. At low magnification, this finding, especially due to prominent acinar architecture and cytoplasmic basophilia, can be misdiagnosed as a low-grade PCa. However, atrophy lacks nuclear and nucleolar enlargement. The spectrum of atrophic conditions can be divided into those conditions related or not to pharmacologic or surgical androgen withdrawal. The first, relatively diffuse throughout the gland, occurs in response to androgen withdrawal or androgen receptor blockade, and is associated to elevated levels of apoptosis (Gleave et al. 1996; Matsushima et al. 1999; Szende et al. 1999). The second, usually focal, is very common and divided into simple atrophy and post-atrophic hyperplasia (Fig. 1.15) (Ruska et al. 1998). It also includes the partial atrophy and the cystic atrophy (simple atrophy with cyst formation) (De Marzo et al. 2006). These lesions usually occur in the setting of chronic and often acute inflammation, and therefore referred as proliferative inflammatory atrophy by some authors that proposed this entity as a lesion from which early prostate carcinogenesis may develop.

1.7.3 Sclerosing Adenosis

The sclerosing adenosis is an incidental, but not rare, finding in TURP or prostatectomy, generally involving the transition zone of the prostate. It consists in a well demarcated proliferation of tightly packed glands and even single cells with signet ring cell-like features in a dense spindle cell stroma (Jones et al. 1991; Sakamoto et al. 1991; Grignon et al. 1992; Luque et al. 2003). It simulates a high grade PCa, but the presence of a nodular circumscription, the lack of cell atypia and the presence of a variably thickened basal cell layer and basal cell markers suggest a diagnosis of sclerosing adenosis. Cheng and Bostwick (Cheng and Bostwick 2010) showed five cases of atypical sclerosing adenosis, differing from normal by the presence of the large nuclear size, macronucleoli and aneuploid DNA in most of cases. Both lesions are considered benign and don't require different therapeutic approach.

1.7.4 Radiation Atypia in Benign Prostatic Glands

Radiation induced atypia is characterized by few small to medium size glands, often atrophic and with nuclear atypia (nuclear enlargement and prominent nucleoli), within a predominant fibrocellular stroma. These are frequently encountered in the peripheral zone of the prostate, in follow-up biopsy or prostatectomy following bachy- or radiotherapy for PCa (Magi-Galluzzi et al. 2003b; Bostwick and Meiers 2007).

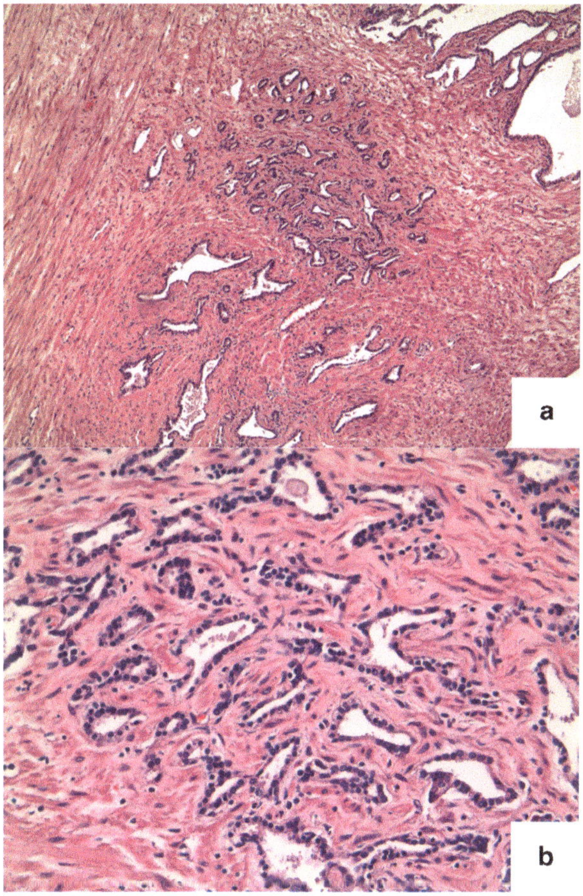

Fig. 1.15 Post-atrophic hyperplasia: (**a**, **b**) haematoxylin eosin staining (**a**: low magnification; **b**: high magnification)

1.8 Conclusion

Despite the advances of molecular pathology, histopathology remains a corner stone of PCa diagnosis and management. Before making a PCa final diagnosis, it is critical to consider the clinic-laboratory and imaging findings, the multiple features on histological examination with haematoxylin and eosin, the wide range of differential diagnosis and the overlapping staining immunohistochemical reactions. In fact, only an appropriate use of clinicopathological features provides a correct diagnostic and therapeutic management of PCa patient, positively contributing to the reduction of PCa mortality rates. The increasing use of novel technologies (imaging, immunohistochemistry, molecular methods) and their incorporation into the clinical practise may led to an improvement of the accuracy of prognostic and predictive systems in order to achieve the goal of individualized therapy making the pathologists the driving force behind the implementation of molecular method in the diagnostic setting.

References

Abdel-Khalek M, El-Baz M, el Ibrahiem H (2004) Predictors of prostate cancer on extended biopsy in patients with high-grade prostatic intraepithelial neoplasia: a multivariate analysis model. BJU Int 94:528–533

Algaba F, Epstein JI, Aldape HC, Farrow GM, Lopez-Beltran A, Maksem J et al (1996) Assessment of prostate carcinoma in core needle biopsy – definition of minimal criteria for the diagnosis of cancer in biopsy material. Cancer 78:376–381

Ali TZ, Epstein JI (2005) Perineural involvement by benign prostatic glands on needle biopsy. Am J Surg Pathol 29:1159–1163

Amin M, Boccon-Gibod L, Egevad L, Epstein JI, Humphrey PA, Mikuz G et al (2005) Prognostic and predictive factors and reporting of prostate carcinoma in prostate needle biopsy specimens. Scand J Urol Nephrol Suppl 20–33

Ananthanarayanan V, Deaton RJ, Yang XJ, Pins MR, Gann PH (2006) Alteration of proliferation and apoptotic markers in normal and premalignant tissue associated with prostate cancer. BMC Cancer 6:73

Andreoiu M, Cheng L (2010) Multifocal prostate cancer: biologic, prognostic, and therapeutic implications. Hum Pathol 41:781–793

Andriole GL, Crawford ED, Grubb RL 3rd, Buys SS, Chia D, Church TR et al (2009) Mortality results from a randomized prostate-cancer screening trial. N Engl J Med 360:1310–1319

Auprich M, Bjartell A, Chun FK, de la Taille A, Freedland SJ, Haese A et al (2011) Contemporary role of prostate cancer antigen 3 in the management of prostate cancer. Eur Urol 60: 1045–1054

Ayala AG, Ro JY (2007) Prostatic intraepithelial neoplasia: recent advances. Arch Pathol Lab Med 131:1257–1266

Aydin H, Zhou M, Herawi M, Epstein JI (2005) Number and location of nucleoli and presence of apoptotic bodies in diagnostically challenging cases of prostate adenocarcinoma on needle biopsy. Hum Pathol 36:1172–1177

Baisden BL, Kahane H, Epstein JI (1999) Perineural invasion, mucinous fibroplasia, and glomerulations: diagnostic features of limited cancer on prostate needle biopsy. Am J Surg Pathol 23:918–924

Basler JW, Thompson IM (1998) Lest we abandon digital rectal examination as a screening test for prostate cancer. J Natl Cancer Inst 90:1761–1763

Bates AW, Baithun SI (2002) Secondary solid neoplasms of the prostate: a clinico-pathological series of 51 cases. Virchows Arch 440:392–396

Beach R, Gown AM, De Peralta-Venturina MN, Folpe AL, Yaziji H, Salles PG et al (2002) P504S immunohistochemical detection in 405 prostatic specimens including 376 18-gauge needle biopsies. Am J Surg Pathol 26:1588–1596

Berruti A, Bollito E, Cracco CM, Volante M, Ciccone G, Porpiglia F et al (2010) The prognostic role of immunohistochemical chromogranin a expression in prostate cancer patients is significantly modified by androgen-deprivation therapy. Prostate 70:718–726

Bethel CR, Faith D, Li X, Guan B, Hicks JL, Lan F et al (2006) Decreased NKX3.1 protein expression in focal prostatic atrophy, prostatic intraepithelial neoplasia, and adenocarcinoma: association with gleason score and chromosome 8p deletion. Cancer Res 66:10683–10690

Bill-Axelson A, Holmberg L, Ruutu M, Garmo H, Stark JR, Busch C et al (2011) Radical prostatectomy versus watchful waiting in early prostate cancer. N Engl J Med 364:1708–1717

Bishara T, Ramnani DM, Epstein JI (2004) High-grade prostatic intraepithelial neoplasia on needle biopsy: risk of cancer on repeat biopsy related to number of involved cores and morphologic pattern. Am J Surg Pathol 28:629–633
Bishop MC, Hardy JG, Taylor MC, Wastie ML, Lemberger RJ (1985) Bone imaging and serum phosphatases in prostatic carcinoma. Br J Urol 57:317–324
Bjartell A, Montironi R, Berney DM, Egevad L (2011) Tumour markers in prostate cancer II: diagnostic and prognostic cellular biomarkers. Acta Oncol 50(Suppl 1):76–84
Borboroglu PG, Comer SW, Riffenburgh RH, Amling CL (2000) Extensive repeat transrectal ultrasound guided prostate biopsy in patients with previous benign sextant biopsies. J Urol 163:158–162
Bostwick DG, Brawer MK (1987) Prostatic intra-epithelial neoplasia and early invasion in prostate cancer. Cancer 59:788–794
Bostwick DG, Cheng L (2012) Precursors of prostate cancer. Histopathology 60:4–27
Bostwick DG, Meiers I (2007) Diagnosis of prostatic carcinoma after therapy. Arch Pathol Lab Med 131:360–371
Bostwick DG, Amin MB, Dundore P, Marsh W, Schultz DS (1993a) Architectural patterns of high-grade prostatic intraepithelial neoplasia. Hum Pathol 24:298–310
Bostwick DG, Srigley J, Grignon D, Maksem J, Humphrey P, van der Kwast TH et al (1993b) Atypical adenomatous hyperplasia of the prostate: morphologic criteria for its distinction from well-differentiated carcinoma. Hum Pathol 24:819–832
Bostwick DG, Pacelli A, Blute M, Roche P, Murphy GP (1998) Prostate specific membrane antigen expression in prostatic intraepithelial neoplasia and adenocarcinoma: a study of 184 cases. Cancer 82:2256–2261
Bostwick DG, Grignon DJ, Hammond ME, Amin MB, Cohen M, Crawford D et al (2000) Prognostic factors in prostate cancer. College of American Pathologists Consensus Statement 1999. Arch Pathol Lab Med 124:995–1000
Bostwick DG, Liu L, Brawer MK, Qian J (2004) High-grade prostatic intraepithelial neoplasia. Rev Urol 6:171–179
Bowen C, Bubendorf L, Voeller HJ, Slack R, Willi N, Sauter G et al (2000) Loss of NKX3.1 expression in human prostate cancers correlates with tumor progression. Cancer Res 60:6111–6115
Brausi M, Castagnetti G, Dotti A, De Luca G, Olmi R, Cesinaro AM (2004) Immediate radical prostatectomy in patients with atypical small acinar proliferation. Over treatment? J Urol 172:906–908; discussion 8–9
Brawer MK, Peehl DM, Stamey TA, Bostwick DG (1985) Keratin immunoreactivity in the benign and neoplastic human prostate. Cancer Res 45:3663–3667
Center MM, Jemal A, Lortet-Tieulent J, Ward E, Ferlay J, Brawley O et al (2012) International variation in prostate cancer incidence and mortality rates. Eur Urol 61:1079–1092, Epub 2012 Mar 8
Chappell B, McLoughlin J (2005a) Technical considerations when obtaining and interpreting prostatic biopsies from men with suspicion of early prostate cancer: part 2. BJU Int 95:1141–1145
Chappell B, McLoughlin J (2005b) Technical considerations when obtaining and interpreting prostatic biopsies from men with suspicion of early prostate cancer: part I. BJU Int 95:1135–1140
Chavin G, Sheinin Y, Crispen PL, Boorjian SA, Roth TJ, Rangel L et al (2009) Expression of immunosuppresive B7-H3 ligand by hormone-treated prostate cancer tumors and metastases. Clin Cancer Res 15:2174–2180
Che M, Sakr W, Grignon D (2003) Pathologic features the urologist should expect on a prostate biopsy. Urol Oncol 21:153–161
Chen H, Toyooka S, Gazdar AF, Hsieh JT (2003) Epigenetic regulation of a novel tumor suppressor gene (hDAB2IP) in prostate cancer cell lines. J Biol Chem 278:3121–3130
Cheng L, Bostwick DG (2010) Atypical sclerosing adenosis of the prostate: a rare mimic of adenocarcinoma. Histopathology 56:627–631
Cheng L, Shan A, Cheville JC, Qian J, Bostwick DG (1998) Atypical adenomatous hyperplasia of the prostate: a premalignant lesion? Cancer Res 58:389–391

Cheng L, Darson MF, Bergstralh EJ, Slezak J, Myers RP, Bostwick DG (1999) Correlation of margin status and extraprostatic extension with progression of prostate carcinoma. Cancer 86:1775–1782

Cheng L, Jones TD, Pan CX, Barbarin A, Eble JN, Koch MO (2005) Anatomic distribution and pathologic characterization of small-volume prostate cancer (<0.5 ml) in whole-mount prostatectomy specimens. Mod Pathol 18:1022–1026

Cheng L, Montironi R, Bostwick DG, Lopez-Beltran A, Berney DM (2012) Staging of prostate cancer. Histopathology 60:87–117

Cheville JC, Reznicek MJ, Bostwick DG (1997) The focus of "atypical glands, suspicious for malignancy" in prostatic needle biopsy specimens: incidence, histologic features, and clinical follow-up of cases diagnosed in a community practice. Am J Clin Pathol 108:633–640

Chin AI, Dave DS, Rajfer J (2007) Is repeat biopsy for isolated high-grade prostatic intraepithelial neoplasia necessary? Rev Urol 9:124–131

Chuang AY, Epstein JI (2008) Positive surgical margins in areas of capsular incision in otherwise organ-confined disease at radical prostatectomy: histologic features and pitfalls. Am J Surg Pathol 32:1201–1206

Chuang AY, DeMarzo AM, Veltri RW, Sharma RB, Bieberich CJ, Epstein JI (2007) Immunohistochemical differentiation of high-grade prostate carcinoma from urothelial carcinoma. Am J Surg Pathol 31:1246–1255

Chun FK, Epstein JI, Ficarra V, Freedland SJ, Montironi R, Montorsi F et al (2010) Optimizing performance and interpretation of prostate biopsy: a critical analysis of the literature. Eur Urol 58:851–864

de Kok JB, Verhaegh GW, Roelofs RW, Hessels D, Kiemeney LA, Aalders TW et al (2002) DD3(PCA3), a very sensitive and specific marker to detect prostate tumors. Cancer Res 62:2695–2698

De Marzo AM, Platz EA, Epstein JI, Ali T, Billis A, Chan TY et al (2006) A working group classification of focal prostate atrophy lesions. Am J Surg Pathol 30:1281–1291

Deftos LJ (1998) Granin-A, parathyroid hormone-related protein, and calcitonin gene products in neuroendocrine prostate cancer. Prostate Suppl 8:23–31

Delahunt B, Miller RJ, Srigley JR, Evans AJ, Samaratunga H (2012) Gleason grading: past, present and future. Histopathology 60:75–86

Derweesh IH, Kupelian PA, Zippe C, Levin HS, Brainard J, Magi-Galluzzi C et al (2004) Continuing trends in pathological stage migration in radical prostatectomy specimens. Urol Oncol 22:300–306

Dovey Z, Corbishley CM, Kirby RS (2005) Prostatic intraepithelial neoplasia: a risk factor for prostate cancer. Can J Urol 12(Suppl 1):49–52; discussion 99–100

Eble JNSG, Epstein JI, Sesterhenn IA (2004) Tumours of the prostate. In: Eble JNSG, Epstein JI, Sester IA (eds) Tumours of the urinary system and male genital organs. IARC, Lyon, pp 159–214

Edge SBD, Compton CC et al (2010) American Joint Committee on cancer staging manual, 7th edn. Springer, New York

Eichler K, Hempel S, Wilby J, Myers L, Bachmann LM, Kleijnen J (2006) Diagnostic value of systematic biopsy methods in the investigation of prostate cancer: a systematic review. J Urol 175:1605–1612

El-Hakim A, Moussa S (2010) CUA guidelines on prostate biopsy methodology. Can Urol Assoc J 4:89–94

Epstein JI (1990) Evaluation of radical prostatectomy capsular margins of resection. The significance of margins designated as negative, closely approaching, and positive. Am J Surg Pathol 14:626–632

Epstein JI (1995a) Adenosis (atypical adenomatous hyperplasia): histopathology and relationship to carcinoma. Pathol Res Pract 191:888–898

Epstein JI (1995b) Diagnostic criteria of limited adenocarcinoma of the prostate on needle biopsy. Hum Pathol 26:223–229

Epstein JI (2004) Diagnosis and reporting of limited adenocarcinoma of the prostate on needle biopsy. Mod Pathol 17:307–315

Epstein JI (2009) Precursor lesions to prostatic adenocarcinoma. Virchows Arch 454:1–16

Epstein JI, Fynheer J (1992) Acidic mucin in the prostate: can it differentiate adenosis from adenocarcinoma? Hum Pathol 23:1321–1325

Epstein JI, Herawi M (2006) Prostate needle biopsies containing prostatic intraepithelial neoplasia or atypical foci suspicious for carcinoma: implications for patient care. J Urol 175:820–834

Epstein JI, Sauvageot J (1997) Do close but negative margins in radical prostatectomy specimens increase the risk of postoperative progression? J Urol 157:241–243

Epstein JI, Partin AW, Sauvageot J, Walsh PC (1996) Prediction of progression following radical prostatectomy. A multivariate analysis of 721 men with long-term follow-up. Am J Surg Pathol 20:286–292

Epstein JI, Allsbrook WC Jr, Amin MB, Egevad LL (2005a) The 2005 International Society of Urological Pathology (ISUP) consensus conference on Gleason grading of prostatic carcinoma. Am J Surg Pathol 29:1228–1242

Epstein JI, Amin M, Boccon-Gibod L, Egevad L, Humphrey PA, Mikuz G et al (2005b) Prognostic factors and reporting of prostate carcinoma in radical prostatectomy and pelvic lymphadenectomy specimens. Scand J Urol Nephrol Suppl 34–63

Fadare O, Wang S, Mariappan MR (2004) Practice patterns of clinicians following isolated diagnoses of atypical small acinar proliferation on prostate biopsy specimens. Arch Pathol Lab Med 128:557–560

Ferlay J, Shin HR, Bray F, Forman D, Mathers C, Parkin DM (2010) Estimates of worldwide burden of cancer in 2008: GLOBOCAN 2008. Int J Cancer 127:2893–2917

Fine SW, Reuter VE (2012) Anatomy of the prostate revisited: implications for prostate biopsy and zonal origins of prostate cancer. Histopathology 60:142–152

Freeman MR, Yang W, Di Vizio D (2012) Caveolin-1 and prostate cancer progression. Adv Exp Med Biol 729:95–110

Gaudin PB, Epstein JI (1995) Adenosis of the prostate. Histologic features in needle biopsy specimens. Am J Surg Pathol 19:737–747

Gelmann EP, Bowen C, Bubendorf L (2003) Expression of NKX3.1 in normal and malignant tissues. Prostate 55:111–117

Girasole CR, Cookson MS, Putzi MJ, Chang SS, Smith JA Jr, Wells N et al (2006) Significance of atypical and suspicious small acinar proliferations, and high grade prostatic intraepithelial neoplasia on prostate biopsy: implications for cancer detection and biopsy strategy. J Urol 175:929–933; discussion 33

Gleave ME, Goldenberg SL, Jones EC, Bruchovsky N, Sullivan LD (1996) Biochemical and pathological effects of 8 months of neoadjuvant androgen withdrawal therapy before radical prostatectomy in patients with clinically confined prostate cancer. J Urol 155:213–219

Godoy G, Taneja SS (2008) Contemporary clinical management of isolated high-grade prostatic intraepithelial neoplasia. Prostate Cancer Prostatic Dis 11:20–31

Gokden N, Roehl KA, Catalona WJ, Humphrey PA (2005) High-grade prostatic intraepithelial neoplasia in needle biopsy as risk factor for detection of adenocarcinoma: current level of risk in screening population. Urology 65:538–542

Grignon DJ, Ro JY, Srigley JR, Troncoso P, Raymond AK, Ayala AG (1992) Sclerosing adenosis of the prostate gland. A lesion showing myoepithelial differentiation. Am J Surg Pathol 16:383–391

Gronberg H (2003) Prostate cancer epidemiology. Lancet 361:859–864

Gupta C, Ren JZ, Wojno KJ (2004) Individual submission and embedding of prostate biopsies decreases rates of equivocal pathology reports. Urology 63:83–86

Gutman AB, Gutman EB (1938) An "acid" phosphatase occurring in the serum of patients with metastasizing carcinoma of the prostate gland. J Clin Invest 17:473–478

Haggman MJ, Macoska JA, Wojno KJ, Oesterling JE (1997) The relationship between prostatic intraepithelial neoplasia and prostate cancer: critical issues. J Urol 158:12–22

Hameed O, Humphrey PA (2010) Pseudoneoplastic mimics of prostate and bladder carcinomas. Arch Pathol Lab Med 134:427–443

Han B, Mehra R, Suleman K, Tomlins SA, Wang L, Singhal N et al (2009) Characterization of ETS gene aberrations in select histologic variants of prostate carcinoma. Mod Pathol 22:1176–1185

Hara R, Jo Y, Fujii T, Kondo N, Yokoyoma T, Miyaji Y et al (2008) Optimal approach for prostate cancer detection as initial biopsy: prospective randomized study comparing transperineal versus transrectal systematic 12-core biopsy. Urology 71:191–195

Hasegawa M, Hagiwara S, Sato T, Jijiwa M, Murakumo Y, Maeda M et al (2007) CD109, a new marker for myoepithelial cells of mammary, salivary, and lacrimal glands and prostate basal cells. Pathol Int 57:245–250

Haussler O, Epstein JI, Amin MB, Heitz PU, Hailemariam S (1999) Cell proliferation, apoptosis, oncogene, and tumor suppressor gene status in adenosis with comparison to benign prostatic hyperplasia, prostatic intraepithelial neoplasia, and cancer. Hum Pathol 30:1077–1086

Heidenreich A, Bellmunt J, Bolla M, Joniau S, Mason M, Matveev V et al (2011) EAU guidelines on prostate cancer. Part I: screening, diagnosis, and treatment of clinically localised disease. Actas Urol Esp 35:501–514

Helpap B (1988) Observations on the number, size and localization of nucleoli in hyperplastic and neoplastic prostatic disease. Histopathology 13:203–211

Helpap B, Bonkhoff H, Cockett A, Montironi R, Troncoso P, Waters D et al (1997) Relationship between atypical adenomatous hyperplasia (AAH), prostatic intraepithelial neoplasia (PIN) and prostatic adenocarcinoma. Pathologica 89:288–300

Henneberry JM, Kahane H, Humphrey PA, Keetch DW, Epstein JI (1997) The significance of intraluminal crystalloids in benign prostatic glands on needle biopsy. Am J Surg Pathol 21:725–728

Henson DE, Hutter RV, Farrow G (1994) Practice protocol for the examination of specimens removed from patients with carcinoma of the prostate gland. A publication of the Cancer Committee, College of American Pathologists. Task force on the examination of specimens removed from patients with prostate cancer. Arch Pathol Lab Med 118:779–783

Herawi M, Parwani AV, Irie J, Epstein JI (2005) Small glandular proliferations on needle biopsies: most common benign mimickers of prostatic adenocarcinoma sent in for expert second opinion. Am J Surg Pathol 29:874–880

Herawi M, Kahane H, Cavallo C, Epstein JI (2006) Risk of prostate cancer on first re-biopsy within 1 year following a diagnosis of high grade prostatic intraepithelial neoplasia is related to the number of cores sampled. J Urol 175:121–124

Hessels D, Schalken JA (2009) The use of PCA3 in the diagnosis of prostate cancer. Nat Rev Urol 6:255–261

Hodge KK, McNeal JE, Terris MK, Stamey TA (1989) Random systematic versus directed ultrasound guided transrectal core biopsies of the prostate. J Urol 142:71–74; discussion 4–5

Humphrey PA (2007) Diagnosis of adenocarcinoma in prostate needle biopsy tissue. J Clin Pathol 60:35–42

Humphrey PA (2012) Histological variants of prostatic carcinoma and their significance. Histopathology 60:59–74

Iczkowski KA (2006) Current prostate biopsy interpretation: criteria for cancer, atypical small acinar proliferation, high-grade prostatic intraepithelial neoplasia, and use of immunostains. Arch Pathol Lab Med 130:835–843

Iczkowski KA, MacLennan GT, Bostwick DG (1997) Atypical small acinar proliferation suspicious for malignancy in prostate needle biopsies: clinical significance in 33 cases. Am J Surg Pathol 21:1489–1495

Iczkowski KA, Casella G, Seppala RJ, Jones GL, Mishler BA, Qian J et al (2002a) Needle core length in sextant biopsy influences prostate cancer detection rate. Urology 59:698–703

Iczkowski KA, Chen HM, Yang XJ, Beach RA (2002b) Prostate cancer diagnosed after initial biopsy with atypical small acinar proliferation suspicious for malignancy is similar to cancer found on initial biopsy. Urology 60:851–854

Jacobsen SJ, Bergstralh EJ, Katusic SK, Guess HA, Darby CH, Silverstein MD et al (1998) Screening digital rectal examination and prostate cancer mortality: a population-based case–control study. Urology 52:173–179

Jani AB, Vaida F, Hanks G, Asbell S, Sartor O, Moul JW et al (2001) Changing face and different countenances of prostate cancer: racial and geographic differences in prostate-specific antigen (PSA), stage, and grade trends in the PSA era. Int J Cancer 96:363–371

Jarrard DF, Paul R, van Bokhoven A, Nguyen SH, Bova GS, Wheelock MJ et al (1997) P-Cadherin is a basal cell-specific epithelial marker that is not expressed in prostate cancer. Clin Cancer Res 3:2121–2128

Jiang Z, Wu CL, Woda BA, Dresser K, Xu J, Fanger GR et al (2002) P504S/alpha-methylacyl-CoA racemase: a useful marker for diagnosis of small foci of prostatic carcinoma on needle biopsy. Am J Surg Pathol 26:1169–1174

Jones EC, Clement PB, Young RH (1991) Sclerosing adenosis of the prostate gland. A clinicopathological and immunohistochemical study of 11 cases. Am J Surg Pathol 15:1171–1180

Joniau S, Goeman L, Pennings J, Van Poppel H (2005) Prostatic intraepithelial neoplasia (PIN): importance and clinical management. Eur Urol 48:379–385

Kahane H, Sharp JW, Shuman GB, Dasilva G, Epstein JI (1995) Utilization of high molecular weight cytokeratin on prostate needle biopsies in an independent laboratory. Urology 45:981–986

Kalos M, Askaa J, Hylander BL, Repasky EA, Cai F, Vedvick T et al (2004) Prostein expression is highly restricted to normal and malignant prostate tissues. Prostate 60:246–256

Kamiya N, Suzuki H, Kawamura K, Imamoto T, Naya Y, Tochigi N et al (2008) Neuroendocrine differentiation in stage D2 prostate cancers. Int J Urol 15:423–428

Karram S, Trock BJ, Netto GJ, Epstein JI (2011) Should intervening benign tissue be included in the measurement of discontinuous foci of cancer on prostate needle biopsy? Correlation with radical prostatectomy findings. Am J Surg Pathol 35:1351–1355

Kausik SJ, Blute ML, Sebo TJ, Leibovich BC, Bergstralh EJ, Slezak J et al (2002) Prognostic significance of positive surgical margins in patients with extraprostatic carcinoma after radical prostatectomy. Cancer 95:1215–1219

Kheirandish P, Chinegwundoh F (2011) Ethnic differences in prostate cancer. Br J Cancer 105:481–485

Kikuchi E, Scardino PT, Wheeler TM, Slawin KM, Ohori M (2004) Is tumor volume an independent prognostic factor in clinically localized prostate cancer? J Urol 172:508–511

Kolonel LN, Altshuler D, Henderson BE (2004) The multiethnic cohort study: exploring genes, lifestyle and cancer risk. Nat Rev Cancer 4:519–527

Kristiansen G (2012) Diagnostic and prognostic molecular biomarkers for prostate cancer. Histopathology 60:125–141

Kristiansen G, Fritzsche FR, Wassermann K, Jager C, Tolls A, Lein M et al (2008) GOLPH2 protein expression as a novel tissue biomarker for prostate cancer: implications for tissue-based diagnostics. Br J Cancer 99:939–948

Kuefer R, Varambally S, Zhou M, Lucas PC, Loeffler M, Wolter H et al (2002) Alpha-methylacyl-CoA racemase: expression levels of this novel cancer biomarker depend on tumor differentiation. Am J Pathol 161:841–848

Kumaresan K, Kakkar N, Verma A, Mandal AK, Singh SK, Joshi K (2010) Diagnostic utility of alpha-methylacyl CoA racemase (P504S) & HMWCK in morphologically difficult prostate cancer. Diagn Pathol 5:83

Kumar-Sinha C, Tomlins SA, Chinnaiyan AM (2008) Recurrent gene fusions in prostate cancer. Nat Rev Cancer 8:497–511

Kunju LP, Rubin MA, Chinnaiyan AM, Shah RB (2003) Diagnostic usefulness of monoclonal antibody P504S in the workup of atypical prostatic glandular proliferations. Am J Clin Pathol 120:737–745

Kunju LP, Mehra R, Snyder M, Shah RB (2006) Prostate-specific antigen, high-molecular-weight cytokeratin (clone 34betaE12), and/or p63: an optimal immunohistochemical panel to distinguish poorly differentiated prostate adenocarcinoma from urothelial carcinoma. Am J Clin Pathol 125:675–681

Kuroda N, Katto K, Tamura M, Shiotsu T, Nakamura S, Ohtsuki Y et al (2010) Immunohistochemical application of D2-40 as basal cell marker in evaluating atypical small acinar proliferation of initial routine prostatic needle biopsy materials. Med Mol Morphol 43:165–169

Lapointe J, Li C, Higgins JP, van de Rijn M, Bair E, Montgomery K et al (2004) Gene expression profiling identifies clinically relevant subtypes of prostate cancer. Proc Natl Acad Sci USA 101:811–816

Laxman B, Morris DS, Yu J, Siddiqui J, Cao J, Mehra R et al (2008) A first-generation multiplex biomarker analysis of urine for the early detection of prostate cancer. Cancer Res 68:645–649

Leite KR, Mitteldorf CA, Srougi M, Dall'oglio MF, Antunes AA, Pontes J Jr et al (2008a) Cdx2, cytokeratin 20, thyroid transcription factor 1, and prostate-specific antigen expression in unusual subtypes of prostate cancer. Ann Diagn Pathol 12:260–266

Leite KR, Srougi M, Dall'Oglio MF, Sanudo A, Camara-Lopes LH (2008b) Histopathological findings in extended prostate biopsy with PSA < or = 4 ng/mL. Int Braz J Urol 34:283–290; discussion 90–92

Lin X, Tascilar M, Lee WH, Vles WJ, Lee BH, Veeraswamy R et al (2001) GSTP1 CpG island hypermethylation is responsible for the absence of GSTP1 expression in human prostate cancer cells. Am J Pathol 159:1815–1826

Loeb S, Roehl KA, Yu X, Han M, Catalona WJ (2007) Use of prostate-specific antigen velocity to follow up patients with isolated high-grade prostatic intraepithelial neoplasia on prostate biopsy. Urology 69:108–112

Lotan TL, Epstein JI (2008) Diffuse adenosis of the peripheral zone in prostate needle biopsy and prostatectomy specimens. Am J Surg Pathol 32:1360–1366

Luo J, Zha S, Gage WR, Dunn TA, Hicks JL, Bennett CJ et al (2002) Alpha-methylacyl-CoA racemase: a new molecular marker for prostate cancer. Cancer Res 62:2220–2226

Luque RJ, Lopez-Beltran A, Perez-Seoane C, Suzigan S (2003) Sclerosing adenosis of the prostate. Histologic features in needle biopsy specimens. Arch Pathol Lab Med 127:e14–e16

Madu CO, Lu Y (2010) Novel diagnostic biomarkers for prostate cancer. J Cancer 1:150–177

Magi-Galluzzi C, Luo J, Isaacs WB, Hicks JL, de Marzo AM, Epstein JI (2003a) Alpha-methylacyl-CoA racemase: a variably sensitive immunohistochemical marker for the diagnosis of small prostate cancer foci on needle biopsy. Am J Surg Pathol 27:1128–1133

Magi-Galluzzi C, Sanderson H, Epstein JI (2003b) Atypia in nonneoplastic prostate glands after radiotherapy for prostate cancer: duration of atypia and relation to type of radiotherapy. Am J Surg Pathol 27:206–212

Mahon SM (2005) Screening for prostate cancer: informing men about their options. Clin J Oncol Nurs 9:625–627

Marks RA, Koch MO, Lopez-Beltran A, Montironi R, Juliar BE, Cheng L (2007) The relationship between the extent of surgical margin positivity and prostate specific antigen recurrence in radical prostatectomy specimens. Hum Pathol 38:1207–1211

Matsushima H, Goto T, Hosaka Y, Kitamura T, Kawabe K (1999) Correlation between proliferation, apoptosis, and angiogenesis in prostate carcinoma and their relation to androgen ablation. Cancer 85:1822–1827

McNeal JE (1969) Origin and development of carcinoma in the prostate. Cancer 23:24–34

McNeal JE, Alroy J, Villers A, Redwine EA, Freiha FS, Stamey TA (1991) Mucinous differentiation in prostatic adenocarcinoma. Hum Pathol 22:979–988

Meyer F, Tetu B, Bairati I, Lacombe L, Fradet Y (2006) Prostatic intraepithelial neoplasia in TURP specimens and subsequent prostate cancer. Can J Urol 13:3255–3260

Miettinen M, Wang ZF, Paetau A, Tan SH, Dobi A, Srivastava S et al (2011) ERG transcription factor as an immunohistochemical marker for vascular endothelial tumors and prostatic carcinoma. Am J Surg Pathol 35:432–441

Miller JS, Chen Y, Ye H, Robinson BD, Brimo F, Epstein JI (2010) Extraprostatic extension of prostatic adenocarcinoma on needle core biopsy: report of 72 cases with clinical follow-up. BJU Int 106:330–333

Montironi R, Mazzucchelli R, Lopez-Beltran A, Cheng L, Scarpelli M (2007) Mechanisms of disease: high-grade prostatic intraepithelial neoplasia and other proposed preneoplastic lesions in the prostate. Nat Clin Pract Urol 4:321–332

Montironi R, Egevad L, Bjartell A, Berney DM (2011a) Role of histopathology and molecular markers in the active surveillance of prostate cancer. Acta Oncol 50(Suppl 1):56–60

Montironi R, Mazzucchelli R, Lopez-Beltran A, Scarpelli M, Cheng L (2011b) Prostatic intraepithelial neoplasia: its morphological and molecular diagnosis and clinical significance. BJU Int 108:1394–1401

Montironi R, Scarpelli M, Mazzucchelli R, Cheng L, Lopez-Beltran A (2012a) The spectrum of morphology in non-neoplastic prostate including cancer mimics. Histopathology 60:41–58

Montironi R, Scarpelli M, Mazzucchelli R, Cheng L, Lopez-Beltran A, Montorsi F (2012b) Extent of cancer of less than 50% in any prostate needle biopsy core: how many millimeters are there? Eur Urol 61:751–756

Moore CK, Karikehalli S, Nazeer T, Fisher HA, Kaufman RP Jr, Mian BM (2005) Prognostic significance of high grade prostatic intraepithelial neoplasia and atypical small acinar proliferation in the contemporary era. J Urol 173:70–72

Moore AL, Dimitropoulou P, Lane A, Powell PH, Greenberg DC, Brown CH et al (2009) Population-based prostate-specific antigen testing in the UK leads to a stage migration of prostate cancer. BJU Int 104:1592–1598

Mosquera JM, Perner S, Genega EM, Sanda M, Hofer MD, Mertz KD et al (2008) Characterization of TMPRSS2-ERG fusion high-grade prostatic intraepithelial neoplasia and potential clinical implications. Clin Cancer Res 14:3380–3385

Naya Y, Ochiai A, Troncoso P, Babaian RJ (2004) A comparison of extended biopsy and sextant biopsy schemes for predicting the pathological stage of prostate cancer. J Urol 171:2203–2208

Netto GJ, Epstein JI (2006) Widespread high-grade prostatic intraepithelial neoplasia on prostatic needle biopsy: a significant likelihood of subsequently diagnosed adenocarcinoma. Am J Surg Pathol 30:1184–1188

Noldus J, Graefen M, Haese A, Henke RP, Hammerer P, Huland H (2000) Stage migration in clinically localized prostate cancer. Eur Urol 38:74–78

Novis DA, Zarbo RJ, Valenstein PA (1999) Diagnostic uncertainty expressed in prostate needle biopsies. A College of American Pathologists Q-probes Study of 15,753 prostate needle biopsies in 332 institutions. Arch Pathol Lab Med 123:687–692

Orandi A (1985) Transurethral incision of prostate (TUIP): 646 cases in 15 years – a chronological appraisal. Br J Urol 57:703–707

Osunkoya AO, Netto GJ, Epstein JI (2007) Colorectal adenocarcinoma involving the prostate: report of 9 cases. Hum Pathol 38:1836–1841

Ouyang RC, Kenwright DN, Nacey JN, Delahunt B (2001) The presence of atypical small acinar proliferation in prostate needle biopsy is predictive of carcinoma on subsequent biopsy. BJU Int 87:70–74

Pacelli A, Lopez-Beltran A, Egan AJ, Bostwick DG (1998) Prostatic adenocarcinoma with glomeruloid features. Hum Pathol 29:543–546

Paner GP, Luthringer DJ, Amin MB (2008) Best practice in diagnostic immunohistochemistry: prostate carcinoma and its mimics in needle core biopsies. Arch Pathol Lab Med 132:1388–1396

Parkin DM, Nambooze S, Wabwire-Mangen F, Wabinga HR (2010) Changing cancer incidence in Kampala, Uganda, 1991–2006. Int J Cancer 126:1187–1195

Partin AW, Yoo J, Carter HB, Pearson JD, Chan DW, Epstein JI et al (1993) The use of prostate specific antigen, clinical stage and Gleason score to predict pathological stage in men with localized prostate cancer. J Urol 150:110–114

Paul B, Dhir R, Landsittel D, Hitchens MR, Getzenberg RH (2005) Detection of prostate cancer with a blood-based assay for early prostate cancer antigen. Cancer Res 65:4097–4100

Pierorazio PM, Lambert SM, Matsukhani M, Sprenkle PC, McCann TR, Katz AE et al (2007) High-grade prostatic intraepithelial neoplasia is an independent predictor of outcome after radical prostatectomy. BJU Int 100:1066–1070

Ploussard G, Durand X, Xylinas E, Moutereau S, Radulescu C, Forgue A et al (2011) Prostate cancer antigen 3 score accurately predicts tumour volume and might help in selecting prostate cancer patients for active surveillance. Eur Urol 59:422–429

Pomerantz MM, Werner L, Xie W, Regan MM, Lee GS, Sun T et al (2011) Association of prostate cancer risk Loci with disease aggressiveness and prostate cancer-specific mortality. Cancer Prev Res (Phila) 4:719–728

Qian J, Jenkins RB, Bostwick DG (1995) Chromosomal anomalies in atypical adenomatous hyperplasia and carcinoma of the prostate using fluorescence in situ hybridization. Urology 46:837–842

Quinn M, Babb P (2002) Patterns and trends in prostate cancer incidence, survival, prevalence and mortality. Part I: international comparisons. BJU Int 90:162–173

Rabets JC, Jones JS, Patel A, Zippe CD (2004) Prostate cancer detection with office based saturation biopsy in a repeat biopsy population. J Urol 172:94–97

Ramos Soler D, Mayordomo Aranda E, Calatayud Blas A, Rubio Briones J, Solsona Narbon E, Llombart Bosch A (2006) Usefulness of bcl-2 expression as a new basal cell marker in prostatic pathology. Actas Urol Esp 30:345–352

Reiter RE, Gu Z, Watabe T, Thomas G, Szigeti K, Davis E et al (1998) Prostate stem cell antigen: a cell surface marker overexpressed in prostate cancer. Proc Natl Acad Sci USA 95:1735–1740

Revelos K, Petraki C, Gregorakis A, Scorilas A, Papanastasiou P, Tenta R et al (2005) p27(kip1) and Ki-67 (MIB1) immunohistochemical expression in radical prostatectomy specimens of patients with clinically localized prostate cancer. In Vivo 19:911–920

Reyes AO, Swanson PE, Carbone JM, Humphrey PA (1997) Unusual histologic types of high-grade prostatic intraepithelial neoplasia. Am J Surg Pathol 21:1215–1222

Rigau M, Morote J, Mir MC, Ballesteros C, Ortega I, Sanchez A et al (2010) PSGR and PCA3 as biomarkers for the detection of prostate cancer in urine. Prostate 70:1760–1767

Ro JY, Ayala AG, Ordonez NG, Cartwright J Jr, Mackay B (1986) Intraluminal crystalloids in prostatic adenocarcinoma. Immunohistochemical, electron microscopic, and x-ray microanalytic studies. Cancer 57:2397–2407

Roscigno M, Scattoni V, Freschi M, Raber M, Colombo R, Bertini R et al (2004) Monofocal and plurifocal high-grade prostatic intraepithelial neoplasia on extended prostate biopsies: factors predicting cancer detection on extended repeat biopsy. Urology 63:1105–1110

Ross HM, Kryvenko ON, Cowan JE, Simko JP, Wheeler TM, Epstein JI (2012) Do adenocarcinomas of the prostate with Gleason score (GS) </=6 have the potential to metastasize to lymph nodes? Am J Surg Pathol 36:1346–1352

Roth TJ, Sheinin Y, Lohse CM, Kuntz SM, Frigola X, Inman BA et al (2007) B7-H3 ligand expression by prostate cancer: a novel marker of prognosis and potential target for therapy. Cancer Res 67:7893–7900

Rubin MA, Zhou M, Dhanasekaran SM, Varambally S, Barrette TR, Sanda MG et al (2002) Alpha-methylacyl coenzyme A racemase as a tissue biomarker for prostate cancer. JAMA 287:1662–1670

Ruska KM, Sauvageot J, Epstein JI (1998) Histology and cellular kinetics of prostatic atrophy. Am J Surg Pathol 22:1073–1077

Sakamoto N, Tsuneyoshi M, Enjoji M (1991) Sclerosing adenosis of the prostate. Histopathologic and immunohistochemical analysis. Am J Surg Pathol 15:660–667

Scattoni V, Roscigno M, Raber M, Deho F, Maga T, Zanoni M et al (2008) Initial extended transrectal prostate biopsy – are more prostate cancers detected with 18 cores than with 12 cores? J Urol 179:1327–1331; discussion 31

Schoenfield L, Jones JS, Zippe CD, Reuther AM, Klein E, Zhou M et al (2007) The incidence of high-grade prostatic intraepithelial neoplasia and atypical glands suspicious for carcinoma on first-time saturation needle biopsy, and the subsequent risk of cancer. BJU Int 99:770–774

Schroder FH, Hugosson J, Roobol MJ, Tammela TL, Ciatto S, Nelen V et al (2009) Screening and prostate-cancer mortality in a randomized European study. N Engl J Med 360:1320–1328

Shah RB, Kunju LP, Shen R, LeBlanc M, Zhou M, Rubin MA (2004) Usefulness of basal cell cocktail (34betaE12 + p63) in the diagnosis of atypical prostate glandular proliferations. Am J Clin Pathol 122:517–523

Shurbaji MS, Kalbfleisch JH, Thurmond TS (1996) Immunohistochemical detection of a fatty acid synthase (OA-519) as a predictor of progression of prostate cancer. Hum Pathol 27:917–921
Siegel R, Naishadham D, Jemal A (2012) Cancer statistics, 2012. CA Cancer J Clin 62:10–29
Singh H, Canto EI, Shariat SF, Kadmon D, Miles BJ, Wheeler TM et al (2004) Predictors of prostate cancer after initial negative systematic 12 core biopsy. J Urol 171:1850–1854
Sreekumar A, Poisson LM, Rajendiran TM, Khan AP, Cao Q, Yu J et al (2009) Metabolomic profiles delineate potential role for sarcosine in prostate cancer progression. Nature 457:910–914
Srigley JR (2004) Benign mimickers of prostatic adenocarcinoma. Mod Pathol 17:328–348
Srigley JR, Amin M, Boccon-Gibod L, Egevad L, Epstein JI, Humphrey PA et al (2005) Prognostic and predictive factors in prostate cancer: historical perspectives and recent international consensus initiatives. Scand J Urol Nephrol Suppl 216:8–19
Stamey TA (1995) Making the most out of six systematic sextant biopsies. Urology 45:2–12
Stamey TA, Yang N, Hay AR, McNeal JE, Freiha FS, Redwine E (1987) Prostate-specific antigen as a serum marker for adenocarcinoma of the prostate. N Engl J Med 317:909–916
Stamey TA, Yemoto CM, McNeal JE, Sigal BM, Johnstone IM (2000) Prostate cancer is highly predictable: a prognostic equation based on all morphological variables in radical prostatectomy specimens. J Urol 163:1155–1160
Szende B, Romics I, Torda I, Bely M, Szegedi Z, Lovasz S (1999) Apoptosis, mitosis, p53, bcl(2), Ki-67 and clinical outcome in prostate carcinoma treated by androgen ablation. Urol Int 63:115–119
Tahir SA, Yang G, Ebara S, Timme TL, Satoh T, Li L et al (2001) Secreted caveolin-1 stimulates cell survival/clonal growth and contributes to metastasis in androgen-insensitive prostate cancer. Cancer Res 61:3882–3885
Takenaka A, Hara R, Ishimura T, Fujii T, Jo Y, Nagai A et al (2008) A prospective randomized comparison of diagnostic efficacy between transperineal and transrectal 12-core prostate biopsy. Prostate Cancer Prostatic Dis 11:134–138
Thorson P, Vollmer RT, Arcangeli C, Keetch DW, Humphrey PA (1998) Minimal carcinoma in prostate needle biopsy specimens: diagnostic features and radical prostatectomy follow-up. Mod Pathol 11:543–551
Tischler V, Fritzsche FR, Gerhardt J, Jager C, Stephan C, Jung K et al (2010) Comparison of the diagnostic value of fatty acid synthase (FASN) with alpha-methylacyl-CoA racemase (AMACR) as prostatic cancer tissue marker. Histopathology 56:811–815
Tomlins SA, Aubin SM, Siddiqui J, Lonigro RJ, Sefton-Miller L, Miick S et al (2011) Urine TMPRSS2:ERG fusion transcript stratifies prostate cancer risk in men with elevated serum PSA. Sci Transl Med 3:94ra72
Totten RS, Heinemann MW, Hudson PB, Sproul EE, Stout AP (1953) Microscopic differential diagnosis of latent carcinoma of prostate. AMA Arch Pathol 55:131–141
Trock BJ, Guo CC, Gonzalgo ML, Magheli A, Loeb S, Epstein JI (2009) Tertiary Gleason patterns and biochemical recurrence after prostatectomy: proposal for a modified Gleason scoring system. J Urol 182:1364–1370
Ung JO, Richie JP, Chen MH, Renshaw AA, D'Amico AV (2002) Evolution of the presentation and pathologic and biochemical outcomes after radical prostatectomy for patients with clinically localized prostate cancer diagnosed during the PSA era. Urology 60:458–463
van Oort IM, Witjes JA, Kok DE, Kiemeney LA, Hulsbergen-vandeKaa CA (2008) Maximum tumor diameter is not an independent prognostic factor in high-risk localized prostate cancer. World J Urol 26:237–241
Varma M, Lee MW, Tamboli P, Zarbo RJ, Jimenez RE, Salles PG et al (2002) Morphologic criteria for the diagnosis of prostatic adenocarcinoma in needle biopsy specimens. A study of 250 consecutive cases in a routine surgical pathology practice. Arch Pathol Lab Med 126: 554–561
Veeramani S, Yuan TC, Chen SJ, Lin FF, Petersen JE, Shaheduzzaman S et al (2005) Cellular prostatic acid phosphatase: a protein tyrosine phosphatase involved in androgen-independent proliferation of prostate cancer. Endocr Relat Cancer 12:805–822
Vesalainen S, Lipponen P, Talja M, Syrjanen K (1995) Mitotic activity and prognosis in prostatic adenocarcinoma. Prostate 26:80–86

Vest D, Schalken JA, Muir G, Dasgupta P (2010) Transmembrane protease serine 2 in prostate cancer. BJU Int 105:1490–1492

Walz J, Graefen M, Chun FK, Erbersdobler A, Haese A, Steuber T et al (2006) High incidence of prostate cancer detected by saturation biopsy after previous negative biopsy series. Eur Urol 50:498–505

Went PT, Sauter G, Oberholzer M, Bubendorf L (2006) Abundant expression of AMACR in many distinct tumour types. Pathology 38:426–432

Wheeler TM, Dillioglugil O, Kattan MW, Arakawa A, Soh S, Suyama K et al (1998) Clinical and pathological significance of the level and extent of capsular invasion in clinical stage T1-2 prostate cancer. Hum Pathol 29:856–862

Williams H, Powell IJ (2009) Epidemiology, pathology, and genetics of prostate cancer among African Americans compared with other ethnicities. Methods Mol Biol 472:439–453

Wojno KJ, Epstein JI (1995) The utility of basal cell-specific anti-cytokeratin antibody (34 beta E12) in the diagnosis of prostate cancer. A review of 228 cases. Am J Surg Pathol 19:251–260

Wolters T, Roobol MJ, van Leeuwen PJ, van den Bergh RC, Hoedemaeker RF, van Leenders GJ et al (2010) Should pathologists routinely report prostate tumour volume? The prognostic value of tumour volume in prostate cancer. Eur Urol 57:821–829

Wright GL Jr, Haley C, Beckett ML, Schellhammer PF (1995) Expression of prostate-specific membrane antigen in normal, benign, and malignant prostate tissues. Urol Oncol 1:18–28

Xu J, Kalos M, Stolk JA, Zasloff EJ, Zhang X, Houghton RL et al (2001) Identification and characterization of prostein, a novel prostate-specific protein. Cancer Res 61:1563–1568

Yadav R, Tu JJ, Jhaveri J, Leung RA, Rao S, Tewari AK (2009) Prostate volume and the incidence of extraprostatic extension: is there a relation? J Endourol 23:383–386

Yang XJ, Wu CL, Woda BA, Dresser K, Tretiakova M, Fanger GR et al (2002) Expression of alpha-methylacyl-CoA racemase (P504S) in atypical adenomatous hyperplasia of the prostate. Am J Surg Pathol 26:921–925

Ye H, Walsh PC, Epstein JI (2010) Skeletal muscle involvement by limited Gleason score 6 adenocarcinoma of the prostate on needle biopsy is not associated with adverse findings at radical prostatectomy. J Urol 184:2308–2312

Yuan H, Wei X, Zhang G, Li C, Zhang X, Hou J (2011) B7-H3 over expression in prostate cancer promotes tumor cell progression. J Urol 186:1093–1099

Zang X, Thompson RH, Al-Ahmadie HA, Serio AM, Reuter VE, Eastham JA et al (2007) B7-H3 and B7x are highly expressed in human prostate cancer and associated with disease spread and poor outcome. Proc Natl Acad Sci USA 104:19458–19463

Zhao Z, Zeng G (2010) Increased serum level of early prostate cancer antigen is associated with subsequent cancer risk in men with high-grade prostatic intraepithelial neoplasia. Endocr Relat Cancer 17:505–512

Zhao Z, Zeng G, Zhong W (2010) Serum early prostate cancer antigen (EPCA) as a significant predictor of incidental prostate cancer in patients undergoing transurethral resection of the prostate for benign prostatic hyperplasia. Prostate 70:1788–1798

Zhao Z, Ma W, Zeng G, Qi D, Ou L, Liang Y (2011) Serum early prostate cancer antigen (EPCA) level and its association with disease progression in prostate cancer in a Chinese population. PLoS One 6:e19284

Zhao Z, Ma W, Zeng G, Qi D, Ou L, Liang Y (2012) Preoperative serum levels of early prostate cancer antigen (EPCA) predict prostate cancer progression in patients undergoing radical prostatectomy. Prostate 72:270–279

Zhigang Z, Wenlv S (2006) Prostate stem cell antigen (PSCA) expression in human prostate cancer tissues: implications for prostate carcinogenesis and progression of prostate cancer. Jpn J Clin Oncol 36:121

Zhou M, Aydin H, Kanane H, Epstein JI (2004) How often does alpha-methylacyl-CoA-racemase contribute to resolving an atypical diagnosis on prostate needle biopsy beyond that provided by basal cell markers? Am J Surg Pathol 28:239–243

Zynger DL, Yang X (2009) High-grade prostatic intraepithelial neoplasia of the prostate: the precursor lesion of prostate cancer. Int J Clin Exp Pathol 2:327–338

Part II
Molecular Pathology

Chapter 2
Molecular Determinants of Cancer-Related Inflammation

Stefania Staibano

Abstract Tumor cells communicate with the cells of their microenvironment via a series of molecular and cellular interactions to aid their progression to a malignant state and ultimately their metastatic spread. Of the cells in the microenvironment with a key role in cancer development, tumor associated macrophages (TAMs) are among the most notable. Tumor cells release a range of chemokines, cytokines and growth factors to attract macrophages, and these in turn release numerous factors (e.g. VEGF, MMP-9 and EGF) that are implicated in invasion-promoting processes such as tumor cell growth, flicking of the angiogenic switch and immunosuppression (Rogers and Holen, J Transl Med 9:177, 2011).

2.1 Background and Aims

A long time ago, Virchow hypothesized the existence of an interplay between inflammation and cancer (Balkwill and Mantovani 2001). Nowadays, i*t has been accepted* that at least 20 % of all human cancers (Vykhovanets et al. 2011) share a common causative *inflammatory* background, and chronic inflammatory states are emerging as having a relevant role also in prostate carcinogenesis. Histology has confirmed the strong association between morphological evidence of chronic inflammation, pre-malignant, and malignant changes in the prostatic epithelium (MacLennan et al. 2006). Chronic inflammatory cells as lymphocytes, tumor-associated macrophages (TAM), mast cells, dendritic cells, natural killer (NK) cells, exert their defensive activity via a plethora of molecules, comprising pro-inflammatory cytokines, growth factors, reactive oxygen species, interferons (IFNs)

S. Staibano (✉)
Department of Advanced Biomedical Sciences, Pathology Section, Faculty of Medicine and Surgery, University of Naples "Federico II", via S. Pansini, n.5, Naples, Italy
e-mail: staibano@unina.it

S. Staibano (ed.), *Prostate Cancer: Shifting from Morphology to Biology*,
DOI 10.1007/978-94-007-7149-9_2, © Springer Science+Business Media Dordrecht 2013

and proteases, membrane perforating agents, matrix metalloproteinase (MMP), and enzymes like as cyclooxygenase-2 (COX-2). They in turn interact with transcription factors as Nuclear Factor κB (NF-κB) (Vendramini-Costa and Carvalho 2012). The hyperactivation of NF-κB maintains an inflammatory status in the prostate(Vykhovanets et al. 2008), and is considered a potential molecular bridge between inflammation and prostate cancer (Karin 2006).

Besides the underlying causative environmental and endogenous factors, the stable umbalance of proinflammatory pathways active in chronic inflammation may lead to the establishment of prostatic regenerative lesions ending to a peculiar type of atrophy, defined as "proliferative inflammatory atrophy (PIA)", which is thought to contribute to the rise of risk for prostate cancer. The relationship between prostate cancer onset and progression and inflammation is being explored at genetic and epigenetic level (Vykhovanets et al. 2011).

Research in this area at present is particularly active, considering its possible beneficial fall-out on population: as an example, we could imagine that the administration of specific anti-inflammatory agents in men may reduce the risk of prostate cancer development.

This chapter will present an overview of the recent knowledge on the role of chronic inflammation in the pathogenesis and/or therapy outcome of prostate preneoplastic lesions and prostate cancer.

Chronic inflammation is associated with the development of several cases of head&neck, esophagus, stomach, colon, liver and urinary bladder cancer (Sugar 2006; Coghill et al. 2011). Besides its initiating causes (either infectious or non-infectious inflammatory diseases, and/or environmental/epigenetic factors), inflammation enhances cellular turnover of the injured cells, as the result of tissue repair processes.

Epidemiological, genetic and molecular findings accumulating from more than a decade, indicate that chronical prostatitis correlate with an increased risk of prostate cancer (PCa), supporting the hypothesis that inflammation may be a cause of neoplastic transformation also for prostatic tissue (Sfanos and De Marzo 2012).

Prostatitis is actually classified into four distinct entities (Murphy et al. 2009).

Category I: acute bacterial prostatitis, due to a uropathogen, often with systemic symptoms.

Category II: chronic bacterial prostatitis due to recurrent episodes of documented infections with the same uropathogen.

Category III: chronic prostatitis/chronic pelvic pain syndrome (CP/CPPS), lacking a documented infection from uropathogens, ending with neurological injury with or without pelvic floor dysfunction.

Category IV: asymptomatic inflammatory prostatitis, of uncertain clinical and biological significance.

Acute prostatitis is infrequent; sometimes (in about 10 % of cases) it may result in a chronic bacterial prostatitis and further 10 % into chronic pelvic pain syndrome (Wagenlehner et al. 2013).

Viruses, fungi, mycobacteria and parasites may cause prostatitis (De Marzo et al. 2007), but the microbial agents responsible of most cases of bacterial prostatitis are represented by *Escherichia coli*, coagulase-negative *Enterococcus* spp. (de Kleijn et al. 1997; Arnow and Flaherty 1997) and Corynebacterium glucuronolyticum.

Chronic non-bacterial prostatitis is much more frequent. They are underdiagnosed with respect to the infectious ones, so that an "histological prostatitis" represents an accidental finding on biopsy for prostate cancer (Stimac et al. 2009; Gui-zhong et al. 2011; Ugurlu et al. 2010; Fujita et al. 2011) or benign prostatic hyperplasia (BPH) (Nickel et al. 1999; De Marzo et al. 2007).

The manifold non-infectious causes of prostate chronic inflammation encompass (De Marzo et al. 2007; Sfanos and De Marzo 2012) hormonal alterations, that may lead to architectural alterations predisposing prostatic tissue to inflammation and physical trauma, particularly related to corpora amylacea, that are thought to represent remnants of past acute inflammatory events, and are composed of organic matrix comprising proteins involved in acute inflammation, such as lactoferrin, myeloperoxidase and α-defensins. These proteins, in turn, may induce the overexpression of stress-proteins by prostate cells (Sfanos et al. 2009).

Other inducers of chronic inflammation are several urine metabolites, which overactivate inflammatory cells in prostate tissue of patients suffering for urine reflux, and dietary and/or environmental carcinogens, reaching the prostate through urine reflux and/or blood.

Interestingly, racial and geographical difference in the incidence of prostatitis have been reported and they paralleled those observed for prostate cancer, with the highest prevalence in African American men and the lesser in Asian men. This, further support the hypothesis of a causative role of inflammation in prostate cancer pathogenesis (Wallace et al. 2008).

Indeed, bacterial and non-infectious chronic prostatitis may lead to prostate cell hyperproliferation, and this event seems to be correlated with the emergence of benign prostatic hyperplasia (BPH) (Nickel 2008).

Chronic inflammation is thought to induce the antigen-presenting capacity of prostatic stromal cells, via the overproduction of the prostate growth-promoting chemokine IL-8 induced by Th1 and Th17 cell-derived inflammatory cytokines (Steiner et al. 2003).

Th17 cells belong to a $CD4^+$ effector T cell lineage which develops through distinct cytokine signals [specifically interleukin (IL)-23] and produce IL-17.

Th17 cells mediate a number of autoimmune diseases, and seem to have a role in inflammation-associated cancer (Weaver et al. 2006; Bettelli et al. 2007).

Long-lasting inflammation induces also the up-regulation of the vitamin D receptor (VDR), which agonizes intra-prostatic androgen signalling by exerting immunostimulating and co-inflammatory effects. BPH stromal cells express high levels of VDR, further supporting the role of chronic inflammation in BPH pathogenesis and their usefulness as therapeutic targets for pharmacological treatment of BPH.

Moreover, chronic inflammation is responsible for the condition termed "Proliferative inflammatory atrophy (PIA)".

Long-standing PIA gradually lost the function of cellular detoxification, by silencing of glutathione-S transferase. This favor an increased susceptibility of prostatic epithelial cells to genomic damage by inflammatory oxidants or nutritional carcinogens, assisted by several other inflammation-induced proteins, as the macrophage scavenger receptor 1 and Toll-like receptor-4.

Prostatitis-derived PIA, then, may gradually transitate to prostatic intraepithelial neoplasia, and take a part in the multifactorial background leading to prostate cancer (Wagenlehner et al. 2007).

PIA, is morphologically characterized by the presence of atrophic-regenerating epithelial cells (De Marzo et al. 1999) that may occupy large regions of the prostate. In these areas, it is frequent the finding of high-grade PIN, (De Marzo et al. 2007; Nelson et al. 2003; Putzi and De Marzo 2000), with variable degree of transitions between PIA, PIN and true prostate cancer (Putzi and De Marzo 2000; Wang et al. 2009b).

PIA has some of the hallmark gene expression changes found in prostate cancer and PIN. For example, two genes which are highly expressed in normal prostate epithelium and frequently down-regulated or absent in PIN and prostate cancer, NKX3.1 and p27, are down-regulated in prostate atrophy (De Marzo et al. 1999, 2007; Bethel et al. 2006), showing, in turn, increased immunostaining for p53, Ki-67, COX-2 and glutathione *S*-transferase-π (GSTP1), particularly in areas adjacent to inflammation (Wang et al. 2009a).

As atrophy/PIA is highly prevalent in the peripheral zone of the prostate, it is possible that a proportion of PIN and/or prostate cancer may originate in these areas (Nakayama et al. 2003).

Recently it has been reported that chronic inflammation in benign tissue was predictive of a higher risk for prostate cancer diagnosis and, specifically, with higher-grade (Gleason score 7–10) disease. The risk of prostate cancer and high-grade prostate cancer also increased with the number of biopsies that were found to contain chronic inflammation

Moreover, several lines of evidence indicate that inflammation in and around prostate cancer is associated with worse disease outcome (Karja et al. 2005; Nonomura et al. 2011).

Among the major effectors of the dangerous potential of inflammation on cancer predisposition are cytokines.

Besides many protean roles in immune system, hematopoiesis, and key biological functions (Sun et al. 2007), these low-molecular weight molecules interact with several types of cells and proteins within the tumor environment (De Marzo et al. 2007; Coussens and Werb 2002), *and* have been associated from long time with the biology and prognosis of several cancer.

As an example, the production of cyclooxygenase (COX) enzymes due to inflammation may alter the environment of precancerous tissues (Mantovani et al. 2008; Coussens and Werb 2002), being important in the pathogenesis of prostate cancer (Wang and Dubois 2006).

Interleukin-4 (IL-4), IL-6 and IL-10 are frequently elevated in blood of prostate cancer patients and, increased levels of transforming growth factor beta (TGFβ), have been detected in serum and in primary and metastatic prostate cancer tissue samples (Perry et al. 1997).

Whereas in normal cells, TGFβ stops cell proliferation, induces differentiation, and/or apoptosis; in cancer cells, mutations of the TGFβ pathway confer resistance to growth inhibition by TGFβ, resulting in uncontrolled cell proliferation. The increase of TGFβ production in cancer cells also stimulates angiogenesis and suppresses the activities of infiltrating immune cells, thereby facilitating the tumor to escape from immunosurveillance. On the other hand, prostate epithelial cancer cells show loss-of-expression of T cell cytolytic promoting IL-7. This causes a severe depletion of prostate-associated lymphocytes (Tang et al. 1997). As an antagonistic relationship has been hypothesized between TGF β and IL-7, it has hypothesized that their level of expression may be used in combination with Gleason score and pre-treatment PSA level to predict prognosis of prostate cancer patients (Tang et al. 1997; Dubinett et al. 1995).

The preliminary results indicate that this addition, at least, doubled the prognostic predictive ability in respect to the use of the sole Gleason score and pre-treatment PSA (Dubinett et al. 1995; Schroten et al. 2012).

Several additional studies have focused, instead, on the specific targeting of the IL-6 as alternative/adjunctive therapy in aggressive, therapy-resistant prostate cancers.

IL-6 is a multifunctional cytokine produced by multiple cell types, including macrophages, endothelial cells and T lymphocytes; it is involved in innate and adaptive inflammatory processes, including acute-phase inflammatory response (Hirano 1992).

When deregulated, IL-6 intervenes in multiple disease processes, including autoimmune disorders, rheumatoid arthritis, osteoporosis, psoriasis, diabetes, atherosclerosis and cancer (Ishihara and Hirano 2002; Kishimoto 2005).

In prostate cells, IL-6 contributes to the activation of androgen receptor (AR) (Culig and Puhr 2012) High-grade prostatic intraepithelial neoplasia (PIN) and prostate cancer cells (Hobisch et al. 2000) overexpress IL-6 and its receptor IL6-R and, patients suffering from metastatic and hormone-refractory prostate cancers, show high IL-6 plasmatic levels (Smith et al. 2001). The linking between IL-6 and prostate cancer morbidity (Twillie et al. 1995), could reside on the stabilization of an 'epigenetic transformed' state of prostate cancer cells, due to the cooperative action of inflammation/IL-6 production, STAT3 and NF-κB activation (Iliopoulos et al. 2009).

Recently, 6 mg/kg anti-IL-6 antibody CNTO328 were administered i.v. every 2 weeks for 12 cycles to 53 patients with castration-resistant prostate cancer pre-treated with taxane chemotherapy. Tumor response was assessed after every three cycles. Primary end- point was PSA response rate defined as a 50 % reduction. Declining C-reactive protein levels during treatment may reflect biological activity.

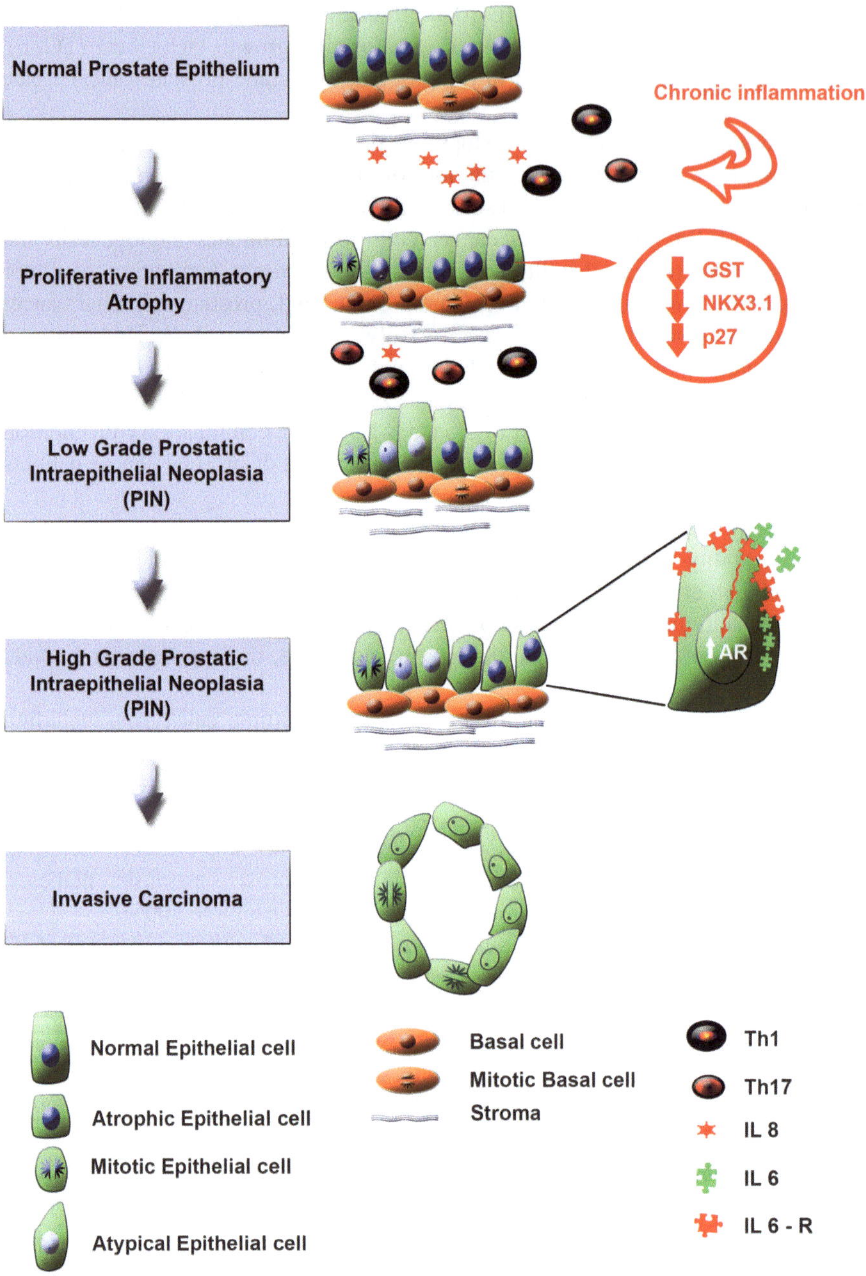

Fig. 2.1 Inflammatory background in prostate carcinogenesis. Chronic inflammation is considered to influence the antigen-presenting capacity of prostatic stromal cells, by the overproduction of the prostate growth promoting chemokine IL8, induced by Th1 and Th17 cells. The stable unbalance of proinflammatory pathways activated in chronic inflammation may lead to the formation of prostatic regenerative lesions, such as a peculiar type of atrophy, defined as

Despite evidence of CNTO-mediated IL-6 inhibition, elevated baseline IL-6 levels portended a poor prognosis (Sfanos and De Marzo 2012). These represent only preliminary data but, considering the potential contribution of IL-6 to therapy for progressing prostate cancers, this cytokine is actually regarded with particular interest for applied prostate cancer research.

Interestingly, it has been found that serum concentrations of IL-6-family cytokines were reduced significantly in animals fed with a tomato-enriched diet. As well, the anti-inflammatory omega 3 PUFA has been associated with a decreased risk of PC (Fradet et al. 2009).

Another inflammatory cytokine of possible significance, as potential mediator between prostatic inflammation pathways and prostate carcinogenesis, is the macrophage inhibitory cytokine 1 (MIC-1), which belongs to the transforming growth factor-β (TGF-β) superfamily (Bootcov et al. 1997).

MIC-1, also known as prostate-derived factor (PDF) or growth differentiation factor-15 (GDF-15), was first identified in activated macrophages (Bootcov et al. 1997) and has been associated with the progression of various types of diseases (Fig. 2.1).

In cancer, macrophages migrate from the circulation into the tissue, and return to the bloodstream or lymphatic system after having phagocytated tumor debris (Faber et al. 2012).

Macrophages are essential in the processes of migration, invasion and tumor metastasis (Stewart et al. 2004; Condeelis and Pollard 2006; Roorda et al. 2009).

As for IL-6, the increased expression of MIC-1 has been associated with a variety of tumors, including breast, gastric, colorectal (Senapati et al. 2010; Breit et al. 2011) and prostate cancer and, high serum levels of MIC-1, have shown to predict poor prognosis of prostate cancer patients (Nakamura et al. 2003; Cheung et al. 2004; Brown et al. 2009).

The MIC-1 gene has emerged as an ideal key candidate to explain the link between macrophage-linked inflammation and prostate cancer pathogenesis (Karan et al. 2009).

The expression of MIC-1 has been reported in conjunction of infiltrating lymphocytes in non-neoplastic human prostate tissues (Paralkar et al. 1998; Bostwick et al. 2003). This has to be considered an early response to inflammation in prostate, which, in the long-time, may enhance cell proliferation (Bootcov et al. 1997; Chen et al. 2007).

Fig. 2.1 (continued) "proliferative inflammatory atrophy (PIA)", that shows an increased susceptibility of prostatic epithelial cells to genomic damage by inflammatory oxidants or nutritional carcinogens, linked to the lost of cellular detoxification, caused by the silencing of glutathione S transferase. Moreover, PIA shows the same gene expression changes found in prostate cancer and PIN, such as downregulation of NKX3.1 and p27. Therefore, PIA gradually transitates to prostatic intraepithelial neoplasia, and take part in the background leading to prostate cancer. Inflammation influences cancer predisposition by production of cytokines. In particular, IL-6 regulates positively the expression of androgen receptor (AR) and is overexpressed in PIN and prostate cancer cells

Macrophages were identified in both epithelium and the stromal area of inflammation-associated to human benign prostatic hyperplasia (BPH) tissues, suggesting that they might play roles in BPH development. Yet the underlying mechanisms remain unclear. New insights for alternative therapeutic approach to counteract BPH via inflammatory signaling pathways (Lu et al. 2012), may thus be provided.

The finding of an autocrine-paracrine positive loop between MIC-1, IL-1β and TNF-α in the human LNCaP prostate cancer cell line, has suggested that MIC-1 overexpression could play a critical role also in the in the early stages of prostate cancer development (Bootcov et al. 1997). The glandular and peri-glandular CD68+ macrophages accumulation in PIA lesions strength the link between the macrophage-rich inflammatory microenvironment and prostate cancer development (Vykhovanets et al. 2008).

This is further supported by the observation that (De Marzo et al. 2007; Platz and De Marzo 2004) prostate cancer is frequently associated with an increased prostate tissue susceptibility to inflammatory injury/infections.

Of additional interest, the overexpression of MIC-1 has been also described during the progression to androgen-independent and metastatic prostate cancers, associated with a poor outcome of patients. In PC3 cells, high levels of MIC-1 were associated with the acquisition of epithelial-mesenchymal transition, higher invasive capacity and docetaxel resistance. These phenomena were in large extent reversed through MIC-1, which proved also effective in promoting the docetaxel-induced cytotoxic effects both on the stem cell-like side population and the non-side population, thus suggesting a promising improving effect of MIC-1 downregulation on the efficacy of current chemotherapies for aggressive prostate cancer (Mimeault et al. 2013).

Moreover, in model studies, macrophages have shown to be sensitive to bisphosphonates, as do osteoclasts, which belong to the same cell lineage, reversing their phenotype from pro-tumoral CD204(+) M2 to tumoricidal CD68(+) M1 upon treatment with zoledronic acid (Rogers and Holen 2011; Fujii et al. 2013).

These exciting results necessitate of further validation on large series of cases. Nevertheless, they indicate that also in prostate cancer, as it is progressively being shown in most of solid cancers (Stewart et al. 2004), stromal cells and their products are determinant for epithelial neoplastic transformation (Pupa et al. 2002). The alteration of the tumor microenvironment homeostasis has a determinant impact on tissue architecture, adhesion, apoptosis and cell proliferation regulation, favoring the shifting toward oncogenic change (Stewart et al. 2004) and conditioning the response to cytotoxic therapies and prognosis of patients.

The reciprocal interaction between the multiple effectors of inflammation could be considered "the epigenetic framework for tumor progression" (Huang and Ingber 2006) and, as such, it represents a potentially modifiable scenario.

Dietary or medicinal intake of anti-inflammatory compounds, as NSAID (Jafari et al. 2009), soy and green tea, are being increasingly proposed to reduce prostate cancer risk by human epidemiology studies and in animal studies (Hsu et al. 2010).

Similarly, in prostate cancer cell lines, the treatment with phytoestrogens genistein and daidzein resulted in demethylation of GSTP1 and ephrin B2 (EPHB2) promoter regions (Vardi et al. 2010) confirming that the protective effects of soy in prostate cancer prevention may involve epigenetic modifications to DNA.

The promise that the malignant phenotype can be reversed through the correction of tumor microenvironment features (Kenny and Bissell 2003), make this one of the most promising and innovative experimental fields on prostate cancer treatment and, the emerging data at this regard, could provide us with a more comprehensive view of prostate cancerogenesis.

References

Arnow PM, Flaherty JP (1997) Fever of unknown origin. Lancet 350:575–580

Balkwill F, Mantovani A (2001) Inflammation and cancer: back to Virchow? Lancet 357:539–545

Bethel CR, Faith D, Li X et al (2006) Decreased NKX3.1 Protein expression in focal prostatic atrophy, prostatic intraepithelial neoplasia, and adenocarcinoma: association with Gleason score and chromosome 8p deletion. Cancer Res 66:10683–10690

Bettelli E, Oukka M, Kuchroo VK (2007) TH-17 cells in the circle of immunity and autoimmunity. Nat Immunol 8:345–350

Bootcov MR, Bauskin AR, Valenzuela SM, Moore AG, Bansal M, He XY, Zhang HP, Donnellan M, Mahler S, Pryor K, Walsh BJ, Nicholson RC, Fairlie WD, Por SB, Robbins JM, Breit SN (1997) MIC-1, a novel macrophage inhibitory cytokine, is a divergent member of the TGF-beta superfamily. Proc Natl Acad Sci USA 94:11514–11519

Bostwick DG, de la Roza G, Dundore P, Corica FA, Iczkowski KA (2003) Intraepithelial and stromal lymphocytes in the normal human prostate. Prostate 55:187–193

Breit SN, Johnen H, Cook AD, Tsai VW, Mohammad MG, Kuffner T, Zhang HP, Marquis CP, Jiang L, Lockwood G, Lee-Ng M, Husaini Y, Wu L, Hamilton JA, Brown DA (2011) The TGF-beta superfamily cytokine, MIC-1/GDF15: a pleotrophic cytokine with roles in inflammation, cancer and metabolism. Growth Factors 29:187–195

Brown DA, Lindmark F, Stattin P et al (2009) Macrophage inhibitory cytokine 1: a new prognostic marker in prostate cancer. Clin Cancer Res 15:6658–6664

Chen SJ, Karan D, Johansson SL, Lin FF, Zeckser J, Singh AP, Batra SK, Lin MF (2007) Prostate-derived factor as a paracrine and autocrine factor for the proliferation of androgen receptor-positive human prostate cancer cells. Prostate 67:557–571

Cheung PK, Woolcock B, Adomat H et al (2004) Protein profiling of microdissected prostate tissue links growth differentiation factor 15 to prostate carcinogenesis. Cancer Res 64:5929–5933

Coghill AE, Newcomb PA, Poole EM et al (2011) Genetic variation in inflammatory pathways is related to colorectal cancer survival. Clin Cancer Res 17:7139–7147

Condeelis J, Pollard JW (2006) Macrophages: obligate partners for tumor cell migration, invasion, and metastasis. Cell 124(2):263–266

Coussens LM, Werb Z (2002) Inflammation and cancer. Nature 420:860–867

Culig Z, Puhr M (2012) Interleukin-6: a multifunctional targetable cytokine in human prostate cancer. Mol Cell Endocrinol 360:52–58

de Kleijn EM, Vandenbroucke JP, van der Meer JW (1997) Fever of unknown origin (FUO): I. A prospective multicenter study of 167 patients with FUO, using fixed epidemiologic entry criteria. The Netherlands FUO Study Group. Medicine (Baltimore) 76:392–400

De Marzo AM, Marchi VL, Epstein JI, Nelson WG (1999) Proliferative inflammatory atrophy of the prostate: implications for prostatic carcinogenesis. Am J Pathol 155:1985–1992

De Marzo AM, Platz EA, Sutcliffe S, Xu J, Grönberg H, Drake CG, Nakai Y, Isaacs WB, Nelson WG (2007) Inflammation in prostate carcinogenesis. Nat Rev Cancer 7(4):256–269

Dubinett SM, Huang M, Dhanani S, Economou JS, Wang J, Lee P, Sharma S, Dougherty GJ, McBride WH (1995) Down-regulation of murine fibrosarcoma transforming growth factor-beta 1 expression by interleukin 7. J Natl Cancer Inst 87:593–597

Faber TJ, Japink D, Leers MP, Sosef MN, von Meyenfeldt MF, Nap M (2012) Activated macrophages containing tumor marker in colon carcinoma: immunohistochemical proof of a concept. Tumour Biol 33(2):435–441

Fradet V, Cheng I, Casey G, Witte JS (2009) Dietary omega-3 fatty acids, cyclooxygenase-2 genetic variation, and aggressive prostate cancer risk. Clin Cancer Res 15:2559–2566

Fujii T, Shimada K, Asai O, Tanaka N, Fujimoto K, Hirao K, Konishi N (2013) Immunohistochemical analysis of inflammatory cells in benign and precancerous lesions and carcinoma of the prostate. Pathobiology 80(3):119–126

Fujita K, Hosomi M, Tanigawa G, Okumi M, Fushimi H, Yamaguchi S (2011) Prostatic inflammation detected in initial biopsy specimens and urinary pyuria are predictors of negative repeat prostate biopsy. J Urol 185:1722–1727

Gui-zhong LI, Libo M, Guanglin H, Jianwei W (2011) The correlation of extent and grade of inflammation with serum PSA levels in patients with IV prostatitis. Int Urol Nephrol 43: 295–301

Hirano T (1992) The biology of interleukin-6. Chem Immunol 51:153–180

Hobisch A, Rogatsch H, Hittmair A et al (2000) Immunohistochemical localization of interleukin-6 and its receptor in benign, premalignant and malignant prostate tissue. J Pathol 191:239–244

Hsu A, Bray TM, Ho E (2010) Anti-inflammatory activity of soy and tea in prostate cancer prevention. Exp Biol Med 235:659–667

Huang S, Ingber DE (2006–2007) A non-genetic basis for cancer progression and metastasis: self-organizing attractors in cell regulatory networks. Breast Dis 26:27–54

Iliopoulos D, Hirsch HA, Struhl K (2009) An epigenetic switch involving NF-[kappa]B, Lin28, Let-7 microRNA, and IL6 links inflammation to cell transformation. Cell 139:693–706

Ishihara K, Hirano T (2002) IL-6 in autoimmune disease and chronic inflammatory proliferative disease. Cytokine Growth Factor Rev 13:357–368

Jafari S, Etminan M, Afshar K (2009) Nonsteroidal anti-inflammatory drugs and prostate cancer: a systematic review of the literature and meta-analysis. Can Urol Assoc J 3:323–330

Karan D, Holzbeierlein J, Thrasher JB (2009) Macrophage inhibitory cytokine-1: possible bridge molecule of inflammation and prostate cancer. Cancer Res 69:2–5

Karin M (2006) Nuclear factor-κB in cancer development and progression. Nature 441:431–436

Karja V, Aaltomaa S, Lipponen P, Isotalo T, Talja M, Mokka R (2005) Tumour-infiltrating lymphocytes: a prognostic factor of PSA-free survival in patients with local prostate carcinoma treated by radical prostatectomy. Anticancer Res 25:4435–4438

Kenny PA, Bissell MJ (2003) Tumor reversion: correction of malignant behavior by microenvironmental cues. Int J Cancer 107:688–695

Kishimoto T (2005) Interleukin-6: from basic science to medicine – 40 years in immunology. Ann Rev Immunol 23:1–21

Lu T, Lin WJ, Izumi K, Wang X, Xu D, Fang LY, Li L, Jiang Q, Jin J, Chang C (2012) Targeting androgen receptor to suppress macrophage-induced EMT and benign prostatic hyperplasia (BPH) development. Mol Endocrinol 26(10):1707–1715

MacLennan GT, Eisenberg R, Fleshman RL, Taylor JM, Fu P, Resnick MI, Gupta S (2006) The influence of chronic inflammation in prostatic carcinogenesis: a 5-year followup study. J Urol 176:1012–1016

Mantovani A, Allavena P, Sica A, Balkwill F (2008) Cancer-related inflammation. Nature 454: 436–444

Mimeault M, Johansson SL, Batra SK (2013) Marked improvement of cytotoxic effects induced by docetaxel on highly metastatic and androgen-independent prostate cancer cells by downregulating macrophage inhibitory cytokine-1. Br J Cancer 108(5):1079–1091

Murphy AB, Macejko A, Taylor A, Nadler RB (2009) Chronic prostatitis: management strategies. Drugs 69(1):71–84
Nakamura T, Scorilas A, Stephan C et al (2003) Quantitative analysis of macrophage inhibitory cytokine-1 (MIC-1) gene expression in human prostatic tissues. Br J Cancer 88:1101–1104
Nakayama M, Bennett CJ, Hicks JL et al (2003) Hypermethylation of the human glutathione S-transferase-[pi] gene (GSTP1) CpG island is present in a subset of proliferative inflammatory atrophy lesions but not in normal or hyperplastic epithelium of the prostate: a detailed study using laser-capture microdissection. Am J Pathol 163:923–933
Nelson WG, De Marzo AM, Isaacs WB (2003) Prostate cancer. N Engl J Med 349:366–381
Nickel JC (2008) Inflammation and benign prostatic hyperplasia. Urol Clin North Am 35(1): 109–115; vii
Nickel JC, Downey J, Young I, Boag S (1999) Asymptomatic inflammation and/or infection in benign prostatic hyperplasia. BJU Int 84:976–981
Nonomura N, Takayama H, Nakayama M et al (2011) Infiltration of tumour-associated macrophages in prostate biopsy specimens is predictive of disease progression after hormonal therapy for prostate cancer. BJU Int 107:1918–1922
Paralkar VM, Vail AL, Grasser WA, Brown TA, Xu H, Vukicevic S, Ke HZ, Qi H, Owen TA, Thompson DD (1998) Cloning and characterization of a novel member of the transforming growth factor-beta/bone morphogenetic protein family. J Biol Chem 273:13760–13767
Perry KT, Anthony CT, Steiner MS (1997) Immunohistochemical localization of TGF beta 1, TGF beta 2, and TGF beta 3 in normal and malignant human prostate. Prostate 33:133–140
Platz EA, De Marzo AM (2004) Epidemiology of inflammation and prostate cancer. J Urol 171:S36–S40
Pupa SM, Ménard S, Forti S, Tagliabue E (2002) New insights into the role of extracellular matrix during tumor onset and progression. J Cell Physiol 192(3):259–267
Putzi MJ, De Marzo AM (2000) Morphologic transitions between proliferative inflammatory atrophy and high-grade prostatic intraepithelial neoplasia. Urology 56:828–832
Rogers TL, Holen I (2011) Tumour macrophages as potential targets of bisphosphonates. J Transl Med 9:177
Roorda BD, ter Elst A, Kamps WA, de Bont ES (2009) Bone marrow-derived cells and tumor growth: contribution of bone marrow-derived cells to tumor micro-environments with special focus on mesenchymal stem cells. Crit Rev Oncol Hematol 69(3):187–198
Schroten C, Dits NF, Steyerberg EW, Kranse R, van Leenders AG, Bangma CH, Kraaij R (2012) The additional value of TGFβ1 and IL-7 to predict the course of prostate cancer progression. Cancer Immunol Immunother 61(6):905–910
Senapati S, Rachagani S, Chaudhary K, Johansson SL, Singh RK, Batra SK (2010) Overexpression of macrophage inhibitory cytokine-1 induces metastasis of human prostate cancer cells through the FAK-RhoA signaling pathway. Oncogene 29:1293–1302
Sfanos KS, De Marzo AM (2012) Prostate cancer and inflammation: the evidence. Histopathology 60:199–215
Sfanos KS, Wilson BA, De Marzo AM, Isaacs WB (2009) Acute inflammatory proteins constitute the organic matrix of prostatic corpora amylacea and calculi in men with prostate cancer. Proc Natl Acad Sci USA 106:3443–3448
Smith PC, Hobisch A, Lin D-L, Culig Z, Keller ET (2001) Interleukin-6 and prostate cancer progression. Cytokine Growth Factor Rev 12:33–40
Steiner GE, Newman ME, Paikl D et al (2003) Expression and function of pro-inflammatory interleukin IL-17 and IL-17 receptor in normal, benign hyperplastic, and malignant prostate. Prostate 56:171–182
Stewart DA, Cooper CR, Sikes RA (2004) Changes in extracellular matrix (ECM) and ECM-associated proteins in the metastatic progression of prostate cancer. Reprod Biol Endocrinol 2:2
Stimac G, Reljic A, Spajic B et al (2009) Aggressiveness of inflammation in histological prostatitis – correlation with total and free prostate specific antigen levels in men with biochemical criteria for prostate biopsy. Scott Med J 54:8–12

Sugar LM (2006) Inflammation and prostate cancer. Can J Urol 13(Suppl 1):46–47
Sun J, Turner A, Xu J, Gronberg H, Isaacs W (2007) Genetic variability in inflammation pathways and prostate cancer risk. Urol Oncol 25:250–259
Tang J, Nuccie BL, Ritterman I, Liesveld JL, Abboud CN, Ryan DH (1997) TGF-beta down-regulates stromal IL-7 secretion and inhibits proliferation of human B cell precursors. J Immunol 159:117–125
Twillie DA, Eisenberger MA, Carducci MA, Hseih W-S, Kim WY, Simons JW (1995) Interleukin-6: a candidate mediator of human prostate cancer morbidity. Urology 45:542–549
Ugurlu O, Yaris M, Oztekin CV, Kosan TM, Adsan O, Cetinkaya M (2010) Impacts of antibiotic and anti-inflammatory therapies on serum prostate-specific antigen levels in the presence of prostatic inflammation: a prospective randomized controlled trial. Urol Int 84:185–190
Vardi A, Bosviel R, Rabiau N et al (2010) Soy phytoestrogens modify DNA methylation of GSTP1, RASSF1A, EPH2 and BRCA1 promoter in prostate cancer cells. In Vivo 24:393–400
Vendramini-Costa DB, Carvalho JE (2012) Molecular link mechanisms between inflammation and cancer. Curr Pharm Des 18(26):3831–3852
Vykhovanets EV, Shukla S, Maclennan GT, Resnick MI, Carlsen H, Blomhoff R, Gupta S (2008) Molecular imaging of NF-kappaB in prostate tissue after systemic administration of IL- 1beta. Prostate 68:34–41
Vykhovanets EV, Maclennan GT, Vykhovanets OV, Gupta S (2011) IL-17 expression by macrophages is associated with proliferative inflammatory atrophy lesions in prostate cancer patients. Int J Clin Exp Pathol 4(6):552–565
Wagenlehner FM, Elkahwaji JE, Algaba F, Bjerklund-Johansen T, Naber KG, Hartung R, Weidner W (2007) The role of inflammation and infection in the pathogenesis of prostate carcinoma. BJU Int 100(4):733–737
Wagenlehner FM, Pilatz A, Bschleipfer T, Diemer T, Linn T, Meinhardt A, Schagdarsurengin U, Dansranjavin T, Schuppe HC, Weidner W (2013) Bacterial prostatitis. World J Urol 22. [Epub ahead of print] PubMed PMID: 23519458
Wallace TA, Prueitt RL, Yi M et al (2008) Tumor immunobiological differences in prostate cancer between African-American and European-American men. Cancer Res 68:927–936
Wang D, Dubois RN (2006) Prostaglandins and cancer. Gut 55:115–122
Wang W, Bergh A, Damber J-E (2009a) Increased p53 immunoreactivity in proliferative inflammatory atrophy of prostate is related to focal acute inflammation. APMIS 117:185–195
Wang W, Bergh A, Damber J-E (2009b) Morphological transition of proliferative inflammatory atrophy to high-grade intraepithelial neoplasia and cancer in human prostate. Prostate 69: 1378–1386
Weaver CT, Harrington LE, Mangan PR, Gavrieli M, Murphy KM (2006) Th17: an effector CD4 T cell lineage with regulatory T cell ties. Immunity 24:677–688

Chapter 3
Apoptosis and Autophagy

Francesco Merolla

Abstract Defects in both apoptotic and non-apoptotic cell-death pathways are strictly associated with tumorigenesis. In particular, resistance to apoptosis is considered to be an hallmark of cancer cells. Defects in apoptosis underlie not only tumorigenesis, but also resistance to cancer treatments.

A better definition of non-apoptotic and apoptotic cell-death pathways interactions is needed. Since the first attempts of cell deaths classification, the caspase-dependent, tolerogenic, programmed and physiological cell death instances have been contrasted to their caspase-independent, immunogenic, accidental and pathological counterparts. However, further investigation of non-apoptotic pathways might provide new therapeutic strategies aimed at inducing the non-apoptotic death of cancer cells.

In the present chapter, apoptotic and non-apoptotic cell death pathways are discussed for what concern neoplastic transformation of prostate gland.

As most human neoplastic diseases, prostate cancers (most of them are adenocarcinomas) develop when the rates of cell division and cell death are no longer equal, leading to uncontrolled tumor growth. Following the initial transformation event, further mutations of a multitude of genes can lead to tumor progression and metastasis. To date, several molecular signalling pathways have been found altered in prostate cancer, such as, to cite some, the Androgen and Estrogen metabolism, the cell cycle progression control, the MAPK signalling pathway, the maintenance of the stability of the genome, the control of Apoptosis and Autophagy.

F. Merolla (✉)
Department of Advanced Biomedical Sciences, Pathology Section, Faculty of Medicine and Surgery, University of Naples "Federico II", via S. Pansini, n.5, Naples, Italy
e-mail: merollafra@gmail.com

S. Staibano (ed.), *Prostate Cancer: Shifting from Morphology to Biology*,
DOI 10.1007/978-94-007-7149-9_3, © Springer Science+Business Media Dordrecht 2013

Both apoptosis and autophagy are stress response mechanisms that have been involved in neoplastic transformation of prostatic gland and that seem to be the most affected especially in the latter stages of prostate cancer progression.

Apoptosis, the most common and well-defined form of programmed cell death (PCD), is a crucial cellular process in normal and pathological conditions: its importance during embryonic development and the maintenance of tissue homeostasis of multicellular organisms has been assessed long ago (Meier et al. 2000). Dysregulation of apoptosis has been implicated in numerous pathological conditions, including neurodegenerative diseases and autoimmunity, moreover its dysfunction is *de facto* accepted as an hallmark of cancer (Hanahan and Weinberg 2000, 2011).

In mammalian cells, the apoptotic process is mediated by a family of cysteine proteases known as the caspases (Alnemri et al. 1996). To keep the apoptotic programme under control, caspases are initially expressed in cells as inactive procaspase precursors. When initiator caspases—such as caspase-8 and caspase-9—are activated by oligomerization, they cleave the precursor forms of effector caspases, such as caspase-3, caspase-6 and caspase-7 (Salvesen and Dixit 1997; Cryns and Yuan 1998; Thornberry and Lazebnik 1998). Activated effector caspases in turn cleave a specific set of cellular substrates, resulting in the well-known constellation of biochemical and morphological changes that are associated with the apoptotic phenotype.

Autophagy is a process in which subcellular membranes undergo dynamic morphological changes that lead to the degradation of cellular proteins and cytoplasmic organelles. This process is an important physiological cellular response to stress or starvation. Many studies have shed light on the involvement of autophagy in cancer, but it is still unclear whether autophagy suppresses tumorigenesis or provides cancer cells with a rescue mechanism under unfavourable conditions. In fact, while apoptosis is clearly a primary cell death mechanism, there is much controversy about the functional role of autophagy in life and death. Depending on the cellular context, the cell line and the stimulus, autophagy either favours or counteracts cell death signalling.

It is believed that multiple connections exist between autophagy and apoptosis, and so the molecular interplay and functional relationship between their pathways have gained considerable interest in normal and neoplastic condition.

In the present chapter, a review of recent literature about the strict relationship between apoptosis, autophagy and prostate cancer is reported, with major emphasis on the role of deregulated apoptosis and autophagy during prostate cancer progression and the therapeutic strategies based on these cellular processes.

3.1 The Biology of Prostate Cancer

Prostate cancer is generally regarded as multifocal, since primary tumors often contain multiple independent histologic foci of cancer, that are often genetically distinct (Aihara et al. 1994; Bostwick et al. 1998; Macintosh et al. 1998; Mehra et al.

2007; Clark et al. 2008). In contrast, molecular and cytogenetic analyses show that multiple metastases in the same patient are clonally related, indicating that advanced prostate cancer is monoclonal (Mehra et al. 2008; Liu et al. 2009).

The prostate gland can be the site of multiple neoplastic transformation events, many of which give rise only to latent prostate cancer that does not progress to clinically detectable disease.

It is widely accepted that PIN represents a precursor for prostate cancer (Bostwick 1989; DeMarzo et al. 2003). PIN is generally characterized at the histological level by the appearance of luminal epithelial hyperplasia, reduction in basal cells, enlargement of nuclei and nucleoli, cytoplasmic hyperchromasia, and nuclear atypia; in addition, high-grade PIN lesions generally display marked elevation of cellular proliferation markers (Bostwick 1989; Shappell et al. 2004). In contrast with prostate cancer, however, basal cells are reduced in number in PIN, but are not absent.

While evidence of major subtypes of prostate cancer is lacking at the histopathological level, recent genomic analyses have provided increasing evidence for molecularly defined subtypes (Tomlins et al. 2008; Palanisamy et al. 2010; Taylor et al. 2010). In particular, expression profiling analyses of prostate cancer specimens have not strictly defined molecular signatures associated with distinct cancer subtypes that specifically correlate with disease outcome (Singh et al. 2002; Lapointe et al. 2004; Tomlins et al. 2007). However, oncogenomic pathway analyses that integrate analyses of gene expression, copy number alterations, and exon resequencing may provide a unified approach for distinguishing prostate cancer subtypes and stratifying patient outcome (Taylor et al. 2010).

Although common sites of secondary metastasis for prostate cancer are lung, liver, and pleura, if prostate cancer metastasizes, it invariably goes to bone, where it forms characteristic osteoblastic lesions (Bubendorf et al. 2000; Logothetis and Lin 2005).

The identification of key molecular alterations in prostate-cancer cells implicates carcinogen defenses (GSTP1), growth-factor-signaling pathways (NKX3.1, PTEN, and p27), and androgens (AR) as critical determinants of the phenotype of prostate-cancer cells. NKX3.1, PTEN, and p27 regulate the growth and survival of prostate cells in the normal prostate. Inadequate levels of PTEN and NKX3.1 lead to a reduction in p27 levels and to increased proliferation and decreased apoptosis. Androgen receptor (AR) is a transcription factor that is normally activated by its androgen ligand. During androgen withdrawal therapy, the AR signal transduction pathway also could be activated by amplification of the AR gene, by AR gene mutations, or by altered activity of AR coactivators. Through these mechanisms, tumor cells lead to the emergence of androgen-independent prostate cancer.

In order to elucidate the relationship between prostate cancer, apoptosis and autophagy, we will focus on the following genes and signalling pathways, often found involved in these tumors:

- *PTEN*
- *AKT/mTOR and MAPK signalling pathways*

- *p53*
- *Bcl2*
- *Beclin I*

3.1.1 PTEN

PTEN (Tumor suppressors phosphatase and tensin-homolog deleted on chromosome 10) was originally identified as a tumor suppressor, is frequently mutated or deleted in many cancers, including prostate (Salmena et al. 2008). The relevance of PTEN loss for prostate cancer was initially inferred from its location on chromosomal region 10q23, which frequently undergoes allelic loss in prostate cancer, as well as by its reduction or loss of expression in prostate tumors (Wang et al. 1998; Whang et al. 1998; McMenamin et al. 1999; Dong et al. 2007). Earlier studies had generated conflicting data regarding whether both alleles of PTEN are deleted in prostate cancer, or, if one allele is deleted, whether the remaining allele is mutated, or if the expression of PTEN protein is reduced, inactivated, or altered in subcellular localization. To resolve these issues, recent studies have investigated PTEN copy number, mutational status, and/or protein expression in primary or castration-resistant tumors using multiple experimental approaches (Verhagen et al. 2006; Schmitz et al. 2007; Sircar et al. 2009; Taylor et al. 2010). In combination with the consensus of previous reports, these studies support the conclusion that PTEN undergoes copy number loss as an early event in prostate carcinogenesis, and is correlated with progression to aggressive, castration-resistant disease. Interestingly, these studies have also suggested that low levels of PTEN activity may be retained in prostate cancer—an observation that parallels the haploinsufficiency of NKX3.1 and the p27 cell cycle regulator (Gao et al. 2004; Abate-Shen et al. 2008), and which may reflect the relative indolence of prostate tumors.

Analyses of Pten deletion in genetically engineered mouse models have uncovered its cooperativity with inactivation of other key genes that are deregulated in prostate tumorigenesis, and have also provided insights into new therapeutic options for the treatment of prostate cancer. Germline loss of Pten in heterozygous mutants or conditional deletion in the prostate epithelium results in PIN and/or adenocarcinoma (Di Cristofano et al. 1998; Podsypanina et al. 1999; Trotman et al. 2003; Wang et al. 2003). Inactivation of Pten has been shown to cooperate with loss of function of the Nkx3.1 homeobox gene, up-regulation of the c-Myc proto-oncogene, or the TMPRSS-ERG fusion (Kim et al. 2002, 2009; Carver et al. 2009; King et al. 2009). Notably, PTEN reduction or loss in prostate cancer predisposes to the emergence of castration-resistant prostate cancer (Mulholland et al. 2006; Shen and Abate-Shen 2007). In particular, perturbation of PTEN expression in human prostate cancer cell lines or targeted deletion of Pten in mouse prostate cancers is sufficient for the development of castration resistance (Lin et al. 2004; Bertram et al. 2006; Gao et al. 2006; Wu et al. 2006). While this may reflect the ability of PTEN to interact directly with AR, the mechanistic details by which PTEN loss promotes castration resistance remain to be resolved.

3.1.2 Akt/mTOR and MAPK Signalling Pathways

Constitutive activation of the PI3K/AKT/mTOR axis is a survival mechanism commonly encountered in human cancer. The abnormal activation of this pathway can be ascribed to diverse cellular events, such as loss of PTEN, loss of tuberous sclerosis complex (TSC) 1 and 2, amplification or mutation of class I PI3K, overexpression of AKT, constitutive activation of tyrosine kinase growth factor receptors and exposure to carcinogens. In prostate cancer, the up-regulation of the Akt/mTOR signaling pathway has been mainly ascribed to the loss of function of Pten gene, primarily through activation of Akt1 (Thomas et al. 2004; Chen et al. 2006b; Mulholland et al. 2006; Shen and Abate-Shen 2007). Nevertheless, up-regulation of this pathway in prostate cancer can also take place through activating mutations of Akt1 (Boormans et al. 2008), or through activation of the p110b isoform of PI3K (Hill et al. 2010; Lee et al. 2010). The functional consequences of Akt/mTOR pathway activation are particularly relevant for castration-resistant prostate cancer, as has been shown in genetically engineered mouse models, in gain-of-function studies with orthotopic grafting or tissue recombination models, as well as in human cell lines (Majumder et al. 2003; Uzgare and Isaacs 2004; Gao et al. 2006; Xin et al. 2006). The consequences of Akt activation are mediated in part by activation of NF-kB signaling via stimulation of IKK (Dan et al. 2008). Conversely, functional studies in mouse models and correlative studies in human prostate cancer have implicated deregulated NF-kB signaling in mediating androgen responsivity, metastasis, and disease outcome (Fradet et al. 2004; Ismail et al. 2004; Lessard et al. 2006; Luo et al. 2007; Zhang et al. 2009).

Constitutive activation of the PI3K/AKT/mTOR axis result in autophagy suppression; the relationship between the PI3K/AKT/mTOR pathway and autophagy is also suggested by the findings that G-protein coupled receptor (GPCR) antagonists to growth factor receptors (GFR), class I PI3K inhibitors such as lithium and carbamazepine, AKT inhibitors such as perifostine and AKT/PKB signaling inhibitor-2 (API-2), and mTOR inhibitors such as rapamycin, RAD-001 and CCI-779, result in autophagy induction (Nicholson and Anderson 2002; Majumder and Sellers 2005; Moretti et al. 2007).

Tumors with high metabolic demands, such as those with constitutively active PI3K mutations, PTEN loss or AKT activation, would be expected to be dependent on autophagy for energy homeostasis and survival. Thus, suppression of autophagy by the PI3K signaling cascade presents a disadvantage that these rapidly proliferating tumor cells may have to overcome to remain viable, and leads to the prediction that compensatory mechanisms, such as deregulated apoptosis and/or metabolism, may be concurrently activated to counteract the negative implications of defective autophagy on tumor cell survival.

In addition to Akt/mTOR signaling, Erk (p42/44) MAPK signaling is also frequently activated in prostate cancer, particularly in advanced disease, and is often coordinately deregulated together with Akt signaling (Abreu-Martin et al. 1999; Gioeli et al. 1999; Paweletz et al. 2001; Malik et al. 2002; Thomas et al. 2004; Kinkade et al. 2008). The mitogen-activated protein kinases (MAPKs) are the family of kinases that transduce signals from the cell membrane to the nucleus in response to a wide range of stimuli, including stress. MAPKs are serine/threonine kinases that, upon stimulation, phosphorylate their specific substrates at serine and/or threonine residues. Such phosphorylation events can either positively or negatively regulate substrate, and thus entire signalling cascade activity. Thus, the MAPK signalling pathways modulate gene expression, mitosis, proliferation, motility, metabolism, and programmed cell death 'apoptosis'. It has been demonstrated that constitutive activation of the MAPK ERK regulates the maturation of autophagosomes (Corcelle et al. 2006); moreover, the oncogenic activation of ras (rasV12), the upstream activator of ERK, has been also reported to induce autophagic vacuolation (Chi et al. 1999; Pattingre et al. 2003).

Simultaneous activation of Akt/mTOR and Erk (p42/44) MAPK signalling pathways promotes tumor progression and castration resistance in prostate cancer cell lines and mouse models (Uzgare and Isaacs 2004; Gao et al. 2006), while combinatorial inhibition of these pathways inhibits castration-resistant prostate cancer in genetically engineered mice (Kinkade et al. 2008). In contrast with Akt/mTOR signalling, the upstream events that lead to activation of Erk MAPK signaling are less well defined, but are thought to be linked to aberrant growth factor signaling (Gioeli 2005). Although mutations of RAS or RAF are rarely found in human prostate cancer, the pathway is frequently perturbed in advanced prostate cancers (Taylor et al. 2010). Notably, expression of activated forms of either Raf or Ras in the mouse prostrate epithelium results in MAPK activation and promotes cancer formation (Jeong et al. 2008; Pearson et al. 2009).

3.1.3 p53

p53 (also known as TP53, for tumor protein p53), is a tumor suppressor protein that is encoded by the *TP53* gene. p53 is crucial in multicellular organisms, where it regulates the cell cycle and functions as a tumor suppressor. Because of its role in conserving genome stability by preventing accumulation of mutations, p53 has been also described as "the guardian of the genome".

p53 plays many roles in anticancer function; among them:

- It can activate DNA repair proteins when DNA has sustained damage.
- It can induce growth arrest by holding the cell cycle at the G_1/S regulation point on DNA damage recognition (allowing for the DNA repair proteins to fix the damage, so to permit the cell to continue the cell cycle).
- It can initiate apoptosis, the programmed cell death, if DNA damage proves to be irreparable.

In unperturbed conditions, the p53 protein is continually produced and degraded in the cell. The degradation of the p53 protein is associated with MDM2 binding. In a negative feedback loop, MDM2 is itself induced by the p53 protein. However, mutant p53 proteins often do not induce MDM2, and are thus able to accumulate at very high concentrations. Worse, mutant p53 protein itself can inhibit normal p53 protein levels.

Patients carrying germline mutations of the *TP53* gene are most likely develop tumors in early adulthood, a disease known as Li-Fraumeni syndrome. Somatic mutations also occurs with a very high rate: more than 50 % of human tumors, in fact, hold p53 mutations (Hollstein et al. 1991).

In prostate cancer, the frequency of p53 mutations seems to be lower than in other cancers. A relatively minor role for p53 in prostate carcinogenesis is consistent with the observation that Li-Fraumeni patients have a low incidence of prostate cancer (Kleihues et al. 1997), although it has been hypothesize that they may die by other carcinomas before they can develop prostate cancer.

Loss of chromosome 17p occurs in advanced stages of prostate cancer and metastatic disease (Cher and Carroll 1994; Cher et al. 1996; Saric et al. 1999), deleting a region that includes the p53 locus, but not BRCA1 (Brooks et al. 1996). It is now generally accepted that mutations of p53 occur infrequently in early invasive carcinoma (Henke et al. 1994; Voeller et al. 1994; Prendergast et al. 1996). In contrast, p53 is mutated in advanced stages of prostate cancer, as well as in recurrent and metastatic disease (Effert et al. 1993; Navone et al. 1993; Aprikian et al. 1994; Eastham et al. 1995; Heidenberg et al. 1995). Moreover, several studies indicate that p53 overexpression is a predictive factor for poor prognosis and disease recurrence, particularly when detected in combination with Bcl2 (Thomas et al. 1993; Shurbaji et al. 1995; Bauer et al. 1996; Moul et al. 1996; Matsushima et al. 1997; Theodorescu et al. 1997; Brewster et al. 1999; Stackhouse et al. 1999).

p53 appears to have a dual role in autophagy regulation. Upon DNA damage, hypoxia and oncogene activation, p53 has been shown to transactivate autophagy-inducing genes and stimulate autophagy by inhibiting mTOR in an AMP-activated protein kinase (AMPK)- and TSC1/TSC2-dependent manner. p53 also induces autophagy via its direct target damage-regulated autophagy modulator (DRAM). At the same time, however, genetic or pharmacologic inactivation of cytoplasmic p53 also triggers autophagy, indicating that the non-nuclear p53 pool is a potent autophagy repressor (Jones et al. 2005; Budanov and Karin 2008; Maiuri et al. 2009a, b; Feng and Levine 2010). Thus, autophagy is activated as a stress-mitigating mechanism by both stress-mediated p53 induction and stress-exacerbating p53 loss. The circumstances and the molecular pathways involved in the decision to use p53 for autophagy activation versus inhibition in cancer cells have not yet been determined. Plausibly, p53 loss, and thus autophagy induction, or negative regulation of autophagy inhibition, may be one of the compensatory mechanisms that tumor cells use to counter-balance the survival-undermining effects of autophagy suppression by an activated PI3K/AKT/mTOR axis.

3.1.4 Bcl2

Bcl-2 (B-cell lymphoma 2) is the founding member of the Bcl-2 family of apoptosis regulator proteins encoded by the BCL2 gene (Tsujimoto et al. 1984; Cleary et al. 1986).

So far, 15 mammalian family members were identified, which were divided into three subfamilies:

1. Bcl-2 subfamily (pro-survival): ***Bcl-2***, Bcl-XL, Bcl-w, Mcl-1 and A1;
2. Bax subfamily (pro-apoptotic): Bax, Bak and Bok;
3. BH3 subfamily (pro-apoptotic): Bad, Bid, Bik, Blk, Hrk, BNIP3 and BimL;

Additionally, several Bcl-2 homologs have been identified in viruses, among others the adenovirus oncoprotein E1B-19 K.

A central checkpoint of apoptosis is the activation of Caspase-9 by mitochondria. Bcl-2, and Bcl-XL, can bind to the C terminal part of Apaf-1 (to the CED-4 like part and the WD-40 domain), thus inhibiting the association of Caspase-9 with Apaf-1. The pro-survival proteins also seem to maintain organelle integrity since Bcl-2 directly or indirectly prevents the release of cytochrome c from mitochondria.

Overexpression of Bcl2 in prostate carcinoma cells is a hallmark of advanced, hormone-refractory disease, and may account for the resistance to apoptosis that is characteristic of late stages (Colombel et al. 1993; McDonnell et al. 1997). Although Bcl2 expression is restricted to basal cells in the normal prostate, forced expression of Bcl2 in LnCAP prostate carcinoma cells protects against apoptosis induced by androgen depletion (Raffo et al. 1995). Moreover, as is the case for p53, Bcl2 expression may provide a prognostic marker that correlates with disease outcome (Mackey et al. 1998). Indeed, several preliminary studies have examined whether Bcl2 inactivation may prevent tumor recurrence (Miyake et al. 1999). Overexpression of Bcl2 has been shown to confer resistance to chemotherapy in prostate carcinoma cell lines (Tu et al. 1995), and current clinical efforts are aimed at modulating the expression of Bcl2 (DiPaola and Aisner 1999).

3.1.5 Beclin-1 and the Crosstalk Between Apoptosis and Autophagy

Apoptosis and autophagy share similarities in that both are self-degradative cellular pathways activated under conditions of stress.

The potential for crosstalk between apoptosis and autophagy was first recognized when Beclin 1 was initially identified as a Bcl-2-interacting protein. Regulators of apoptosis, such as Bcl-2/Bcl-xL and the BH3-only proteins, interact with Beclin 1 and can modulate autophagy (Wang 2008). The anti-apoptotic protein Bcl-2 binds to Beclin 1 under non-stress conditions and inhibits autophagy in the ER, whereas the BH3-only protein Bad, BNIP3, and BH3 mimetics, such as ABT737, competitively inhibit the interaction between Beclin 1 and Bcl-2/BclxL and stimulate autophagy (Wang 2008). We can conclude that, up to our knowledge, positive regulators of apoptosis also induce autophagy, which is reasonable given that both pathways are activated under similar stress conditions. The cell fate, in response to metabolic stress, is determined by the functional status and the interaction between the stress-mitigating pathways of apoptosis and autophagy.

In prostate cancer some data are available from cultured cell lines experiments. It has been recently demonstrated that in PC3, prostate cancer cell lines, the Ursolic Acid-induced autophagy is mediated through the Beclin-1 and Akt/mTOR pathways. (Ursolic acid is a pentacyclic triterpenoid, that inhibit the growth of cancer cells by cell cycle arrest and the stimulation of apoptosis). Inhibition of autophagy by either 3-methyladenine or Beclin-1/Atg5 small interfering RNA enhanced UA-induced apoptosis (Shin et al. 2012). Moreover, it has been shown that autophagy is elevated in LNCaP cells under androgen deprivation conditions, which results in increased cell viability (Li et al. 2008) (Fig. 3.1).

3.2 Non-apoptotic/Non-autophagic Cell Death Pathways in Prostate Cancer

3.2.1 Anoikis

Anoikis (from a Greek word meaning "homelessness") is defined as anchorage-dependent programmed cell death. It can be considered an apoptotic process that is induced by inadequate or inappropriate cell–matrix interactions. Anokis is used to describe the apoptotic response elicited by the absence of cell-matrix interactions (Frisch and Screaton 2001).

In prostate carcinoma cell lines, anoikis has been reported to be regulated by Bcl2-independent pathways; mitochondrial DNA depletion in prostate epithelial cells promotes anoikis resistance and invasion through activation of PI3K/Akt2. Several papers propose Anoikis as a novel therapeutic target for prostate cancer (Bondar and McConkey 2002; Garrison and Kyprianou 2004; Hasanuzzaman et al. 2007; Moro et al. 2009; Sakamoto and Kyprianou 2010).

3.2.2 Autoschizis

Autoschizis is a term derived from the Greek roots "auto" meaning self, and "skhizein" to split. It indicates a recently described form of cancer cell death characterized by a reduction in cell size due to the loss of cytoplasm through self-excision. This process occurs without cell organelles loss, in absence of morphologic degradation of the cells nucleus and nucleolus and without the formation of apoptotic bodies and destruction of the cell membrane. The cell death results from karyorrhexis and karyolysis. Autoschizis can be initiated via in vivo treatment with Vitamin C (VC), synthetic Vitamin K (VK3) or a combination of both. The treatment has been tested on various types of cancers with positive results (Jamison et al. 2002)

A combination of vitamin C/K(3) has been reported to induce cell death by autoschizis in prostate carcinoma cell lines (Taper et al. 2001; Lasalvia-Prisco et al. 2003; Gilloteaux et al. 2005; Tomasetti et al. 2010).

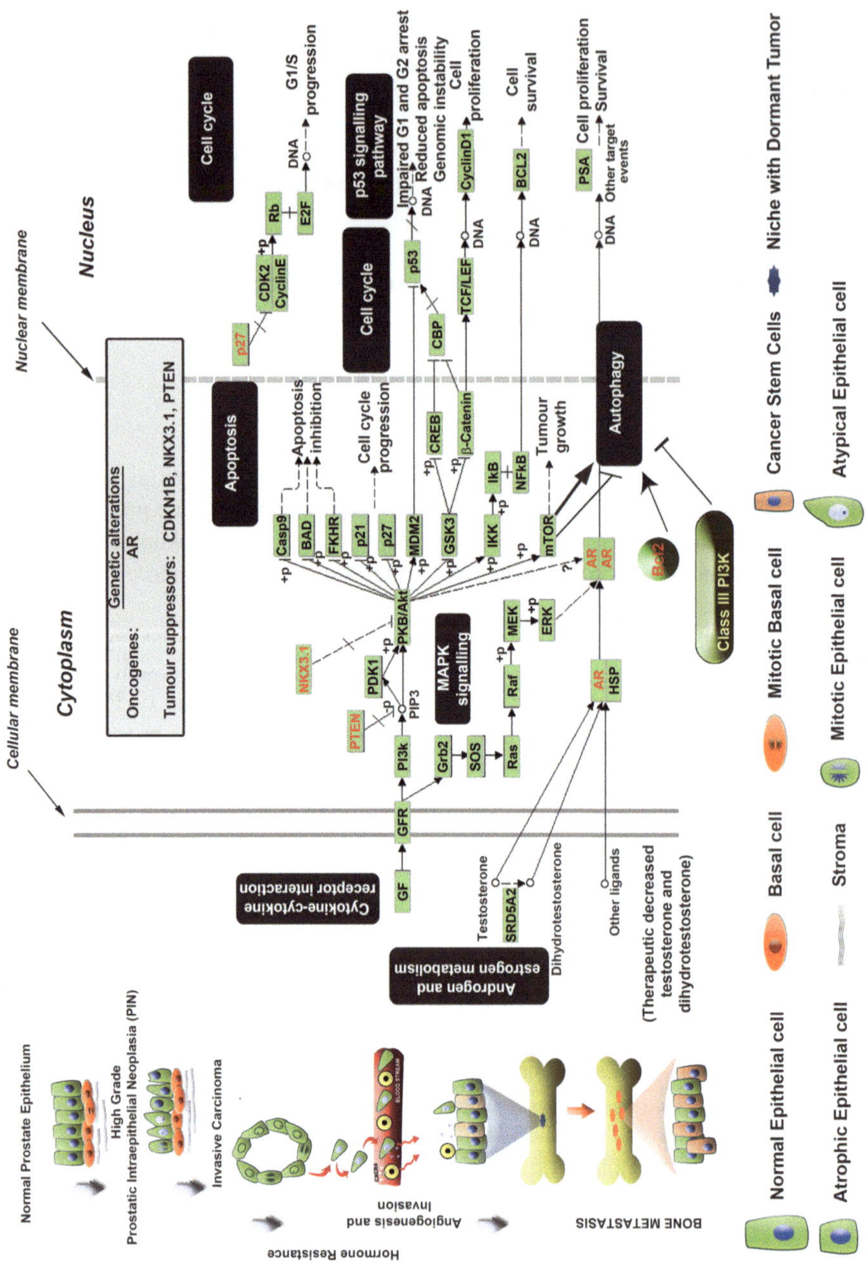

Cellular membrane
Nuclear membrane
Cytoplasm
Nucleus
Genetic alterations
Oncogenes: AR
Tumour suppressors: CDKN1B, NKX3.1, PTEN
Cytokine-cytokine receptor interaction
Androgen and estrogen metabolism
GF
GFR
PI3k
PIP3
PTEN
PDK1
NKX3.1
PKB/Akt
Grb2
SOS
Ras
Raf
MEK
ERK
MAPK signalling
Testosterone
SRD5A2
Dihydrotestosterone
Other ligands
(Therapeutic decreased testosterone and dihydrotestosterone)
AR HSP
AR AR
Bcl2
Class III PI3K
Casp9
BAD
FKHR
p21
p27
MDM2
GSK3
IKK
IkB
NFkB
mTOR
CREB
β-Catenin
Apoptosis
Apoptosis inhibition
Cell cycle progression
Tumour growth
Autophagy
p27
CDK2 CyclinE
Rb
E2F
DNA
G1/S progression
Cell cycle
p53
CBP
TCF/LEF
CyclinD1
BCL2
PSA
p53 signalling pathway
Impaired G1 and G2 arrest
Reduced apoptosis
Genomic instability
Cell proliferation
Cell survival
Cell proliferation
Survival
Other target events
Normal Prostate Epithelium
High Grade Prostatic Intraepithelial Neoplasia (PIN)
Invasive Carcinoma
Hormone Resistance
Angiogenesis and Invasion
BLOOD STREAM
BONE METASTASIS
Normal Epithelial cell
Basal cell
Mitotic Basal cell
Cancer Stem Cells
Niche with Dormant Tumor
Atrophic Epithelial cell
Stroma
Mitotic Epithelial cell
Atypical Epithelial cell

Fig. 3.1 Mechanism of cellular death in the genesis and progression of PCa. Apoptosis and autophagy are both self-degradative cellular pathways activated under conditions of stress to eliminate unwanted or irreparably damaged cells and as a stress adaptive response to prolong cell survival, respectively. In this figure is reported the complex ridge of molecular pathway activated in cancerogenesis and progression of PCa, in which are comprised those regulating apoptosis and autophagy.
Loss of PTEN is an early event in prostate carcinogenesis and correlates with progression to castration-resistant disease and parallels with loss of function of the Nkx3.1 homeobox gene, up-regulation of the c-Myc proto-oncogene, or the TMPRSS-ERG fusion, involved in cellular death. Moreover, loss of function of PTEN causes the abnormal activation of the PI3K/AKT/mTOR axis and result in autophagy suppression. The latter event leads to compensatory mechanisms, such as deregulated apoptosis and/or metabolis to counteract the negative implications of defective autophagy on tumor cell survival.
The constitutive activation of the MAPK ERK regulates the maturation of autophagosomes and induce autophagic vacuolation.
p53 is a tumor suppressor protein that regulates the cell cycle and functions as a tumor suppressor. It regulate autophagy in a double manner: by inhibiting mTOR in an AMP-activated protein kinase (AMPK)- and TSC1/TSC2-dependent manner and via its direct target damage-regulated autophagy modulator (DRAM).
Bcl-2 inhibits the association of Caspase-9 with Apaf-1, and, therefore, its activation, so regulating negatively the apoptotic mechanism.
A potential for crosstalk between apoptosis and autophagy is hypothesizable because Beclin 1 is a Bcl-2-interacting protein and can modulate autophagy. The anti-apoptotic protein Bcl-2 binds to Beclin 1 under non-stress conditions and inhibits autophagy in the ER

3.2.3 *Entosis*

Entosis is a form of cell death that occurs when a cell dyes being engulfed by a neighboring cell. The process was discovered by Overholtzer, et al. as reported in *Cell* (Overholtzer et al. 2007).

Several works indicate entosis as a non-genetic cause of aneuploidy. Aneuploidy is common in human tumors and is often indicative of aggressive disease. Aneuploidy can result from cytokinesis failure, which produces binucleate cells that generate aneuploid offspring with subsequent divisions. In cancers, disruption of cytokinesis is known to result from genetic perturbations to mitotic pathways or checkpoints. It has been described a non-genetic mechanism of cytokinesis failure that occurs as a direct result of cell-in-cell formation by entosis. Live cells internalized by entosis, which can persist through the cell cycle of host cells, disrupt formation of the contractile ring during host cell division. As a result, cytokinesis frequently fails, generating binucleate cells that produce aneuploid cell lineages (White 2007; Janssen and Medema 2011; Krajcovic et al. 2011).

3.2.4 *Excitotoxicity*

The overactivation of receptors for the excitatory neurotransmitter glutamate (glutamate receptors) such as the NMDA receptor and AMPA receptor can lead to the so called Excitotoxicity, that is the pathological process by which nerve cells are damaged and killed by excessive stimulation.

Excitotoxins like NMDA and kainic acid which bind to these receptors, as well as pathologically high levels of glutamate, can cause excitotoxicity by allowing high levels of calcium ions ($Ca2+$) to enter the cell. $Ca2+$ influx into cells activates a number of enzymes, including phospholipases, endonucleases, and proteases such as calpain. These enzymes go on to damage cell structures such as components of the cytoskeleton, membrane, and DNA.

Glutamate carboxypeptidase II (GCPII) is a membrane responsible for the cleavage of N-acetyl-L-aspartyl-L-glutamate (NAAG) yielding free glutamate in the synaptic cleft, and is implicated in various pathologic conditions associated with glutamate excitotoxicity. The prostate form of GCPII, termed prostate-specific membrane antigen (PSMA), is up-regulated in cancer and used as an effective prostate cancer marker (Barinka et al. 2004; Ding et al. 2007).

3.2.5 *Mitotic Catastrophe*

Mitotic catastrophe is an event in which a cell is destroyed during mitosis. This is believed by some to occur as a result of an attempt at aberrant chromosome segregation early in mitosis, or as a result of DNA damage later. Cells which fail to go through a mitotic catastrophe after a mitotic failure are likely to create aneuploid cells when they later reproduce, posing a risk of oncogenesis, potentially leading to cancer.

Mitotic catastrophe thus may be conceived as a molecular device that prevents aneuploidization, which may participate in oncogenesis. Mitotic catastrophe is controlled by numerous molecular players, in particular, cell-cycle-specific kinases (such as the cyclin B1-dependent kinase Cdk1, polo-like kinases and Aurora kinases), cell-cycle checkpoint proteins, survivin, p53, caspases and members of the Bcl-2 family (Castedo et al. 2004).

A large body of works show the correlation between mitotic catastrophe and prostate cancer. Taxols, such as docetaxel, has been shown to induce cell death by mitotic catastrophe, in prostate cancer cells, by concomitant activation of caspase and lysosomal pathways; resistance to docetaxel has been proven as a complex mechanism involving several genes, as shown by several genomic and proteomic approaches (Fabbri et al. 2008; Mediavilla-Varela et al. 2009; Balasubramani et al. 2011; Desarnaud et al. 2011).

3.2.6 *Necrosis and Oncosis*

Necrosis does not indicate a form of cell death but refers to changes secondary to cell death by any mechanism, including apoptosis. The term oncosis (derived from ónkos, meaning swelling) was proposed in 1910 by von Reckling-hausen precisely to mean cell death with swelling. Oncosis leads to necrosis with karyolysis and stands in contrast to apoptosis, which leads to necrosis with karyorhexis and cell shrinkage (Majno and Joris 1995).

Some compounds, such as Kahalalide F, a marine-derived compound, have been reported to induce oncosis in human prostate cancer cells (Suarez et al. 2003).

3.2.7 *Paraptosis*

Paraptosis, which has been observed in a variety of cell types in response to insulin derived growth factor 1 receptor, differs from apoptosis because of the lack of fragmentation of the cell, its nucleus, and its DNA, and from necrosis due to its requirement for new RNA and protein synthesis.

Paraptosis, like apoptosis, does indeed involve a caspase, caspase-9. Compounds able to modulate the proteasome along with Hsp90 protein, were also able to induce prostate carcinoma cell lines death by paraptosis (Wang et al. 2012).

3.3 Other Cell Death Pathways

3.3.1 Parthanatos

Parthanatos is a form of cell death that often occurs as a result of ischemia reperfusion injury. This form of cell death is distinct from apoptosis, necrosis, or autophagy and is being referred to also as PARP1-dependent cell death [poly(ADP-ribose) polymerase 1-dependent cell death]. Although it shows some features of cell death by apoptosis, it is not associated with the formation of apoptotic bodies. Parthanatos also differs from autophagy in that it does not involve the formation of autophagic vacuoles and lysosomal degradation (David et al. 2009).

3.4 Therapeutic Implications

The requirement of the mTORC2 complex as well as the p110b isoform of PI3K for tumor formation following Pten loss suggests that these signaling components may provide additional and/or alternative targets for therapeutic intervention (Jia et al. 2008; Guertin et al. 2009). Moreover, the observation that complete inactivation of Pten in mouse prostate tumors leads to cellular senescence (Chen et al. 2006a) has led to the idea that novel therapeutic approaches might promote senescence for selective targeting of prostate tumor cells through knockdown of Pten function (Alimonti et al. 2010) or targeting of Skp2 (Lin et al. 2010). Furthermore, a small percentage of aggressive prostate tumors contains a translocation of B-RAF or C-RAF that results in activation (Palanisamy et al. 2010). This let envisage a further therapeutical approach based on the pharmacological inhibition of RAF kinase.

Finally, several novel prostate cancer cells killing strategies are based on different cell death pathways, especially apoptosis and autophagy, that have been proved to be an interesting field of investigation in order to find efficient therapeutic approaches for androgen-resistant and metastatic prostate carcinomas.

References

Abate-Shen C, Shen MM, Gelmann E (2008) Integrating differentiation and cancer: the Nkx3.1 homeobox gene in prostate organogenesis and carcinogenesis. Differentiation 76:717–727

Abreu-Martin MT, Chari A, Palladino AA, Craft NA, Sawyers CL (1999) Mitogen-activated protein kinase kinase kinase 1 activates androgen receptor-dependent transcription and apoptosis in prostate cancer. Mol Cell Biol 19:5143–5154

Aihara M, Wheeler TM, Ohori M, Scardino PT (1994) Heterogeneity of prostate cancer in radical prostatectomy specimens. Urology 43:60–66; discussion 6–7

Alimonti A, Nardella C, Chen Z, Clohessy JG, Carracedo A, Trotman LC et al (2010) A novel type of cellular senescence that can be enhanced in mouse models and human tumor xenografts to suppress prostate tumorigenesis. J Clin Invest 120:681–693

Alnemri ES, Livingston DJ, Nicholson DW, Salvesen G, Thornberry NA, Wong WW et al (1996) Human ICE/CED-3 protease nomenclature. Cell 87:171
Aprikian AG, Zhang ZF, Fair WR (1994) Prostate adenocarcinoma in men younger than 50 years. A retrospective review of 151 patients. Cancer 74:1768–1777
Balasubramani M, Nakao C, Uechi GT, Cardamone J, Kamath K, Leslie KL et al (2011) Characterization and detection of cellular and proteomic alterations in stable stathmin-overexpressing, taxol-resistant BT549 breast cancer cells using offgel IEF/PAGE difference gel electrophoresis. Mutat Res 722:154–164
Barinka C, Sacha P, Sklenar J, Man P, Bezouska K, Slusher BS et al (2004) Identification of the N-glycosylation sites on glutamate carboxypeptidase II necessary for proteolytic activity. Protein Sci 13:1627–1635
Bauer JJ, Sesterhenn IA, Mostofi FK, McLeod DG, Srivastava S, Moul JW (1996) Elevated levels of apoptosis regulator proteins p53 and bcl-2 are independent prognostic biomarkers in surgically treated clinically localized prostate cancer. J Urol 156:1511–1516
Bertram J, Peacock JW, Tan C, Mui AL, Chung SW, Gleave ME et al (2006) Inhibition of the phosphatidylinositol 3′-kinase pathway promotes autocrine Fas-induced death of phosphatase and tensin homologue-deficient prostate cancer cells. Cancer Res 66:4781–4788
Bondar VM, McConkey DJ (2002) Anoikis is regulated by BCL-2-independent pathways in human prostate carcinoma cells. Prostate 51:42–49
Boormans JL, Hermans KG, van Leenders GJ, Trapman J, Verhagen PC (2008) An activating mutation in AKT1 in human prostate cancer. Int J Cancer 123:2725–2726
Bostwick DG (1989) Prostatic intraepithelial neoplasia (PIN). Urology 34:16–22
Bostwick DG, Shan A, Qian J, Darson M, Maihle NJ, Jenkins RB et al (1998) Independent origin of multiple foci of prostatic intraepithelial neoplasia: comparison with matched foci of prostate carcinoma. Cancer 83:1995–2002
Brewster SF, Oxley JD, Trivella M, Abbott CD, Gillatt DA (1999) Preoperative p53, bcl-2, CD44 and E-cadherin immunohistochemistry as predictors of biochemical relapse after radical prostatectomy. J Urol 161:1238–1243
Brooks JD, Bova GS, Ewing CM, Piantadosi S, Carter BS, Robinson JC et al (1996) An uncertain role for p53 gene alterations in human prostate cancers. Cancer Res 56:3814–3822
Bubendorf L, Schopfer A, Wagner U, Sauter G, Moch H, Willi N et al (2000) Metastatic patterns of prostate cancer: an autopsy study of 1,589 patients. Hum Pathol 31:578–583
Budanov AV, Karin M (2008) p53 target genes sestrin1 and sestrin2 connect genotoxic stress and mTOR signaling. Cell 134:451–460
Carver BS, Tran J, Gopalan A, Chen Z, Shaikh S, Carracedo A et al (2009) Aberrant ERG expression cooperates with loss of PTEN to promote cancer progression in the prostate. Nat Genet 41:619–624
Castedo M, Perfettini JL, Roumier T, Andreau K, Medema R, Kroemer G (2004) Cell death by mitotic catastrophe: a molecular definition. Oncogene 23:2825–2837
Chen BY, Lin DP, Liu JY, Chang H, Huang PH, Chen YL et al (2006a) A mouse prostate cancer model induced by Hedgehog overexpression. J Biomed Sci 13:373–384
Chen ML, Xu PZ, Peng XD, Chen WS, Guzman G, Yang X et al (2006b) The deficiency of Akt1 is sufficient to suppress tumor development in Pten+/− mice. Genes Dev 20:1569–1574
Cher ML, Carroll PR (1994) Screening for prostate cancer. West J Med 160:250
Cher ML, Bova GS, Moore DH, Small EJ, Carroll PR, Pin SS et al (1996) Genetic alterations in untreated metastases and androgen-independent prostate cancer detected by comparative genomic hybridization and allelotyping. Cancer Res 56:3091–3102
Chi S, Kitanaka C, Noguchi K, Mochizuki T, Nagashima Y, Shirouzu M et al (1999) Oncogenic Ras triggers cell suicide through the activation of a caspase-independent cell death program in human cancer cells. Oncogene 18:2281–2290
Clark J, Attard G, Jhavar S, Flohr P, Reid A, De-Bono J et al (2008) Complex patterns of ETS gene alteration arise during cancer development in the human prostate. Oncogene 27:1993–2003

Cleary ML, Smith SD, Sklar J (1986) Cloning and structural analysis of cDNAs for bcl-2 and a hybrid bcl-2/immunoglobulin transcript resulting from the t(14;18) translocation. Cell 47:19–28

Colombel M, Symmans F, Gil S, O'Toole KM, Chopin D, Benson M et al (1993) Detection of the apoptosis-suppressing oncoprotein bc1-2 in hormone-refractory human prostate cancers. Am J Pathol 143:390–400

Corcelle E, Nebout M, Bekri S, Gauthier N, Hofman P, Poujeol P et al (2006) Disruption of autophagy at the maturation step by the carcinogen lindane is associated with the sustained mitogen-activated protein kinase/extracellular signal-regulated kinase activity. Cancer Res 66:6861–6870

Cryns V, Yuan J (1998) Proteases to die for. Genes Dev 12:1551–1570

Dan HC, Cooper MJ, Cogswell PC, Duncan JA, Ting JP, Baldwin AS (2008) Akt-dependent regulation of NF-{kappa}B is controlled by mTOR and Raptor in association with IKK. Genes Dev 22:1490–1500

David KK, Andrabi SA, Dawson TM, Dawson VL (2009) Parthanatos, a messenger of death. Front Biosci 14:1116–1128

DeMarzo AM, Nelson WG, Isaacs WB, Epstein JI (2003) Pathological and molecular aspects of prostate cancer. Lancet 361:955–964

Desarnaud F, Geck P, Parkin C, Carpinito G, Makarovskiy AN (2011) Gene expression profiling of the androgen independent prostate cancer cells demonstrates complex mechanisms mediating resistance to docetaxel. Cancer Biol Ther 11:204–212

Di Cristofano A, Pesce B, Cordon-Cardo C, Pandolfi PP (1998) Pten is essential for embryonic development and tumour suppression. Nat Genet 19:348–355

Ding P, Helquist P, Miller MJ (2007) Design, synthesis and pharmacological activity of novel enantiomerically pure phosphonic acid-based NAALADase inhibitors. Org Biomol Chem 5:826–831

DiPaola RS, Aisner J (1999) Overcoming bcl-2- and p53-mediated resistance in prostate cancer. Semin Oncol 26:112–116

Dong B, Kim S, Hong S, Das Gupta J, Malathi K, Klein EA et al (2007) An infectious retrovirus susceptible to an IFN antiviral pathway from human prostate tumors. Proc Natl Acad Sci USA 104:1655–1660

Eastham JA, Stapleton AM, Gousse AE, Timme TL, Yang G, Slawin KM et al (1995) Association of p53 mutations with metastatic prostate cancer. Clin Cancer Res 1:1111–1118

Effert PJ, McCoy RH, Walther PJ, Liu ET (1993) p53 gene alterations in human prostate carcinoma. J Urol 150:257–261

Fabbri F, Amadori D, Carloni S, Brigliadori G, Tesei A, Ulivi P et al (2008) Mitotic catastrophe and apoptosis induced by docetaxel in hormone-refractory prostate cancer cells. J Cell Physiol 217:494–501

Feng Z, Levine AJ (2010) The regulation of energy metabolism and the IGF-1/mTOR pathways by the p53 protein. Trends Cell Biol 20:427–434

Fradet V, Lessard L, Begin LR, Karakiewicz P, Masson AM, Saad F (2004) Nuclear factor-kappaB nuclear localization is predictive of biochemical recurrence in patients with positive margin prostate cancer. Clin Cancer Res 10:8460–8464

Frisch SM, Screaton RA (2001) Anoikis mechanisms. Curr Opin Cell Biol 13:555–562

Gao H, Ouyang X, Banach-Petrosky W, Borowsky AD, Lin Y, Kim M et al (2004) A critical role for p27kip1 gene dosage in a mouse model of prostate carcinogenesis. Proc Natl Acad Sci USA 101:17204–17209

Gao H, Ouyang X, Banach-Petrosky WA, Shen MM, Abate-Shen C (2006) Emergence of androgen independence at early stages of prostate cancer progression in Nkx3.1; Pten mice. Cancer Res 66:7929–7933

Garrison JB, Kyprianou N (2004) Novel targeting of apoptosis pathways for prostate cancer therapy. Curr Cancer Drug Targets 4:85–95

Gilloteaux J, Jamison JM, Neal DR, Summers JL (2005) Cell death by autoschizis in TRAMP prostate carcinoma cells as a result of treatment by ascorbate: menadione combination. Ultrastruct Pathol 29:221–235

Gioeli D (2005) Signal transduction in prostate cancer progression. Clin Sci (Lond) 108:293–308

Gioeli D, Mandell JW, Petroni GR, Frierson HF Jr, Weber MJ (1999) Activation of mitogen-activated protein kinase associated with prostate cancer progression. Cancer Res 59: 279–284

Guertin DA, Stevens DM, Saitoh M, Kinkel S, Crosby K, Sheen JH et al (2009) mTOR complex 2 is required for the development of prostate cancer induced by Pten loss in mice. Cancer Cell 15:148–159

Hanahan D, Weinberg RA (2000) The hallmarks of cancer. Cell 100:57–70

Hanahan D, Weinberg RA (2011) Hallmarks of cancer: the next generation. Cell 144:646–674

Hasanuzzaman M, Kutner R, Agha-Mohammadi S, Reiser J, Sehgal I (2007) A doxycycline-inducible urokinase receptor (uPAR) upregulates uPAR activities including resistance to anoikis in human prostate cancer cell lines. Mol Cancer 6:34

Heidenberg HB, Sesterhenn IA, Gaddipati JP, Weghorst CM, Buzard GS, Moul JW et al (1995) Alteration of the tumor suppressor gene p53 in a high fraction of hormone refractory prostate cancer. J Urol 154:414–421

Henke RP, Kruger E, Ayhan N, Hubner D, Hammerer P, Huland H (1994) Immunohistochemical detection of p53 protein in human prostatic cancer. J Urol 152:1297–1301

Hill KM, Kalifa S, Das JR, Bhatti T, Gay M, Williams D et al (2010) The role of PI 3-kinase p110beta in AKT signally, cell survival, and proliferation in human prostate cancer cells. Prostate 70:755–764

Hollstein M, Sidransky D, Vogelstein B, Harris CC (1991) p53 mutations in human cancers. Science 253:49–53

Ismail HA, Lessard L, Mes-Masson AM, Saad F (2004) Expression of NF-kappaB in prostate cancer lymph node metastases. Prostate 58:308–313

Jamison JM, Gilloteaux J, Taper HS, Calderon PB, Summers JL (2002) Autoschizis: a novel cell death. Biochem Pharmacol 63:1773–1783

Janssen A, Medema RH (2011) Entosis: aneuploidy by invasion. Nat Cell Biol 13:199–201

Jeong JH, Wang Z, Guimaraes AS, Ouyang X, Figueiredo JL, Ding Z et al (2008) BRAF activation initiates but does not maintain invasive prostate adenocarcinoma. PLoS One 3:e3949

Jia L, Yu W, Wang P, Sanders BG, Kline K (2008) In vivo and in vitro studies of anticancer actions of alpha-TEA for human prostate cancer cells. Prostate 68:849–860

Jones RG, Plas DR, Kubek S, Buzzai M, Mu J, Xu Y et al (2005) AMP-activated protein kinase induces a p53-dependent metabolic checkpoint. Mol Cell 18:283–293

Kim MJ, Cardiff RD, Desai N, Banach-Petrosky WA, Parsons R, Shen MM et al (2002) Cooperativity of Nkx3.1 and Pten loss of function in a mouse model of prostate carcinogenesis. Proc Natl Acad Sci USA 99:2884–2889

Kim J, Eltoum IE, Roh M, Wang J, Abdulkadir SA (2009) Interactions between cells with distinct mutations in c-MYC and Pten in prostate cancer. PLoS Genet 5:e1000542

King JC, Xu J, Wongvipat J, Hieronymus H, Carver BS, Leung DH et al (2009) Cooperativity of TMPRSS2-ERG with PI3-kinase pathway activation in prostate oncogenesis. Nat Genet 41:524–526

Kinkade CW, Castillo-Martin M, Puzio-Kuter A, Yan J, Foster TH, Gao H et al (2008) Targeting AKT/mTOR and ERK MAPK signaling inhibits hormone-refractory prostate cancer in a preclinical mouse model. J Clin Invest 118:3051–3064

Kleihues P, Schauble B, zur Hausen A, Esteve J, Ohgaki H (1997) Tumors associated with p53 germline mutations: a synopsis of 91 families. Am J Pathol 150:1–13

Krajcovic M, Johnson NB, Sun Q, Normand G, Hoover N, Yao E et al (2011) A non-genetic route to aneuploidy in human cancers. Nat Cell Biol 13:324–330

Lapointe J, Li C, Higgins JP, van de Rijn M, Bair E, Montgomery K et al (2004) Gene expression profiling identifies clinically relevant subtypes of prostate cancer. Proc Natl Acad Sci USA 101:811–816

Lasalvia-Prisco E, Cucchi S, Vazquez J, Lasalvia-Galante E, Golomar W, Gordon W (2003) Serum markers variation consistent with autoschizis induced by ascorbic acid-menadione in patients with prostate cancer. Med Oncol 20:45–52
Lee KN, Seo MC, Bae IH, Oh SH, Jang WG, Jeong BC et al (2010) COMP-Ang1, a variant of angiopoietin 1, inhibits serum-deprived apoptosis of mesenchymal cells via PI3K/Akt and mitogen-activated protein kinase pathways. Pharmacology 86:327–335
Lessard L, Karakiewicz PI, Bellon-Gagnon P, Alam-Fahmy M, Ismail HA, Mes-Masson AM et al (2006) Nuclear localization of nuclear factor-kappaB p65 in primary prostate tumors is highly predictive of pelvic lymph node metastases. Clin Cancer Res 12:5741–5745
Li M, Jiang X, Liu D, Na Y, Gao GF, Xi Z (2008) Autophagy protects LNCaP cells under androgen deprivation conditions. Autophagy 4:54–60
Lin HK, Hu YC, Lee DK, Chang C (2004) Regulation of androgen receptor signaling by PTEN (phosphatase and tensin homolog deleted on chromosome 10) tumor suppressor through distinct mechanisms in prostate cancer cells. Mol Endocrinol 18:2409–2423
Lin HK, Chen Z, Wang G, Nardella C, Lee SW, Chan CH et al (2010) Skp2 targeting suppresses tumorigenesis by Arf-p53-independent cellular senescence. Nature 464:374–379
Liu W, Laitinen S, Khan S, Vihinen M, Kowalski J, Yu G et al (2009) Copy number analysis indicates monoclonal origin of lethal metastatic prostate cancer. Nat Med 15:559–565
Logothetis CJ, Lin SH (2005) Osteoblasts in prostate cancer metastasis to bone. Nat Rev Cancer 5:21–28
Luo JL, Tan W, Ricono JM, Korchynskyi O, Zhang M, Gonias SL et al (2007) Nuclear cytokine-activated IKKalpha controls prostate cancer metastasis by repressing Maspin. Nature 446: 690–694
Macintosh CA, Stower M, Reid N, Maitland NJ (1998) Precise microdissection of human prostate cancers reveals genotypic heterogeneity. Cancer Res 58:23–28
Mackey TJ, Borkowski A, Amin P, Jacobs SC, Kyprianou N (1998) bcl-2/bax ratio as a predictive marker for therapeutic response to radiotherapy in patients with prostate cancer. Urology 52:1085–1090
Maiuri MC, Malik SA, Morselli E, Kepp O, Criollo A, Mouchel PL et al (2009a) Stimulation of autophagy by the p53 target gene Sestrin2. Cell Cycle 8:1571–1576
Maiuri MC, Tasdemir E, Criollo A, Morselli E, Vicencio JM, Carnuccio R et al (2009b) Control of autophagy by oncogenes and tumor suppressor genes. Cell Death Differ 16:87–93
Majno G, Joris I (1995) Apoptosis, oncosis, and necrosis. An overview of cell death. Am J Pathol 146:3–15
Majumder PK, Sellers WR (2005) Akt-regulated pathways in prostate cancer. Oncogene 24: 7465–7474
Majumder PK, Yeh JJ, George DJ, Febbo PG, Kum J, Xue Q et al (2003) Prostate intraepithelial neoplasia induced by prostate restricted Akt activation: the MPAKT model. Proc Natl Acad Sci USA 100:7841–7846
Malik SN, Brattain M, Ghosh PM, Troyer DA, Prihoda T, Bedolla R et al (2002) Immunohistochemical demonstration of phospho-Akt in high Gleason grade prostate cancer. Clin Cancer Res 8:1168–1171
Matsushima H, Kitamura T, Goto T, Hosaka Y, Homma Y, Kawabe K (1997) Combined analysis with Bcl-2 and P53 immunostaining predicts poorer prognosis in prostatic carcinoma. J Urol 158:2278–2283
McDonnell TJ, Navone NM, Troncoso P, Pisters LL, Conti C, von Eschenbach AC et al (1997) Expression of bcl-2 oncoprotein and p53 protein accumulation in bone marrow metastases of androgen independent prostate cancer. J Urol 157:569–574
McMenamin ME, Soung P, Perera S, Kaplan I, Loda M, Sellers WR (1999) Loss of PTEN expression in paraffin-embedded primary prostate cancer correlates with high Gleason score and advanced stage. Cancer Res 59:4291–4296
Mediavilla-Varela M, Pacheco FJ, Almaguel F, Perez J, Sahakian E, Daniels TR et al (2009) Docetaxel-induced prostate cancer cell death involves concomitant activation of caspase and lysosomal pathways and is attenuated by LEDGF/p75. Mol Cancer 8:68

Mehra R, Han B, Tomlins SA, Wang L, Menon A, Wasco MJ et al (2007) Heterogeneity of TMPRSS2 gene rearrangements in multifocal prostate adenocarcinoma: molecular evidence for an independent group of diseases. Cancer Res 67:7991–7995
Mehra R, Tomlins SA, Yu J, Cao X, Wang L, Menon A et al (2008) Characterization of TMPRSS2-ETS gene aberrations in androgen-independent metastatic prostate cancer. Cancer Res 68:3584–3590
Meier P, Finch A, Evan G (2000) Apoptosis in development. Nature 407:796–801
Miyake H, Hara I, Yamanaka K, Gohji K, Arakawa S, Kamidono S (1999) Overexpression of Bcl-2 enhances metastatic potential of human bladder cancer cells. Br J Cancer 79:1651–1656
Moretti L, Cha YI, Niermann KJ, Lu B (2007) Switch between apoptosis and autophagy: radiation-induced endoplasmic reticulum stress? Cell Cycle 6:793–798
Moro L, Arbini AA, Yao JL, di Sant'Agnese PA, Marra E, Greco M (2009) Mitochondrial DNA depletion in prostate epithelial cells promotes anoikis resistance and invasion through activation of PI3K/Akt2. Cell Death Differ 16:571–583
Moul JW, Bettencourt MC, Sesterhenn IA, Mostofi FK, McLeod DG, Srivastava S et al (1996) Protein expression of p53, bcl-2, and KI-67 (MIB-1) as prognostic biomarkers in patients with surgically treated, clinically localized prostate cancer. Surgery 120:159–166; discussion 66–67
Mulholland DJ, Dedhar S, Wu H, Nelson CC (2006) PTEN and GSK3beta: key regulators of progression to androgen-independent prostate cancer. Oncogene 25:329–337
Navone NM, Troncoso P, Pisters LL, Goodrow TL, Palmer JL, Nichols WW et al (1993) p53 protein accumulation and gene mutation in the progression of human prostate carcinoma. J Natl Cancer Inst 85:1657–1669
Nicholson KM, Anderson NG (2002) The protein kinase B/Akt signalling pathway in human malignancy. Cell Signal 14:381–395
Overholtzer M, Mailleux AA, Mouneimne G, Normand G, Schnitt SJ, King RW et al (2007) A nonapoptotic cell death process, entosis, that occurs by cell-in-cell invasion. Cell 131:966–979
Palanisamy N, Ateeq B, Kalyana-Sundaram S, Pflueger D, Ramnarayanan K, Shankar S et al (2010) Rearrangements of the RAF kinase pathway in prostate cancer, gastric cancer and melanoma. Nat Med 16:793–798
Pattingre S, Bauvy C, Codogno P (2003) Amino acids interfere with the ERK1/2-dependent control of macroautophagy by controlling the activation of Raf-1 in human colon cancer HT-29 cells. J Biol Chem 278:16667–16674
Paweletz CP, Liotta LA, Petricoin EF 3rd (2001) New technologies for biomarker analysis of prostate cancer progression: laser capture microdissection and tissue proteomics. Urology 57:160–163
Pearson JF, Hughes S, Chambers K, Lang SH (2009) Polarized fluid movement and not cell death, creates luminal spaces in adult prostate epithelium. Cell Death Differ 16:475–482
Podsypanina K, Ellenson LH, Nemes A, Gu J, Tamura M, Yamada KM et al (1999) Mutation of Pten/Mmac1 in mice causes neoplasia in multiple organ systems. Proc Natl Acad Sci USA 96:1563–1568
Prendergast NJ, Atkins MR, Schatte EC, Paulson DF, Walther PJ (1996) p53 immunohistochemical and genetic alterations are associated at high incidence with post-irradiated locally persistent prostate carcinoma. J Urol 155:1685–1692
Raffo AJ, Perlman H, Chen MW, Day ML, Streitman JS, Buttyan R (1995) Overexpression of bcl-2 protects prostate cancer cells from apoptosis in vitro and confers resistance to androgen depletion in vivo. Cancer Res 55:4438–4445
Sakamoto S, Kyprianou N (2010) Targeting anoikis resistance in prostate cancer metastasis. Mol Aspects Med 31:205–214
Salmena L, Carracedo A, Pandolfi PP (2008) Tenets of PTEN tumor suppression. Cell 133:403–414
Salvesen GS, Dixit VM (1997) Caspases: intracellular signaling by proteolysis. Cell 91:443–446
Saric T, Brkanac Z, Troyer DA, Padalecki SS, Sarosdy M, Williams K et al (1999) Genetic pattern of prostate cancer progression. Int J Cancer 81:219–224

Schmitz M, Grignard G, Margue C, Dippel W, Capesius C, Mossong J et al (2007) Complete loss of PTEN expression as a possible early prognostic marker for prostate cancer metastasis. Int J Cancer 120:1284–1292

Shappell SB, Thomas GV, Roberts RL, Herbert R, Ittmann MM, Rubin MA et al (2004) Prostate pathology of genetically engineered mice: definitions and classification. The consensus report from the Bar Harbor meeting of the Mouse Models of Human Cancer Consortium Prostate Pathology Committee. Cancer Res 64:2270–2305

Shen MM, Abate-Shen C (2007) Pten inactivation and the emergence of androgen-independent prostate cancer. Cancer Res 67:6535–6538

Shin SW, Kim SY, Park JW (2012) Autophagy inhibition enhances ursolic acid-induced apoptosis in PC3 cells. Biochim Biophys Acta 1823:451–457

Shurbaji MS, Kalbfleisch JH, Thurmond TS (1995) Immunohistochemical detection of p53 protein as a prognostic indicator in prostate cancer. Hum Pathol 26:106–109

Singh J, Young L, Handelsman DJ, Dong Q (2002) Prostate epithelial expression of a novel androgen target gene. J Androl 23:652–660

Sircar K, Yoshimoto M, Monzon FA, Koumakpayi IH, Katz RL, Khanna A et al (2009) PTEN genomic deletion is associated with p-Akt and AR signalling in poorer outcome, hormone refractory prostate cancer. J Pathol 218:505–513

Stackhouse GB, Sesterhenn IA, Bauer JJ, Mostofi FK, Connelly RR, Srivastava SK et al (1999) p53 and bcl-2 immunohistochemistry in pretreatment prostate needle biopsies to predict recurrence of prostate cancer after radical prostatectomy. J Urol 162:2040–2045

Suarez Y, Gonzalez L, Cuadrado A, Berciano M, Lafarga M, Munoz A (2003) Kahalalide F, a new marine-derived compound, induces oncosis in human prostate and breast cancer cells. Mol Cancer Ther 2:863–872

Taper HS, Jamison JM, Gilloteaux J, Gwin CA, Gordon T, Summers JL (2001) In vivo reactivation of DNases in implanted human prostate tumors after administration of a vitamin C/K(3) combination. J Histochem Cytochem 49:109–120

Taylor BS, Schultz N, Hieronymus H, Gopalan A, Xiao Y, Carver BS et al (2010) Integrative genomic profiling of human prostate cancer. Cancer Cell 18:11–22

Theodorescu D, Broder SR, Boyd JC, Mills SE, Frierson HF Jr (1997) p53, bcl-2 and retinoblastoma proteins as long-term prognostic markers in localized carcinoma of the prostate. J Urol 158:131–137

Thomas DJ, Robinson M, King P, Hasan T, Charlton R, Martin J et al (1993) p53 expression and clinical outcome in prostate cancer. Br J Urol 72:778–781

Thomas GV, Horvath S, Smith BL, Crosby K, Lebel LA, Schrage M et al (2004) Antibody-based profiling of the phosphoinositide 3-kinase pathway in clinical prostate cancer. Clin Cancer Res 10:8351–8356

Thornberry NA, Lazebnik Y (1998) Caspases: enemies within. Science 281:1312–1316

Tomasetti M, Strafella E, Staffolani S, Santarelli L, Neuzil J, Guerrieri R (2010) Alpha-Tocopheryl succinate promotes selective cell death induced by vitamin K3 in combination with ascorbate. Br J Cancer 102:1224–1234

Tomlins SA, Mehra R, Rhodes DR, Cao X, Wang L, Dhanasekaran SM et al (2007) Integrative molecular concept modeling of prostate cancer progression. Nat Genet 39:41–51

Tomlins SA, Laxman B, Varambally S, Cao X, Yu J, Helgeson BE et al (2008) Role of the TMPRSS2-ERG gene fusion in prostate cancer. Neoplasia 10:177–188

Trotman LC, Niki M, Dotan ZA, Koutcher JA, Di Cristofano A, Xiao A et al (2003) Pten dose dictates cancer progression in the prostate. PLoS Biol 1:E59

Tsujimoto Y, Finger LR, Yunis J, Nowell PC, Croce CM (1984) Cloning of the chromosome breakpoint of neoplastic B cells with the t(14;18) chromosome translocation. Science 226:1097–1099

Tu SM, McConnell K, Marin MC, Campbell ML, Fernandez A, von Eschenbach AC et al (1995) Combination adriamycin and suramin induces apoptosis in bcl-2 expressing prostate carcinoma cells. Cancer Lett 93:147–155

Uzgare AR, Isaacs JT (2004) Enhanced redundancy in Akt and mitogen-activated protein kinase-induced survival of malignant versus normal prostate epithelial cells. Cancer Res 64:6190–6199

Verhagen PC, van Duijn PW, Hermans KG, Looijenga LH, van Gurp RJ, Stoop H et al (2006) The PTEN gene in locally progressive prostate cancer is preferentially inactivated by bi-allelic gene deletion. J Pathol 208:699–707
Voeller HJ, Sugars LY, Pretlow T, Gelmann EP (1994) p53 oncogene mutations in human prostate cancer specimens. J Urol 151:492–495
Wang J (2008) Beclin 1 bridges autophagy, apoptosis and differentiation. Autophagy 4:947–948
Wang SI, Parsons R, Ittmann M (1998) Homozygous deletion of the PTEN tumor suppressor gene in a subset of prostate adenocarcinomas. Clin Cancer Res 4:811–815
Wang S, Gao J, Lei Q, Rozengurt N, Pritchard C, Jiao J et al (2003) Prostate-specific deletion of the murine Pten tumor suppressor gene leads to metastatic prostate cancer. Cancer Cell 4:209–221
Wang WB, Feng LX, Yue QX, Wu WY, Guan SH, Jiang BH et al (2012) Paraptosis accompanied by autophagy and apoptosis was induced by celastrol, a natural compound with influence on proteasome, ER stress and Hsp90. J Cell Physiol 227:2196–2206
Whang YE, Wu X, Suzuki H, Reiter RE, Tran C, Vessella RL et al (1998) Inactivation of the tumor suppressor PTEN/MMAC1 in advanced human prostate cancer through loss of expression. Proc Natl Acad Sci USA 95:5246–5250
White E (2007) Entosis: it's a cell-eat-cell world. Cell 131:840–842
Wu Z, Conaway M, Gioeli D, Weber MJ, Theodorescu D (2006) Conditional expression of PTEN alters the androgen responsiveness of prostate cancer cells. Prostate 66:1114–1123
Xin L, Teitell MA, Lawson DA, Kwon A, Mellinghoff IK, Witte ON (2006) Progression of prostate cancer by synergy of AKT with genotropic and nongenotropic actions of the androgen receptor. Proc Natl Acad Sci USA 103:7789–7794
Zhang L, Altuwaijri S, Deng F, Chen L, Lal P, Bhanot UK et al (2009) NF-kappaB regulates androgen receptor expression and prostate cancer growth. Am J Pathol 175:489–499

Chapter 4
Androgen Receptor and Steroidogenesis Pathway Control

Simona Romano, Rita Bisogni, and Maria Fiammetta Romano

Abstract Prostate cancer is the most common male malignancy and the leading cause of mortality in western countries. Androgens, the hormones that regulate prostate development and physiology, play a pivotal role also in the maintenance and progression of prostate cancer (Chen et al, Curr Opin Pharmacol, 8:440–448, 2008). Approximately 80–90 % of these tumors are dependent on androgen at initial diagnosis. Therapies that counteract androgen, by reducing its levels, blocking it, or antagonizing the androgen receptor (AR) and its target genes, represent the mainstay of treatment for prostate cancer (Chen et al, Curr Opin Pharmacol, 8:440–448, 2008). However, androgen ablation therapy ultimately fails because prostate cancer progresses to a hormone refractory state.

This chapter focuses on the role of AR-coactivators in prostate cancerogenesis.

AR is expressed throughout prostate cancer progression, and it persists in the majority of patients with hormone refractory disease (Chen et al. 2008). The AR gene is a member of the steroid hormone receptor family of genes and is located on the long (q) arm of the X chromosome at position 12 (Xq11-12) (McEwan 2004). The eight exons of the AR gene code for functionally distinct regions of the protein are similar to the modular structure of other steroid hormone receptor genes. The first exon codes for the N-terminal domain (NTD), that is the transcriptional regulatory region of the protein; exons 2 and 3 code for the central DNA-binding domain (DBD); and exons 4 through 8 code for the C-terminal LBD (ligand binding domain) (McEwan 2004). At least 85 mutations in the AR gene have been associated with prostate cancer (McEwan 2004). Almost all of these mutations are somatic, which means they develop during a person's life and occur only in certain cells

S. Romano (✉) • R. Bisogni • M.F. Romano
Department of Biochemistry and Medical Biotechnology, Federico II University, via S. Pansini, n.5, Naples, Italy
e-mail: romanos@dbbm.unina.it; bisogni@dbbm.unina.it; mariafiammetta.romano@unina.it

S. Staibano (ed.), *Prostate Cancer: Shifting from Morphology to Biology*,
DOI 10.1007/978-94-007-7149-9_4, © Springer Science+Business Media Dordrecht 2013

(in this case, cells in the prostate). Somatic mutations are not inherited and are not passed to future generations. Most AR mutations are capable of transcriptional activity, thereby excluding that loss of AR function is a major cause of androgen ablation failure in prostate cancer therapy. Instead, increasing evidence suggests that a disregulation of AR activity, due to alteration in co-regulators and/or mutations of AR, can enable the receptor to become transcriptionally active in the absence of ligands. This article presents an overview of such co-regulators, with particular focus on the FK506 binding protein (FKBP51), an important factor in the androgen superchaperone receptor complex, which according to recent studies, is a crucial element in prostate cancer biology and progression.

4.1 The Superchaperone Receptor Complex of Steroid Hormone

Similar to all steroid hormones, androgens derive from the precursor pregnenolone, which is produced directly from cholesterol. In humans and other vertebrates, androgens are made primarily in the male testes, female ovaries, and adrenal glands. All the steroid hormones exert their action by passing through the plasma membrane and binding to intracellular receptors. The steroid hormone-receptor (SR) complexes exert their action by binding to specific nucleotide sequences in the DNA of responsive genes. These DNA sequences are identified as hormone response elements. The interaction of steroid-receptor complexes with DNA leads to gene transcription.

In absence of the ligand hormone, inactive SR is maintained in a heterocomplex with several molecular chaperones and co-chaperones, including Hsp90, p23, FKBP51 and FKBP52, in a step-wise manner. The receptor is first delivered by Hsp70 to Hsp90 through the Hsp organizing protein (Hop), which binds both chaperones through two separate tetratricopeptide repeat (TPR) motifs. Notably, TPR motif is a flexible, mutable domain of fundamental biological importance in coordinating interactions among proteins. The Hsp70 interacting protein, Hip, is also involved in the dynamics of formation of this intermediate receptor complex. Hsp90 has ATPase activity that is stimulated by binding with the steroid receptor and inhibited by Hop (McLaughlin et al. 2002). ATP-binding weakens the affinity of Hsp90 for Hop. The Hsp90-ATP-bound form is stabilized by p23, a small acidic protein generally involved in the stabilization of client protein -Hsp90 complexes. Detachment of Hop (and HSP70) allows for the recruitment of other TPR proteins: the immunophilins FKBP52, FKBP51, and Cyclophilin 40 (Cyp40), or the serine-threonine protein phosphatase 5 (PP5). Finally, this assembly and maturation process produces a large, oligomeric steroid hormone-Hsp90 complex with high hormone binding affinity. Hsp90 and co-chaperones maintain the SR in a particular structural conformation that is highly responsive to hormones. Hsp90 is required for high affinity ligand binding because it contacts directly with the hormone-binding domain of the receptor. In the absence of this chaperone, the SR affinity for hormones is reduced by 100-fold (Fang et al. 1996). The activity of the folding complex is also determined by the TPR protein occupying the acceptor site on Hsp90. Binding of different immunophilins (FKBP51, FKBP52, or Cyp40) diversifies the response mediated by the receptor and affects the receptor's ability to bind ligand. In addition, immunophilins have a variable ability to bind to the motor protein dynein, thus inducing transport to the nucleus and modifying transcription of target genes. FKBP52 and Cyp-40 bind dynein, whereas FKBP51 does not bind or binds it very weakly (Pratta et al. 2004).

In the glucocorticoid receptor (GR), the FKBP52-containing superchaperone complexes facilitate dimerization and nuclear transport of the GR. In a different way, the FKBP51-containing complexes attenuate the glucocorticoid binding, resulting in attenuation of the glucocorticoid action. Glucocorticoids up-regulate the gene for FKBP51, which provides a mechanism for desensitization of cells after an initial exposure to the hormone (Cheung and Smith 2000). FKBP51 reduces estrogen-, progesterone-, and mineralcorticoid-receptor transcriptional activities, but for AR, the situation is quite different.

In the AR superchaperone complexes, the FKBP51 increases the amount of hormone-bound receptor (Ni et al. 2010), which strengthens the androgenic signals. Androgens cause upregulation of FKBP51 expression through a direct binding of AR to enhancer elements in the FKBP51 gene, creating an auto-regulatory pathway designed to increase androgen sensitivity. These facts position FKBP51 as an interesting and potentially important candidate in the etiology of prostate cancer. Elevated levels of FKBP51 in human prostate cancer samples correspond with this concept (Fig. 4.1).

4.2 The FK506 Binding Protein 51 (FKBP51)

FKBP51 is a member of the FK506 binding proteins (FKBP), which belongs to the protein family of immunophilins (Dornan et al. 2003; Fischer and Aumüller 2003; Somarelli et al. 2008). This protein family also includes cyclophilins, the intracellular targets of cyclosporin A. FKBPs derive their name from their ability to bind immunosuppressant agents like rapamycin and FK506 (Fischer and Aumüller 2003). FKBP51 maps to chromosome 6 (6p21.31). This gene, isolated from human and mouse genomic DNA in 2003 (Scammell et al. 2001; Hubler et al. 2003; Cioffi et al. 2011), has 13 exons and 12 introns spanning more than 150 kb. FKBP51 contains an N-terminal FK1 domain responsible for peptidyl-prolyl cis-trans isomerase activity (PPIase), catalyzing the isomerization of peptidyl-prolyl imide bonds, from cis to trans, in protein substrates (Dornan et al. 2003; Fischer and Aumüller 2003). FKBP51 also has a PPIase-like FK2 domain, which shares 32 % sequence homology with FK1 and exhibits no PPIase activity. At the C-terminal, FKBP51 contains three TPR units of a 34-amino acids, for protein/protein interaction. FKBP51 expression appeared first to be restricted to T lymphocytes (Baughman et al. 1995). However, subsequent studies in humans

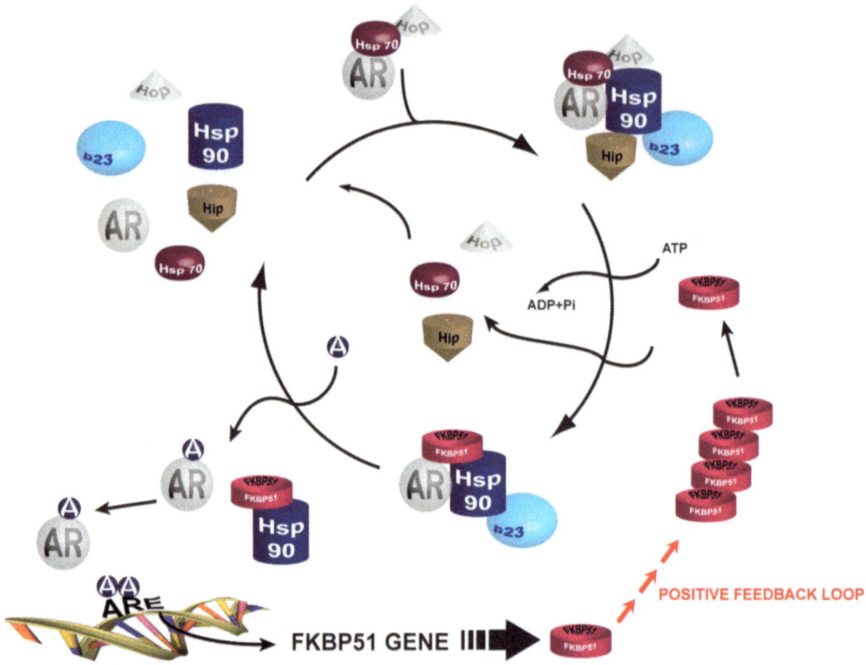

Fig. 4.1 Chaperones and co-chaperones in AR activity control. In the absence of the ligand hormone, inactive AR is maintained in a heterocomplex with several molecular chaperones and co-chaperones, including Hsp90, p23, FKBP51 and FKBP52, in a step-wise manner. The AR receptor is first delivered by Hsp70 to Hsp90 through the Hsp organizing protein (Hop). The Hsp90-ATP-bound form is stabilized by p23, a small acidic protein generally involved in the stabilization of client protein-Hsp90 complexes. Detachment of Hop (and HSP70) allows for the recruitment of FKBP51. Hsp90 and FKBP51 maintain the SR in a particular structural conformation that is highly responsive to hormones. Hsp90 is required for high affinity ligand binding

confirmed that while FKBP51 is abundantly expressed in T lymphocytes, it is also expressed in several other tissues (Baughman et al. 1997). Thanks to its multiple domains, FKBP51 is able to exert a wide variety of cellular functions, including protein folding, improvement of kinase performance, steroid receptor signaling, and transcription (Romano et al. 2011). Increasing evidence indicate that enhanced expression of FKBP51 sustains cell survival and growth in both non-neoplastic and neoplastic conditions (Vittorioso et al. 1998; Liu et al. 2007; Menicanin et al. 2009). FKBP51 has a specialized role and is preferentially expressed in mitotically active cells in the very early phases of differentiation (Liu et al. 2007; Menicanin et al. 2009; Yeh et al. 1995). FKBP51 is among the top gene candidates expressed during early mesenchymal differentiation into the three mesodermal lineages, namely, osteogenesis, chondrogenesis, and adipogenesis (Menicanin et al. 2009). At this stage, FKBP51 is co-expressed with the zinc-finger protein, ZNF145, which regulates cell cycle progression (Menicanin et al. 2009; Yeh et al. 1995; Zhang et al. 1999). The concept that FKBP51 is an essential factor for cell proliferation is also supported by studies on myeloproliferative disorders (Komura et al. 2003, 2005; Periyasamy et al. 2007). Overexpression of FKBP51 in this disorder regulates the growth factor independence of megakaryocyte progenitors (Komura et al. 2003). Finally, several lines of evidence involve FKBP51 hyperexpression in carcinogenesis and tumor growth (reviewed in Romano et al. 2010b, 2011).

4.3 FKBP51 and Prostate Cancer

Recent studies (Ni et al. 2010; Febbo et al. 2005; Makkonen et al. 2009; Periyasamy et al. 2010) point to deregulated FKBP51 as a prime factor in the etiology of androgen-dependent prostate cancer. Febbo et al. found that FKBP51 is regulated by androgens and physically associates with the AR before ligand binding in the androgen-dependent prostate cancer cell line LNCaP and prostate tumor tissue (Febbo et al. 2005). According to Makkonen et al. AR induces FKBP51 more rapidly and more strongly than does PSA, the classical AR target in prostate (Makkonen et al. 2009). Periyasamy et al. found FKBP51 upregulated in association with cyclophilin Cyp40 in prostate cancer (Periyasamy et al. 2010). In androgen-dependent tumor cell lines, FKBP51 hyperexpression increased androgen receptor transcriptional activity in the presence and absence of androgens, while knockdown of FKBP51 dramatically decreased androgen dependent gene transcription and proliferation (Periyasamy et al. 2010). FK506, the immune suppressant macrolide produced by Streptomyces tsukubaensis (Dornan et al. 2003), showed similar inhibitory effects on androgen-induced growth of prostate cancer cells (Periyasamy et al. 2010). In a study that compared AR-positive prostate cancer cells to two AR-negative prostate cancer cell lines (DU145 and PC-3), an inhibitory effect of FK506 on growth was only found in the androgen-dependent cells treated with androgens (Periyasamy et al. 2007), suggesting that the effect of FK506 requires hormone binding.

Notably, FK506 typically binds to the FK domain of FKBP51 and inhibits peptidyl prolyl isomerase (PPIase) activity (Dornan et al. 2003). Although the role of the FKBP51 PPIase function has not yet been determined in AR signaling, evidence suggests that FK506 inhibited the ligand-induced activity of AR (Periyasamy et al. 2010). The fact that hyperexpression of FKBP51 increases AR transcriptional activity even in the absence of hormone ligand (Ni et al. 2010; Periyasamy et al. 2010) suggests a role for FKBP51 in enhancing AR activity, which is not affected by FK506.

Though androgen ablation remains the backbone of prostate cancer treatment, patients develop resistance to this therapy and progress to castration-resistant prostate cancer with an attendant poor prognosis. Mechanisms that drive from androgen-dependent prostate cancer to castration-resistant prostate cancer are currently unknown. AR is expressed in castration-resistant prostate cancer and may function in an androgen-independent manner through autocrine signaling or crosstalk with other prosurvival and proliferative pathways (Attard et al. 2009;

Montgomery et al. 2008). According to (Periyasamy et al. 2010), androgen-bound AR regulates FKBP51 expression, creating a feed-forward mechanism that could amplify AR signaling under the low-hormone conditions that occur during androgen ablation. A study employing xenograph animal models, androgen-independent tumors, which arose from androgen-dependent tumors, had higher levels of FKBP51, suggesting that continued activation of AR despite androgen deprivation may be sustained, at least partly, by the continued FKBP51 expression in some prostate cancers following castration (Velasco et al. 2004).

Available immunohistochemistry data shows an appreciable homogeneous expression of FKBP51 in normal prostates and in benign prostatic hyperplasia (Staibano et al. 2011), but to a lesser extent in prostate cancer specimens (Febbo et al. 2005; Staibano et al. 2011; Romano et al. 2010a). According to Febbo et al. (2005), any significant difference in immunohistochemical staining for FKBP51 of malignant prostate epithelium between samples from individuals with or without previous androgen ablation was found, whereas the protein expression decreased in normal prostate epithelial cells following castration. A more intense signal for FKBP51 in tumors with a high Gleason grade, in respect to the well-differentiated ones, was instead described by (Staibano et al. 2011). The intracellular localization of the protein resulted in both cytoplasmatic and nuclear signal, with a prevalent nuclear shifting in more aggressive tumors (Staibano et al. 2001; Romano et al. 2010a). Taken together, these findings support a pathogenetic role for FKBP51 in tumor progression and suggest this protein can be a promising prognostic marker for prostate cancer.

In conclusion, the overall data underlines a crucial role for FKBP51, a protein that enhances androgen signaling, in the pathogenesis and progression of prostate cancer. The central role of this molecule in tumor growth has been defined clearly in androgen-sensitive prostate cancer (Ni et al. 2010; Febbo et al. 2005; Makkonen et al. 2009; Periyasamy et al. 2010; Stechschulte and Sanchez 2011). Increasing evidence points to FKBP51 as a useful target for innovative therapies that block the androgen-receptor signaling axis in endocrine-dependent prostate cancer (reviewed in Stechschulte and Sanchez 2011). A deregulated AR signal can persist also in androgen-insensitive tumors (Amler et al. 2000; Madan et al. 2011; Chmelar et al. 2007; Hobisch et al. 1995). However, even if FKBP51 appears to increase the transcriptional activity of AR in the absence of hormones (Febbo et al. 2005; Makkonen et al. 2009; Periyasamy et al. 2010), further studies are needed to clarify the efficacy of effectively blocking FKBP51 in castration resistant prostate cancers.

References

Amler LC, Agus DB, LeDuc C, Sapinoso ML, Fox WD, Kern S, Lee D, Wang V, Leysens M, Higgins B, Martin J, Gerald W, Dracopoli N, Cordon-Cardo C, Scher HI, Hampton GM (2000) Dysregulated expression of androgen-responsive and nonresponsive genes in the androgen-independent prostate cancer xenograft model CWR22-R1. Cancer Res 60:6134–6141

Attard G, Cooper CS, de Bono JS (2009) Steroid hormone receptors in prostate cancer: a hard habit to break? Cancer Cell 16:458–462

Baughman G, Wiederrecht GJ, Faith Campbell N, Martin MM, Bourgeois S (1995) FKBP51, a novel T-cell specific immunophilin capable of calcineurin inhibition. Mol Cell Biol 15: 4395–4440
Baughman G, Wiederrecht GJ, Chang F, Martin MM, Bourgeois S (1997) Tissue distribution and abundance of human FKBP51, an FK506-binding protein that can mediate calcineurin inhibition. Biochem Biophys Res Commun 232:437–443
Chen Y, Sawyers CL, Scher HI (2008) Targeting the androgen receptor pathway in prostate cancer. Curr Opin Pharmacol 8:440–448
Cheung J, Smith DF (2000) Molecular chaperone interactions with steroid receptors: an update. Mol Endocrinol 14:939–946
Chmelar R, Buchanan G, Need EF, Tilley W, Greenberg NM (2007) Androgen receptor coregulators and their involvement in the development and progression of prostate cancer. Int J Cancer 120:719–733
Cioffi DL, Hubler TR, Scammell JG (2011) Organization and function of the FKBP52 and FKBP51 genes. Curr Opin Pharmacol 11:308–313
Dornan J, Taylor P, Walkinshaw MD (2003) Structures of immunophilins and their ligand complexes. Curr Top Med Chem 3:1392–1409. Review
Fang Y, Fliss AE, Robins DM, Caplan AJ (1996) Hsp90 regulates androgen receptor hormone binding affinity in vivo. J Biol Chem 271:28697–28702
Febbo PG, Lowenberg M, Thorner AR, Brown M, Loda M, Golub TR (2005) Androgen mediated regulation and functional implications of FKBP51 expression in prostate cancer. J Urol 173:1772–1777
Fischer G, Aumüller T (2003) Regulation of peptide bond cis/trans isomerization by enzyme catalysis and its implication in physiological processes. Rev Physiol Biochem Pharmacol 148:105–150
Hobisch A, Culig Z, Radmayr C, Bartsch G, Klocker H, Hittmair A (1995) Distant metastases from prostatic carcinoma express androgen receptor protein. Cancer Res 55:3068–3072
Hubler TR, Denny WB, Valentine DL, Cheung-Flynn J, Smith DF, Scammell JG (2003) The FK506-binding immunophilin FKBP51 is transcriptionally regulated by progestin and attenuates progestin responsiveness. Endocrinology 144:2380–2387
Komura E, Chagraoui H, Mansat de Mas V, Blanchet B, de Sepulveda P, Larbret F, Larghero J, Tulliez M, Debili N, Vainchenker W, Giraudier S (2003) Spontaneous STAT5 activation induces growth factor independence in idiopathic myelofibrosis: possible relationship with FKBP51 overexpression. Exp Hematol 31:622–630
Komura E, Tonetti C, Penard-Lacronique V, Chagraoui H, Lacout C, Lecouédic JP, Rameau P, Debili N, Vainchenker W, Giraudier S (2005) Role for the nuclear factor kappaB pathway in transforming growth factor-beta1 production in idiopathic myelofibrosis: possible relationship with FK506 binding protein 51 overexpression. Cancer Res 65:3281–3289
Liu TM, Martina M, Hutmacher DW, Hui JH, Lee EH, Lim B (2007) Identification of common pathways mediating differentiation of bone marrow- and adipose tissue- derived human mesenchymal stem cells into three mesenchymal lineages. Stem Cells 25:750–760
Madan RA, Pal SK, Sartor O, Dahut WL (2011) Overcoming chemotherapy resistance in prostate cancer. Clin Cancer Res 17:3892–3902
Makkonen H, Kauhanen M, Paakinaho V, Jääskeläinen T, Palvimo JJ (2009) Long-range activation of FKBP51 transcription by the androgen receptor via distal intronic enhancers. Nucleic Acids Res 37:4135–4148
McEwan IJ (2004) Molecular mechanisms of androgen receptor-mediated gene regulation: structure–function analysis of the AF-1 domain. Endocr Relat Cancer 11:281–293
McLaughlin SH, Smith HW, Jackson SE (2002) Stimulation of the weak ATPase activity of human Hsp90 by a client protein. J Mol Biol 315:787–798
Menicanin D, Bartold PM, Zannettino AC, Gronthos S (2009) Genomic profiling of mesenchymal stem cells. Stem Cell Rev 5:36–50
Montgomery RB, Mostaghel EA, Vessella R, Hess DL, Kalhorn TF, Higano CS, True LD, Nelson PS (2008) Maintenance of intratumoral androgens in metastatic prostate cancer: a mechanism for castration-resistant tumor growth. Cancer Res 68:4447–4454

Ni L, Yang CS, Gioeli D, Frierson H, Toft DO, Paschal BM (2010) FKBP51 promotes assembly of the Hsp90 chaperone complex and regulates androgen receptor signaling in prostate cancer cells. Mol Cell Biol 30:1243–1253

Periyasamy S, Warrier M, Tillekeratne MP, Shou W, Sanchez ER (2007) The immunophilin ligands cyclosporin a and fk506 suppress prostate cancer cell growth by androgen receptor-dependent and -independent mechanisms. Endocrinology 148:4716–4726

Periyasamy S, Hinds T Jr, Shemshedini L, Shou W, Sanchez ER (2010) FKBP51 and Cyp40 are positive regulators of androgen-dependent prostate cancer cell growth and the targets of FK506 and cyclosporin A. Oncogene 29:1691–1701

Pratta WB, Galignianaa MD, Harrella JM, DeFranco DB (2004) Role of hsp90 and the hsp90-binding immunophilins in signalling protein movement. Cell Signal 16:857–872

Romano S, D'Angelillo A, Staibano S, Ilardi G, Romano MF (2010a) FK506-binding protein 51 is a possible novel tumoral marker. Cell Death Dis 1:55

Romano S, Di Pace A, Sorrentino A, Bisogni R, Sivero L, Romano MF (2010b) FK506 binding proteins as targets in anticancer therapy. Anticancer Agents Med Chem 10:651–656 (review)

Romano S, Sorrentino A, Di Pace A, Nappo G, Mercogliano C, Romano MF (2011) The emerging role of large immunophlin FK506 binding protein 51 in cancer. Curr Med Chem 18:5424–5429 (review)

Scammell JG, Denny WB, Valentine DL, Smith DF (2001) Overexpression of the FK506-binding immunophilin FKBP51 is the common cause of glucocorticoid resistance in three New World primates. Gen Comp Endocrinol 124:152–165

Somarelli JA, Lee SY, Skolnick J, Herrera RJ (2008) Structure-based classification of 45 FK506-binding proteins. Proteins 72:197–208

Staibano S, Mascolo M, Ilardi G, Siano M, De Rosa G (2011) Immunohistochemical analysis of FKBP51 in human cancers. Curr Opin Pharmacol 11:338–347

Stechschulte LA, Sanchez ER (2011) FKBP51-a selective modulator of glucocorticoid and androgen sensitivity. Curr Opin Pharmacol 11:332–337

Velasco AM, Gillis KA, Li Y, Brown EL, Sadler TM, Achilleos M, Greenberger LM, Frost P, Bai W, Zhang Y (2004) Identification and validation of novel androgen-regulated genes in prostate cancer. Endocrinology 145:3913–3924

Vittorioso P, Cowling R, Faure JD, Caboche M, Bellini C (1998) Mutation in the Arabidopsis PASTICCINO1 gene, which encodes a new FK506-binding protein-like protein, has a dramatic effect on plant development. Mol Cell Biol 18:3034–3043

Yeh WC, Li TK, Bierer BE, McKnight SL (1995) Identification and characterization of an immunophilin expressed during the clonal expansion phase of adipocyte differentiation. Proc Natl Acad Sci U S A 92:11081–11085

Zhang T, Xiong H, Kan LX, Zhang CK, Jiao XF, Fu G, Zhang QH, Lu L, Tong JH, Gu BW, Yu M, Liu JX, Licht J, Waxman S, Zelent A, Chen E, Chen SJ (1999) Genomic sequence, structural organization, molecular evolution and abberrant rearrangement of promyelocytic leukemia zinc finger gene. Proc Natl Acad Sci U S A 96:11422–11427

Chapter 5
Neuroendocrine Differentiation in Prostate Cancer

Renato Franco, Paolo Chieffi, Sisto Perdonà, Gaetano Facchini, and Michele Caraglia

Abstract Neuroendocrine differentiation (ND) is widely observed in prostate cancer (PC). Its role in clinical practice is controversial, but preclinical and clinical evidences underline the association of ND with poor prognosis in PC patients. Neuroendocrine (NE) cells could condition the PC progression, mainly stimulating the PC exocrine neoplastic cells proliferation through the production of paracrine growth factors. Thus, the castrated adapted neoplastic cells are favored to outgrowth through an androgen receptor independent mechanism. Moreover proportion of NE cells in PC increases because of tumor treatment, mainly androgen deprivation therapy, enormously amplifying the promotion of the PC exocrine component growth stimulated by neuroendocrine paracrine growth factors.

This chapter provides an overview of the most relevant clinical studies demonstrating a significant correlation between ND and PC behavior, indicating that ND could represent a prognostic parameter in PC, and strongly suggesting that NE cells in a castrate resistant patients could be targeted through specific treatment.

R. Franco (✉)
Department of Laboratory Medicine, National Institute of Tumors Fondazione "G. Pascale", Via Mariano Semmola 1, 80131 Naples, Italy
e-mail: r.franco@istitutotumori.na.it

P. Chieffi
Department of Experimental Medicin, Second University of Naples, Via Costantinopoli 16, 80138 Naples, Italy
e-mail: paolo.chieffi@unina2.it

S. Perdonà • G. Facchini
Department of Urogynecologic Oncology, National Institute of Tumors Fondazione "G. Pascale", Via Mariano Semmola 1, 80131 Naples, Italy
e-mail: s.perdona@istitutotumori.na.it; g.facchini@istitutotumori.na.it

M. Caraglia
Department of Biochemistry and Biophysics, Second University of Naples, Via Costantinopoli, 16, 80138 Naples, Italy
e-mail: michele.caraglia@unina2.it

S. Staibano (ed.), *Prostate Cancer: Shifting from Morphology to Biology*,
DOI 10.1007/978-94-007-7149-9_5, © Springer Science+Business Media Dordrecht 2013

Abbreviations

ND	Neuroendocrine differentiation
NE	Neuroendocrine
PC	Prostate cancer
PSA	The prostate specific antigen
CGA	ChromograninA
NSE	Neuron-specific enolase
VIP	Vasoactive intestinal peptide
GRP	Bombesin/gastrin releasing peptide
aHCG	Alpha-human chorionic gonadotropin
PTHrP	Parathyroid hormonerelated protein
VEGF	Vascular endothelial growth factor
SCC	Small cell carcinomas
PIN	Prostatic intraepithelial neoplasia
AMACR	Alpha-methylacyl-CoA racemase
ADT	Androgen deprivation therapy
uPA	Urokinase-type plasminogen activator
PAI-1	Plasminogen activator inhibitor-1
MMP	Metalloprotease
MDV	Microvascular density
MAPKs	Mitogen activated protein kinases
PKA	Cyclic AMP-dependent protein kinase
PI3K	Phosphatidylinositol 3-kinase
CDK	Cyclin-dependent kinase
CGB	Chromogranin B
CGC	Chromogranin C
ProGRP	Progastrin-releasing peptide
BPH	Prostatic hyperplasia
PS	Performance status
PET	Positron emission tomography
FDG	F18-fluorodeoxyglucose
DTPA	Diethylenetriaminepentaacetic acid
DTX	Docetaxel
OS	Overall survival
TTP	Time to progression
DOTA	1,4,7,10-Tetraazacyclododecane-1,4,7,10-tetraacetic acid
PTP	Protein tyrosine phosphatase

5.1 Distribution and Origin of NE Cells in Normal Gland

The normal human prostate ducts and acini epithelium is composed of two layers cells, including luminal secretory, basal and neuroendocrine (NE) cells (di Sant'Agnese 1992, 1998; Cohen et al. 1994).

Specificity of NE cells has been systematically defined during the 1980s. Thus NE cells have been described as an important component of the prostate (di Sant'Agnese 1992; Berruti et al. 2000).

The origin of these cells in normal prostate is not completely known. In fact different theories about their ontogenesis have been proposed. The most recent studies have postulated that NE, basal and secretory luminal cells originate from a common endodermal pluripotent stem cell (Huss et al. 2004). Thus, it has been hypotized that stem cells in multiple stem cell units (Isaacs and Coffey 1989; Bonkhoff et al. 1994; Bonkhoff 1996) with an unlimited self-renewal ability provide, with an unlimited self- renewal ability, progeny able to differentiate into either transit-amplifying or NE cells (Bonkhoff et al. 1994; Bonkhoff 1996). The luminal secretory cells or basal cells derive subsequently from the transit-amplifying cells. According to this theory, the prostate specific antigen (PSA) has been demonstrated also in NE cells, suggesting the same common precursor of secretory cell (Aprikian et al. 1993). Moreover, the expression of CD44, a marker of lymphocytes and cancer stem cells, has been observed in cells of NE phenotype (Palapattu et al. 2009). On the other hand NE cells could be represent a specific cell lineage of neural crest origin, substantially distinct from urogenital sinus- derived prostate secretory and basal cells (Pearse and Takor 1979).

NE cells are not homogenously distributed in the prostate glands, being consistently found in the periurethral ducts and *veru montanum* regions. NE cells are generally more numerous in the transition and the peripheral zones than in the central zone (Sant'Agnese 1998).

NE cells are normally not identifiable on standard stained histological sections. They are only recognized by electron microscopy or immunohistochemical staining for NE markers, being chromogranin A (CGA) and synaptophysin the most common.

As scattered or small groups, the distribution of NE cells according to electron microscopy studies could be in two different manner: (i) *open*, with cells directly extended to gland lumen; and (ii) *closed*, with cell dendritic-like processes extending between adjacent cells, on the basal lamina and strictly related to stromal nerves (Kamiya et al. 2008).

Intracinar and intraductal prostatic NE cells secrete serotonin and many other neuropeptides. Prostatic NE cells contain dense-core cytoplasmic granules with variable sizes and morphologies, storing peptide hormones and pro-hormones and recognizable by electron microscopy. The products contained in neurosecretory granules are either as a single typology or as mix of different peptides such as CGA, chromogranin B (CGB), somatostatin, neuron-specific enolase (NSE), parathyroid hormone-related protein (PTHrP), bombesin, thyroid-stimulating hormone and calcitonin gene family (calcitonin, katacalcin, and calcitonin gene-related peptide), vasoactive intestinal peptide (VIP), neuropeptide Y, alpha-human chorionic gonadotropin (aHCG), bombesin/gastrin releasing peptide (GRP), thyroid stimulating hormone-like peptide, adrenomedullin, cholecystokinin, and vascular endothelial growth factor (VEGF) (Huang and di Sant'Agnese 2002) (Table 5.1).

Table 5.1 Main characteristics of neuroendocrine cell in normal prostate

General characteristics
Androgen-receptor negative
Non-proliferating
PSA-negative
Bcl-2-negative
Express intermediate and luminal cytokeratins
Functional roles
Regulation of cell growth and differentiation
Regulation of homeostasis
Regulation of prostatic secretion
Products
Calcitonin gene family
Chromogranin A
Chromogranin B
Cholecystokinin (CCK)
Gastrin-releasing peptide
Histamine
Neuron-specific enolase
Neuropeptide Y
Parathyroid hormone-related protein
Proadrenomedullin N-terminal peptide
Serotonin
Somatostatin
TSH-like peptide
Vascular endothelial growth factor
Receptors
Gastrin releasing peptide (GRPR)
Serotonin (5HTR1A,B)
Somatostatin (SSTR 1–5)
Calcitonin (hCTR-2)
Cholecystokinin
Neuropeptide Y
Vasoactive intestinal peptide
PTHrp receptor (highly expressed in bone metastases from prostate)

Also NE peptides receptors have been identified on NE prostate cells, including receptors for serotonin (5HT1a) (Abdul et al. 1994), bombesin/GRP (GRPR) (Markwalder and Reubi 1999), neurotensin (Seethalakshmi et al. 1997), somatostatin (SSTR1-5) (Dizeyi et al. 2002), cholecystokinin, Neuropeptide Y and calcitonin (Wu et al. 1996). Thus, through a complex network of NE peptides and respective receptors on exocrine cells in prostate gland, NE cells may control the growth, differentiation and secretory activity of the prostatic epithelium through a paracrine mechanism. In addition, the activity of the NE cells activity may be regulated by the neural network.

5.2 Physiopathology of ND in PC

NE cells are also present in PC. Rarely, NE cells are the only neoplastic component in prostatic carcinoma, such as carcinoid tumors and small cell carcinomas (SCC). A component of SCC or carcinoid could be also associated to conventional adenocarcinoma. More commonly, however, conventional adenocarcinoma are associated by scattered NE cells and distributed as single or grouped elements.

5.2.1 *SCC and Carcinoid Tumors*

Pure SCC of the prostate represent no more than 1 % of all prostate carcinomas. They have a very aggressive behavior, being often locally advanced or metastatic at diagnosis (Erasmus et al. 2002). It is occasionally associated with paraneoplastic syndromes (Kawai et al. 2003). Rarely, SCC has been observed in conventional adenocarcinoma patients treated with hormonal therapy (Tanaka et al. 2001). Indeed SCC is more commonly a component of mixed tumors with conventional adenocarcinoma. Histologically, prostate SCC is similar to lung SCC, being constituted both by neoplastic small cells in a solid growth pattern, with fine cromatin and common nuclear crushing. Necrosis and high mitotic index is commonly observed (Yao et al. 2006). Immunohistochemical study can greatly help in diagnosis of this rare tumor. In fact neoplastic cells are commonly positive for one or more neuroendocrine markers, such as chromogranin, synaptophysin, CD56 or NSE. Dot-like positivity for Cytokeratin is commonly observed (Kawai et al. 2003). Differently from conventional prostate adenocarcinoma, prostate SCC does not express Androgen Receptor. Thus hormonal therapy is not adequate in the treatment of such tumors while chemotherapy may be responsible of initial clinical response (Helpap 2002).

Carcinoid is a very rare tumor in prostate. Microscopically all neoplastic cells show complete ND and the organoid pattern of growth is similar to neuroendocrine tumors of other districts. Also carcinoid could be part of mixed tumor with conventional adenocarcinoma. (Ghannoum et al. 2004).

5.2.2 *Focal ND in PC*

Focal ND occurs in conventional prostatic adenocarcinomas. As in normal prostate, NE cells are not distinguishable from other cancer cells, unless immunohistochemical staining for NE markers are used. Also in PC CGA is usually considered sensitive and specific. Some NE cells seem to be present in all conventional adenocarcinoma, but only about 5–10 % contain a large number of NE cells (Abrahamsson et al. 1987). They are described in a large spectrum of prostatic neoplasia, from prostatic intraepithelial neoplasia (PIN) (Bostwick et al. 1994) to metastatic disease (Bostwick et al. 2002).

5.3 The Origin of NE Cells in PC

The biological features of PC NE cells suggest that these cells are different from normal prostate NE cells.

Some observational evidences suggest that the cells are more similar to secretory cells that normal NE cells. As normal NE cells, PC NE cells do not express Androgen Receptor (AR) and PSA, normally present in secretory component (Huang et al. 2006). Also neuron-like processes of normal NE are often not observed in PC NE cells (Xing et al. 2001). Immunohistochemical studies have demonstrated that normally expressed basal cell markers of normal NE cells are not expressed by PC NE cells, such as p63 and cytokeratin 5 (Huang et al. 2006). On the contrary, they are positive for luminal secretory cells markers such as cytokeratin 18 (van Bokhoven et al. 2003). Also alpha-methylacyl-CoA racemase (AMACR), an enzyme involved in the ß-oxidation of fatty acids, expressed in PC non NE cells rather than in normal prostatic secretory cells, has been demonstrated in PC NE cells (Huang et al. 2006). In addition anti-apoptotic Bcl-2 protein is normally expressed by PC NE cells but not by normal NE (Segal et al. 1994). Finally genetic analysis support that PC NE cells are similar to PC secretive cells and not to normal NE (Sauer et al. 2006).

All these considerations have led to the frequent use of the term 'NE-like' PC cell to define this neoplastic population (Yuan et al. 2007). Thus the theory that PC NE cells derive from neural crest cells has been finally abandoned (Schron et al. 1984). Two main hypotheses justifying neuroendocrine differentiation in PC are taken into account, transdifferentiation and origin from a common neoplastic stem cell (Bonkhoff et al. 1995).

The *transdifferentiation* model seems to occur as in response to hormonal and growth factor microenvironmental changes. In fact it has been supported by *in vitro* experiments on LNCaP cells, an androgen-dependent cell line, developing ND through induction by androgen deprivation or other agents increasing intracellular levels of cAMP such as epinephrine (Cox et al. 2000) interleukin-6, (Deeble et al. 2001) and genistein (Pinski et al. 2006). Moreover, *in vivo* androgen deprivation can stimulate PC ND either in the animal model or in humans. In fact, an increase percentage of NE cells has been demonstrated in matched tumor samples collected from patients with early PC before and after androgen deprivation therapy (ADT) (Ahlgren et al. 2000). In patients with advanced PC a relative increase of CgA serum levels has been documented after ADT (Berruti et al. 2005). Antineoplastic treatment, not only androgen deprivation, seems to be responsible of ND in PC. In fact, fractionated ionizing radiation (IR) can stimulate ND *in vitro* in LNCaP (Deng et al. 2008) and docetaxel (DTX) can induce in animal model ND with the same relevance of the androgen deprivation. It has to be emphasized that both therapies are used in the treatment of castration resistant patients (Tang et al. 2009).

Recently, it has been demonstrated that human PC cell lines express the stem cell marker CD44 (Palapattu et al. 2009) as well as human PC tissues were highly positive for CD44 both in secretive and NE component. These evidences could suggest that PC NE cell and PCA secretive cells share the *same potential progenitor*.

Moreover, the ND in PC is not a stable phenotype. In fact NE trans-differentiation *in vitro* is a reversible phenomenon, as the neoplastic population could lose its neuroendocrine phenotype after the removal of the inducing agents (Palapattu et al. 2009).

Different signaling pathways are involved in the ND of PC. PI3K-AKT-mTOR pathway seem the main involved pathway (Wu and Huang 2007). The Notch signaling pathway, and specifically of hASH1 (human achaete-scute homolog-1 transcription factor) have been also demonstrated to be as responsible of ND setting (Guillemot et al. 1993). Finally, all treatments responsible of *in vitro* NE trans-differentiation in LNCaP cells caused deregulated expression of a unique set of genes (Mori et al. 2009).

5.4 The Function of NE Cells in PC

As NE tumor cells do not express AR, their growth is androgen independent. Therefore, they continue to survive and perform their functions in a *milieu* devoid of androgens, building a fine network of autocrine and paracrine relationships with the rest of the tumor, providing androgen-independent growth. In fact LNCaP xenografts, being androgen-dependent, cannot normally grow in castrated mice, while they can grow when also NE cells from a mouse NE tumor (NE-10) are transplanted in the castrated hosts (Jin et al. 2004). *In vitro* studies have demonstrated that NE cells products could favor growth and invasiveness of PC cell lines (Jongsma et al. 2000). Neuropeptides are active on different pathways, up-regulating proteins critical for tumor growth, invasiveness, angiogenesis and metastasis. PC NE cells may promote androgen-independent PC growth through the production of paracrine signals interacting with the secretory PC cells by AR dependent or independent mechanisms. In fact, one of the main pathway activated by neuropeptides is the same activated by androgens. It has been demonstrated that NF-kB pathway activated by neuropeptides could promote cancer growth AR pathway (Jin et al. 2008). Neuropeptides could act on G protein-coupled receptors, overexpressed in PC, and then aberrantly activate AR pathway even in the absence of androgens. The same authors through the development of gastrin related peptide (GRP) overexpressing LNCaP model demonstrated that this cell line showed androgen independent growth and an enhanced motility *in vitro*. Furthermore, in castrated nude mice, LNCaP-GRP derived tumors were very aggressive, and produced GRP, PSA and nuclear AR. Chromatin immune-precipitation studies of LNCaP-GRP suggested AR recruitment to the cognate promoter also in the absence of androgens (Jin et al. 2008).

The network of protein related to local invasion proteolytic enzyme urokinase-type plasminogen activator (uPA), plasminogen activator inhibitor-1 (PAI-1) and metalloprotease (MMP) are stimulated by bombesin (Festuccia et al. 1998). Bombesin and MMP-9 are highly expressed in high grade carcinoma (Ishimaru et al. 2002). Other MMPs are activated by neuropeptides (Sehgal and Thompson 1999). GRP increases MT1-MMP in androgen independent cell lines DU-145, amplifying their ability of Matrigel invasion (Nagakawa et al. 1999). Finally MMPs are highly expressed by androgen independent respect than androgen dependent cell lines.

Cytokines and their own receptors could contribute to androgen independent PC growth in androgen independent manner. IL-8 is normally overexpressed by PCa NE cells and its receptors CXCR1 has been demonstrated on PCa secretory cells (Huang et al. 2005). IL-8 promotes androgen independent growth and migration of LNCaP cells, a model of the *in vivo* paracrine mechanism of androgen independent growth and invasion (Lee et al. 2004).

Resistance to apoptosis of PCa is partially conferred by PC NE cells. Survivin is highly expressed by PC NE cells, subsequently more resistant to stress and then to apoptosis (Xing et al. 2001). Bcl-2, the most relevant anti-apoptotic agent, is not expressed by PC NE cells, but it is demonstrated in PC non NE cells, above all in those closest to PC NSE positive NE cells suggesting that apoptosis resistance in PC cells could be induced by NE cells neuropeptides (Segal et al. 1994). Finally bombesin and calcitonin are able to prevent apoptosis in PC cell lines (Salido et al. 2000, 2004; Vilches et al. 2004).

Neovascularization is also promoted by PC NE cells, mainly through VEGF and IL-8 (Chevalier et al. 2002) In particular, VEGF and neovascularization have been significantly reduced in prostate surgical sample of patients previously treated with complete androgen blockade, except in the areas with PC NE cells (Mazzucchelli et al. 2002). Moreover, the number of NE cells are predictor of neovascularization (Grobholz et al. 2000).

Finally VEGF expression seems to be significantly related to microvascular density (MVD), high tumor stage, high Gleason grade and shorter disease free survival (Borre et al. 2000). Some neuropeptides are involved in stimulation of proangiogenic factors VEGF and IL-8 *in vitro* model. Bombesin activates the expression of such factors in PC-3 cells, probably through NF- kB pathway (Levine et al. 2003). Instead calcitonin appears to stimulate vessel formation by acting directly on endothelial cells. CGA fragment 286–301, representing the C-terminal of pancreastatin, induced the invasive ability of PC-3 and DU-145 PC cell lines. CGA (286–301) also increased the haptotactic migration of these cells and the production of urokinase type plasminogen activator (Nagakawa et al. 1999). Through invasion assay, it has been demonstrated that gastrin-releasing peptide, calcitonin gene-related peptide, and parathyroid hormone-related protein increased invasive ability of PCa cells (Nagakawa et al. 2001).

5.5 Molecular Mechanisms of ND

NE phenotype is inversely correlated to active AR signaling (Wright et al. 2003). These data could explain increased ND, during inhibition of AR signaling, both in patients with androgen blockade and in LNCaP cell line in androgen-deprived media, through different mechanism, requiring activation of multiple protein, such as ERK (Wu et al. 2006). Thus, androgen deprivation is the main stimulator of ND, but it is not the only one. In fact, each factor determining increase of intracytoplasmic cAMP and then of cAMP-dependent kinase activity could favour ND *in vitro* model, such as epinephrine and forskolin, IL-6, IL-1 (Albrecht et al. 2004). IL-6 induced ND involving the protein tyrosine kinase pathway (Chung et al. 2000), JAK-STAT signaling, mitogen activated protein kinases (MAPKs), cyclic AMP-dependent protein kinase (PKA) phosphatidylinositol 3-kinase (PI3K) induction of cyclin-dependent kinase (CDK) inhibitor p27 (Kip1) and inhibition of CDKs (Mori et al. 1999). IL-6 induced ND in PC is irreversible respect to epinephrine induced ND and the aggressive phenotype *in vitro* and *in vivo* model seems to be related to IL-6 concentration used to treat LNCaP cells (Wang et al. 2004). Also induction of NFkB signaling, as inhibition of proinflammatoy enzyme COX-2 does, could promote ND (Meyer-Siegler 2001). Moreover inhibition of FGF signaling could induce ND in PC cells of transgenic mice promotes ND.

Recently, the activation of phosphatidylinositol 3-kinase-AKT-mammalian target of rapamycin (PI3K-AKT-mTOR) pathway has been demonstrated as to be essential for ND in PC (Wu et al. 2006). In fact ND of LNCaP is induced also through activation of the PI3K-AKT-mTOR signaling pathways. Moreover, Rapamycin, an inhibitor of mTOR, significantly inhibited the expression of NSE in LNCaP cells induced by androgen withdrawal. Recently, it has been demonstrated that also irradiation of LNCaP can promote ND, through nuclear increasing of CREB, a transcription factor potentially enhancer of ND, and cytoplasmic accumulation of the transcription factor potentially suppressor of NED ATF2 (Deng et al. 2008). Protocadherin-PC, a member of protocadherin gene family, has been shown to be as critical for ND of PC through Wnt signaling activation (Yang et al. 2009).

Heparin binding epidermal growth factor (HB-EGF) activates ND through the involvement of mitogen-activated protein kinase (MAPK) signaling pathway (Kim et al. 2002).

Also autocrine neuropeptides themselves could be directly involved in ND. In fact, it has been documented that VIP, commonly secreted by PC NE cells, induces ND through activation of ERK, PKA and PI-3-kinase pathways (Collado et al. 2004, 2005).

5.6 NE Differentiation and Hormone-Refractory Prostate Cancer

AR is normally present in both prostate stromal and secretory cells and, the androgen, acts in different manner in the two compartment. In fact, on stromal cells, it favors andromedins secretion, supporting the survival and proliferation of luminal secretary epithelial cells (Isaacs 2008), while in luminal secretory cells the androgen suppresses the cell growth through expression of p27Kip1 expression (Waltregny et al. 2001). Clinically, the long-term treatment of patients with androgen deprivation is responsible of the castration-resistant state. This condition has to be partially

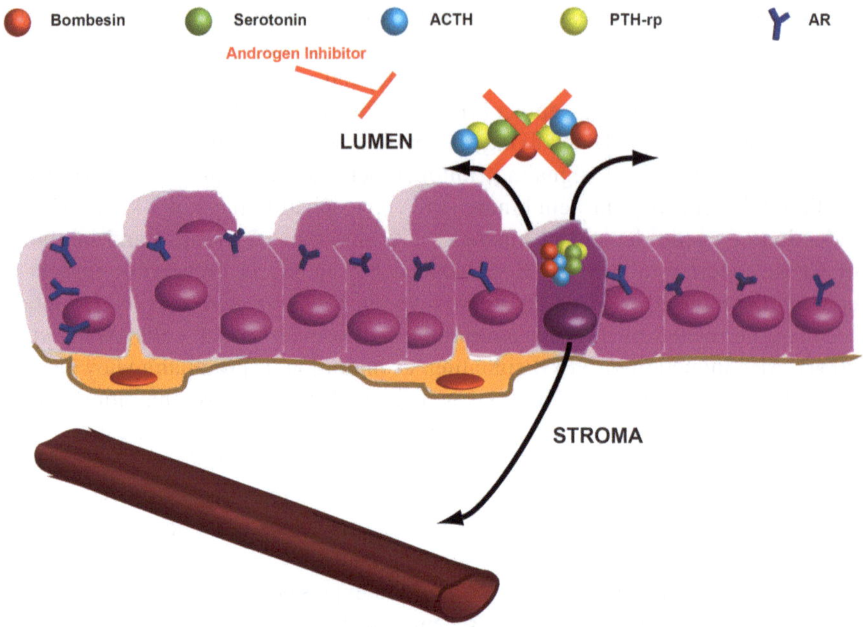

Fig. 5.1 Neuroendocrine cells in PCa. NE phenotype is inversely correlated to active AR signaling and, therefore, androgen deprivation is the main stimulator of NED that play a pivotal role in conversion to androgen independent growth of Pca. Thus since androgen deprivation depresses the growth of cancer cells, the main effect of treatment occurs on the stromal cells. The paracrine role of stromal cells, sensible to androgen deprivation, could be gradually substituted by NE insensible to androgen deprivation and able to secrete a large variety of neuropeptides and cytokines promoting survival and proliferation of neoplastic epithelial cells. Neuropeptides are active on different pathways, upregulating proteins critical for tumor growth, invasiveness, angiogenesis and metastasis

attributable to the alteration of AR signaling, including AR gene amplification, responsible of a response to low levels of circulating androgens, AR mutations, causing AR pathway activation through antiandrogens or weak androgens, the local synthesis/concentration of androgens, AR activation through growth factors/kinase pathways, and/or changes in AR coregulators (Scher and Sawyers 2005). The most relevant role in the conversion to androgen independent growth of PC seems to be played by ND. Thus, since androgen deprivation depresses the growth of cancer cells growth, translating the normal model to PC model, it can be assumed that the main effect of treatment occurs on the stromal cells, rather than the neoplastic secretory compartment. Neither AR mutation in secretory neoplastic cells, recorded in only 10 % of cases, could be justify a real change of activity in those cells (Buchanan et al. 2001). The paracrine role of stromal cells, sensible to Androgen deprivation, could be gradually substituted by NE cells insensible to androgen deprivation and able to secrete a large variety of neuropeptides and cytokines promoting survival and proliferation of neoplastic epithelial cells (Fig. 5.1).

5.7 The Diagnosis of Neuroendocrine Prostate Cancer

5.7.1 Serological Markers

The identification of neuroendocrine markers in the PC patients serum represents a more complete indicator and more objective quantification of ND of tumors, because it corresponds to the entire primary tumor and its associated metastases.

Herein, we list the most relevant serological markers associate to ND:

CGA. This marker is a tumor cell population product, and sometimes, the clue to diagnose special subtypes, such as pure small cell prostate cancer. CGA is an excellent marker for ND in tumors and its measurement is also useful to identify prostatic carcinoma in patients with not elevated PSA. Angelsen and co-workers (1997) showed that the number of CGA-positive NE cells in tumoral tissue significantly related with serum CGA levels in patients with prostatic carcinoma. However elevated serum CGA level could be observed in unrelated tumor conditions, such as in patients with impaired renal function or in those receiving omeprazole treatment for peptic ulcer. Kadmon and co-workers (1991) reported elevated plasma CGA levels in 48 % patients with metastatic PC. Similarly, (Logothetis and Hoosi 1992) found also elevated plasma bombesin levels in 47 % of PC patients treated for androgen-independent growth.

Kadmon observed among advanced stage of PC patients a relative low frequency of high CGA levels (17 %) respect to 33 % of patients with normal conventional PSA and PAP markers, suggesting that high CGA patients was not representative of usual progressive tumor status (Kadmon et al. 1991). The fact that prostatic carcinomas expressed neuropeptide markers (CgA, 17 %; NSE, 15 %) before any endocrine therapy suggests that neuroendocrine products may induce PC progression independently of androgen withdrawal. Data by Deftos et al. (1996), and a more recent by Kimura and co-workers (1997), demonstrated that CGA could be an useful marker in advanced disease. Tarle and Rados (1991) found that elevated plasma NSE levels was observed more frequently in untreated PC patients with localized tumors (28.6 %) than in untreated subjects with disseminated disease (10.7 %).

In a comparison study of serum levels of CGA, pancreastatin, a breakdown product of CGA, c CBG (Angelsen et al. 1997) and chromogranin C (CGC) (Schmid et al. 1994), CGA appears to be the best marker of neuroendocrine prostate tumor activity.

Recently, plasma CGA was assessed by ELISA in 14 patients with Castration resistant PC (CRPC) receiving 3-weekly docetaxel. Increased plasma CgA was observed in 64.3 % of patients. No correlation between baseline CGA and PSA has been observed. Two patients with PSA < 10 ng/ml had elevated CGA. Baseline CGA was not conditioned by clinical parameters such as presence of metastasis, metastasis sites and time to develop CRPC status. Seven patients (50 %) showed PSA-response and five (36 %) CGA-response. In two patients PSA response and CGA response were discordant. Compared to men with normal baseline CGA, a higher proportion of those with elevated baseline CGA had PSA response (55 % vs 40 %), symptomatic response (66 % vs 40 %) and radiological response (55 % vs 20 %). Two patients with symptomatic response had only CGA response. Three patients with disease progression, despite PSA response, had increasing CGA. On the basis of these results, CGA and PSA have been proposed as complementary tumor biomarkers in castration resistant prostate cancer and CGA may be useful to predict the response to therapy with DTX. Increased serum CGA during therapy may be associated with poor prognosis, whereas CGA response is likely to be associated with clinical response (Sarkar et al. 2010). Moreover, in a series of 135 patients with prostatic carcinoma and 28 with benign prostatic hyperplasia plasma CGA, NSE and other neuroendocrine biomarkers have been analyzed and compared to clinical and pathological stages of disease. Particularly elevated levels of CGA were detected in 15 % of patients with PC, before any treatment, but elevation of plasma CGA and NSE levels was observed respectively in 55 and 30 % of the patients during hormone resistant prostate cancer progression. In addition log-rank analysis in stage D3 patients revealed a statistically significant difference between positive and negative CGA groups These data confirmed the role of CGA and, at a lesser extent, of NSE in predicting the prognosis of PC patients even if their role as markers of ND still need additional investigations. Finally, serum NSE and plasma CGA were evaluated in 141 patients with prostatic hyperplasia (BPH), 54 patients with PIN, and 159 patients with Pca; 119 patients were bearing hormone-naive disease and 40 were bearing hormone-refractory disease. Supernormal CGA was observed more frequently in Stage D2 disease patients (45.5 %) compared with Stage D1 (33.3 %), Stage C disease (16.7 %), Stage A/B disease (18.8 %), PIN (25.9 %), and BPH (17.0 %) patients ($P < 0.02$). Supernormal NSE did not show differences in any of the patient subgroups stages. Elevated CGA was observed in 36.0 % of metastatic patients with hormone-naive disease and in 45.0 % of metastatic patients with hormone-refractory disease, although without statistical significance. Supernormal NSE and CgA values were observed as predictors for poor prognosis in patients with hormone-refractory disease. Significant decreased baseline CgA values has been observed in 1 of 12 patients who received luteinizing hormone-releasing hormone analogs and in 2 of 12 patients receiving chemotherapy. Elevated CGA levels correlated with poor prognosis and were scarcely influenced by either endocrine therapy or chemotherapy (Berruti et al. 2000).

Gastrin-releasing peptide (GRP), a 27 amino acids neuropeptides, seems to play a role as an autocrine/paracrine growth factor in several cancers. Progastrin-releasing peptide (ProGRP), comprehending three different subtypes of precursors for GRP, has a longer half-life than GRP, with levels similar to GRP itself. The values show excellent sensitivity and correlation with the therapeutic response in neuroendocrine tumors. Indeed ProGRP is mainly used in clinical diagnosis and follow-up of small cell lung cancer. ProGRP levels have also been observed to be elevated in other tumors with neuroendocrine features, such as colorectal, thyroid, and breast cancer. In 60 patients with benign BPH and 200 with PCa, increased ProGRP value was significantly observed in the androgen-independent group ($P < 0.0001$). In a subset of patients, the ProGRP levels increased transiently when the cancer shifts to became androgen independent status, but remained unchanged or decreased at the androgen-dependent stage. In addition positive ProGRP immunostaining occurred in a different distribution in tumoral tissues when comparing to CGA immunostaining. The clinical results confirm the existence of a regulatory mechanism for GRP, demonstrated in cell lines. These findings suggest that GRP is a growth factor potentially upregulated by androgens but it does not rely principally on androgen modulation (Yashi et al. 2003). Subsequently, serum ProGRP status was determined in 460 men with benign and malignant prostatic diseases, chronic renal failure, and healthy controls. The increased serum of ProGRP was observed in patients with the progression of PC into metastatic and androgen-independent stages. In addition multivariate analysis demonstrated that Performance Status (PS), serum ProGRP, and nadir PSA held an independent predictive value for Progression Free Survival ($P < 0.05$). Finally Serum ProGRP was the most significant predictor among pre-treatment factors in this model ($P = 0.0094$) (Yashi et al. 2003).

5.7.2 *Immunohistochemistry*

The role of immunohistochemistry in the last decades has been directed to the definition of potential prediction of androgen resistant status related to ND. In this setting, however, no univoque method of interpretation has been used and multiple neuroendocrine markers have been proposed. The methods of evaluation of ND in PC are mainly based of definition of percentage of positive cells (Berruti et al. 2010) The definition of cut off also remain less characterized. An interesting paper defined as how critical for prognosis, the number of NE cells per hot spot area and the pattern of CGA positivity are. In fact, tumors with more than 30 CGA positive cells for hot spot area have a significant worse prognosis than those tumors with less than 30 CGA positive cells for host spot area. In addition, cases with large clusters of NE cells were significantly more aggressive compared to tumors with no cluster or small clusters of NE cells (Grobholz et al. 2005). Although many immunohistochemical markers have been proposed in order to correctly define the ND in PC, CGA remains the most reliable. NE cells were are more common in higher grade and stage disease, but no difference of 5-year survival between patients with NE cell-positive and -negative tumors have been recorded (Allen et al. 1995). However, McWilliam et al. (1997) found that ND correlates with high grade tumor, bone metastasis and shorter patient survival. In addition, Weinstein et al. (1996) reported that ND determined through immunohistochemistry using CGA represented an independent prognostic factor for biochemical progression in clinical organ-confined PC treated by radical prostatectomy. Moreover Kokubo et al. (2005) demonstrated that 22 % of stage D2 PC showed immunohistochemical CGA overexpression and that CGA staining significantly related to shorter time to recurrence after hormone therapy. Finally, Kamiya and co-workers (2008) demonstrated that positive staining for independent

CGA and combined CGA with NSE after hormone therapy in stage D2 PC was significantly related to shorter overall survival (OS) representing, also, independent factors in multivariate analysis. Also when the CGA expression was assessed on PC biopsies seems to be a new predictive marker of early resistance to ADT (Berruti et al. 2010).

5.8 Imaging of Neuroendocrine Prostate Cancer

5.8.1 PET

Positron emission tomography (PET) is a new imaging modality which has been widely used for the detection of metastasis in various malignancies. F18-fluorodeoxyglucose (FDG), the most common radiotracer, is used for glycolysis evaluation and glucose transporter expression, because most of malignant tumors show increased glucose metabolism. Unfortunately, the use of FDG-PET is not common PCa because of low rate of glycolysis of the tumor cells. In addition, physiologic urinary excretion of FDG does not allow a good visualization of the pelvis (Powles et al. 2007). Indeed Liu et al. found only 4 % sensitivity for identification of primary PCa though FDG-PET (Liu et al. 2001).

But an increased sensitivity for detecting Pca has been obtained through using continuous bladder irrigation (Oyama et al. 2001). FDG-PET could be used to identify local recurrence and distant metastases in patients with increasing PSA after definite local therapy for PCa (Schoder et al. 2005).

In addition, Morris et al. reported that using PSA levels, bone scintigraphy and soft tissue imaging as references, FDG-PET might be a promising outcome measure after chemotherapy in prostate cancer (Morris et al. 2005). Recently, a case was reported showing FDG PET-CT intense uptake in neuroendocrine tumor of the prostate with multiple metastases (Liu 2008). In fact, the cells with ND secrete a variety of factors that can influence growth patterns and metabolic pathways involved in this process and the tumor has different biological behaviour.

5.8.2 Peptide Imaging (PET or Scintigraphic Detection)

Alternatively, neuroendocrine tumors of the prostate can be imaged through the use of probes of the receptors specifically expressed by prostate cancer cells behaving neuroendocrine phenotype. These receptors bind peptides that play a modulator role also in numerous cancers (Reubi 2003; Reubi et al. 2005). This is the key reason or the use of regulatory peptide receptors in cancer imaging in recent times. The first, and currently best, example of targeted peptide receptors is represented by the somatostatin receptors, discovered to be overexpressed in mot neuroendocrine tumors (Reubi 2003).

Somatostatin. In somatostatin-based cancer imaging, a stable somatostatin analog linked to a chelator that can bind radioactive metals such as ^{111}In, ^{99}mTc, or ^{68}Ga, is injected intravenously. The tracer will selectively bind to somatostatin receptors if the patient cancer contains somatostatin receptors in large amounts. The internalization of ligands lead to a radioactivity accumulation in the tumor, compared with the rest of the organs. Normally, rapid and specific uptake is observed in the tumor, and concomitantly in the kidney and bladder, because of predominant urinary radioligands excretion.

The first commercially available agent was 111 the In- diethylenetriaminepentaacetic acid (DTPA)0-octreotide, but its binding affinity to sst2 is moderate and it is not a suitable chelator for b-emitters such as ^{90}Y and ^{177}Lu. For these radiometals, it is better to use the macrocyclic chelator 1, 4, 7, 10-tetraazacyclododecane-1,4,7,10-tetraacetic acid (DOTA), because of the formation of stable metal complexes. The most frequently used DOTA-coupled, somatostatin-based radiopeptides are [DOTA0, Tyr3]-octreotide (DOTATOC) and [DOTA0, Tyr3, Thr8]-octreotide (DOTATATE) (Rufini et al. 2006). In order to improve tracer pharmacokinetics coupling octreotide and octreotate has been developed (Schottelius et al. 2004). 6-Hydrazinopyridine-3-carboxylic acid-TATE (HYNIC-TATE), HYNIC-TOC, and N4-TATE were designed for high-specific-activity labeling with ^{99}mTc, demonstrating relevant additional compounds in octreotide backbone clinical practice (Rufini et al. 2006).

Bombesin. Bombesin-based ligands with high affinity for gastrin-releasing peptide (GRP) receptors have been developed. An early report used a ^{99}mTc-based ligand, RP 527, an N3S chelator coupled to bombesin demonstrated identification of primary prostate cancers and their metastases (Van de Wiele et al. 2008). ^{177}Lu-AMBA is a more recently developed analog, with possible diagnostic and therapeutic applications (Lantry et al. 2006). Another bombesin agonist has been developed with high affinity to all 3 bombesin receptors with possibly broader indications (Zhang et al. 2004). ^{99}mTc-demobesin is an interesting compound, that does not internalize significantly into PC-3 tumor cells but ableto label *in vivo* GRP-R–expressing PC-3 tumors more intensely and for a longer time than the best available GRP-R agonists (Cescato et al. 2008). This extends the paradigm shift on tumor imaging observed earlier with somatostatin antagonists to GRP-R.

Several of these new bombesin-based radiopeptides are conjugated to DOTA and can be labeled with ^{68}Ga. PET studies with a ^{68}Ga-labeled bombesin analog were performed in 11 patients with prostate cancer (Hofmann et al. 2004). Primary tumors were visible in all patients, being the smallest tumor size 5 mm and a plateau of tumor uptake at 15–25 min after injection. Lymph node metastases have been found in three of these patients. However, in four patients a significant nonspecific enrichment has been observed in the upper abdomen, probably due to the pancreas uptake.

5.9 Therapy of Neuroendocrine-Differentiated Prostate Cancers

5.9.1 *ND and New Treatment Modalities*

Despite the initial efficacy of androgen ablation therapies, hormone refractory stage is the normal evolution of PC. Nowadays, a truly hormone-refractory condition is considered when the patient no longer responds to any of the second-line hormonal alternatives. This disease phase often parallels ND in PC. Currently chemotherapy represents the only non-experimental option available. Recently, it was demonstrated that 3-weekly schedule of DTX and prednisone is the standard first-line chemotherapy hormone-refractory PC phase, with a positive impact on OS and time to progression (TTP) (Facchini et al. 2010). But response duration with current chemotherapies is often short. Therefore, novel therapeutic options are needed.

Thus novel approaches currently being tested in early clinical trials include angiogenesis inhibitors, immunological therapies, gene therapy, differentiation therapies and interference in growth-factor-mediated pathways.

5.9.2 *Somatostatin Analogues*

Newly developed somatostatin analogues could be useful in the treatment of PC (Hansson and Abrahamsson 2003). Potential mechanisms of antitumor action include the suppression of circulating levels of trophic hormones and growth factors as well as direct effect on tumoral growth, involving autocrine/paracrine mechanisms.

Somatostatin family include regulatory peptides produced by normal neuroendocrine, inflammatory and immune cells, but also by many tumor activated cells.

Exogenously administration of somatostatin induces a wide range of effects on multiple target sites. Thus selective non-peptide agonists have been developed for four of the somatostatin receptors (SSTR) subtypes. Main somatostatin effect is the prevention of cell proliferation through inducing cell cycle arrest and apoptosis. These effects are mediated by SSTR expressed by tumor cells and by non-tumor-cells, secreting hormones and growth factors promoting tumor cell growth. Four SSTRs induce cell cycle arrest via protein tyrosine phosphatase (PTP)-dependent modulation of MAPK, associated with induction of retinoblastoma tumor suppressor protein and p21. In addition SSTR3 triggers apoptosis, through activation of p53 and the pro-apoptotic protein BAX. Thus it seems that somatostatin plays an important role in tumor development and in the future there may be a potential role for somatostatin analogues in the treatment of the PC (Hansson and Abrahamsson 2003). In this view, recently 38 stage D3 PC patients (mean age 71.8 ± 5.9 years) have continued to receive androgen ablation therapy in combination with oral dexamethasone (4 mg daily for the first month of treatment, tapered down to 1 mg daily by the fourth month, with 1 mg daily maintenance dose thereafter) and somatostatin analog (20 mg octreotide i.m. injections every 28 days). Twenty-three of those thirty-eight patients (60.5 %) had partial responses (PR, ≥50 % PSA decline), 9 (21.1 %) had stable disease and 7 (18.4 %) had disease progression. In 47.7 % (18 of 38) of patients, serum PSA levels decreased with treatment but did not return to their respective baselines until the end of follow-up (or death from non-prostate cancer-related causes). All patients reported significant and durable improvement of bone pain and PS (for a median duration of 14 months; 95 % CI, 9–19 months). In addition a statistically significant ($P < 0.01$) reduction of serum insulin-like growth factor-1 levels was recorded in patients with response to the combination therapy (Koutsilieris et al. 2004). On the basis of this latter study, a randomized controlled clinical trial of 38 stage D3 patients (mean age 72.8 ± 6.8 years) has been performed in order to compare the combination of somatostatin analog (octreotide 20 mg i.m. every 28 days) and oral dexamethasone (4 mg daily for 1 month, gradually reduced to 1 mg daily by the fourth month, with a 1 mg daily maintenance dose thereafter) plus zoledronate (4 mg i.v. every 4 weeks) vs. zoledronate only. All patients in both arms remained in basic androgen blockade. Partial responses (PR, > or =50 % PSA decline) was recorded in 13 out of 20 patients with combination therapy vs. none with zoledronate (Mitsiades et al. 2006).

It was recently proposed as therapy of ND in hormone-independent PC, a combination of oestrogens and somatostatin analogues. The combination of ethinyl estradiol and the somatostatin analogue lanreotide, binding 3/5 SSTRs, showed a favourable toxicity profile and offered objective and symptomatic responses in patients with refractoriness to conventional hormonal therapy strategies. In addition a higher median OS was observed (Sciarra et al. 2003).

5.9.3 Serotonin Antagonists

NE cells produce and secrete 5-HT, a biogenic amine, neurotransmitter and potent mitogen associated with tumor growth. 5-HT receptors (5-HTR), such as 5-HTR1 and 5-HTR4, are overexpressed in hormone refractory PC tissues and in PC cell lines. Recently promising results have been demonstrated with the use of 5-HTR antagonists (Abrahamsson et al. 1986).

5.9.4 Bombesin Antagonists

Bombesin produces androgen-dependent growth and invasiveness of PC cells. Bombesin also carries metastatic potential in androgen-insensitive PC. Therefore, bombesin-like antagonists could become an effective treatment option in the future (Hansson and Abrahamsson 2003; Levine et al. 2003).

5.9.5 Cytokines

Recently, IL-6, an inflammatory cytokine that not only regulates the immune response, but also modulates cancer cell growth, differentiation and survival has been proposed as a possible target in the treatment of androgen resistant PC patients. Recently 53 patients with castration-resistant PC pre- treated with taxane chemotherapy were treated with 6 mg/kg anti-IL-6 antibody, CNTO328 i.v. every 2 weeks for 12 cycles. Two patients (3.8 %; 95 % CI, 0.5–13.0 %) had PSA response. None of the 31 patients with measurable disease had a RECIST (Response Evaluation Criteria in Solid Tumors) response but 7 (23 %) had stable disease. After a median follow-up of 14.8 months, the median progression-free survival (PFS) was 1.6 months (95 % CI, 1.6–1.7) and median OS was 11.6 months (95 % CI, 7.5–19.0). Thirty-two out of thirty-eight patients had C-reactive protein plasma levels decline at 6 weeks. In conclusion, CNTO328 resulted in a PSA response rate of 3.8 % and a RECIST stable disease rate of 23 %, while declining C- reactive protein levels during treatment may reflect biological activity. Despite evidence of CNTO-mediated IL-6 inhibition, elevated baseline IL-6 levels portended a poor prognosis.

In another open-label phase II trial mitoxantrone/prednisone (M/P) with and without CNTO328 was performed in metastatic patients with castration-resistant PC who have had received DTX- based chemotherapy. This trial concluded that while CNTO328 plus M/P appeared well tolerated, improvement in outcomes was not demonstrable (Fizazi et al. 2012).

In conclusion, recent progress in terms of PC research, especially the role of ND in PC has lead to the development of entirely new therapeutic modalities for hormone-refractory PC.

References

Abdul M, Anezinis PE, Logothetis CJ et al (1994) Growth inhibition of human prostatic carcinoma cell lines by serotonin antagonists. Anticancer Res 14:1215–1220

Abrahamsson PA, Wadstrom LB, Alumets J et al (1986) Peptide hormone- and serotonin-immunoreactive cells in normal and hyperplastic prostate glands. Pathol Res Prac 181:675–683

Abrahamsson PA, Wadstrom LB, Alumets J et al (1987) Peptide-hormone and serotonin-immunoreactive tumour cells in carcinoma of the prostate. Pathol Res Pract 182:298–307

Ahlgren G, Pedersen K, Lundberg S et al (2000) Regressive changes and neuroendocrine differentiation in prostate cancer after neoadjuvant hormonal treatment. Prostate 42:274–279

Albrecht M, Doroszewicz J, Gillen S et al (2004) Proliferation of prostate cancer cells and activity of neutral endopeptidase is regulated by bombesin and IL-1beta with IL-1beta acting as a modulator of cellular differentiation. Prostate 58:82–94

Allen FJ, Van Velden DJ, Heyns CF (1995) Are neuroendocrine cells of practical value as an independent prognostic parameter in prostate cancer? Br J Urol 75:751–754

Angelsen A, Syversen U, Stridsberg M et al (1997) Use of neuroendocrine serum markers in the follow-up of patients with cancer of the prostate. Prostate 31:110–117

Aprikian AG, Cordon-Cardo C, Fair WR et al (1993) Characterization of neuroendocrine differentiation in human benign prostate and prostatic adenocarcinoma. Cancer 71:3952–3965

Berruti A, Dogliotti L, Mosca A et al (2000) Circulating neuroendocrine markers in patients with prostate carcinoma. Cancer 88:2590–2597

Berruti A, Mosca A, Tucci M et al (2005) Independent prognostic role of circulating chromogranin a in prostate cancer patients with hormonerefractory disease. Endocr Relat Cancer 12:109–117

Berruti A, Bollito E, Cracco CM et al (2010) The prognostic role of immunohistochemical chromogranin a expression in prostate cancer patients is significantly modified by androgen-deprivation therapy. Prostate 70:718–726

Bonkhoff H (1996) Role of the basal cells in premalignant changes of the human prostate: a stem cell concept for the development of prostate cancer. Eur Urol 30:201–205

Bonkhoff H, Stein U, Remberger K (1994) Multidirectional differentiation in the normal, hyperplastic, and neoplastic human prostate: simultaneous demonstration of cell-specific epithelial markers. Hum Pathol 25:42–46

Bonkhoff H, Stein U, Remberger K (1995) Endocrine-paracrine cell types in the prostate and prostatic adenocarcinoma are postmitotic cells. Hum Pathol 26:167–170

Borre M, Nerstrom B, Overgaard J (2000) Association between immunohistochemical expression of vascular endothelial growth factor (VEGF), VEGF-expressing neuroendocrine-differentiated tumor cells, and outcome in prostate cancer patients subjected to watchful waiting. Clin Cancer Res 6:1882–1890

Bostwick DG, Dousa MK, Crawford BG et al (1994) Neuroendocrine differentiation in prostatic intraepithelial neoplasia and adenocarcinoma. Am J Surg Pathol 18:1240–1246

Bostwick DG, Qian J, Pacelli A et al (2002) Neuroendocrine expression in node positive prostate cancer: correlation with systemic progression and patient survival. J Urol 168:1204–1211

Buchanan G, Greenberg NM, Scher HI et al (2001) Collocation of androgen receptor gene mutations in prostate cancer. Clin Cancer Res 7:1273–1281

Cescato R, Maina T, Nock B et al (2008) Bombesin receptor antagonists may be preferable to agonists for tumor targeting. J Nucl Med 49:318–326

Chevalier S, Defoy I, Lacoste J et al (2002) Vascular endothelial growth factor and signaling in the prostate: more than angiogenesis. Mol Cell Endocrinol 189:169–179

Chung TD, Yu JJ, Kong TA et al (2000) Interleukin-6 activates phosphatidylinositol-3 kinase, which inhibits apoptosis in human prostate cancer cell lines. Prostate 42:1–7

Cohen MK, Arber DA, Coffield KS et al (1994) Neuroendocrine differentiation in prostatic adenocarcinoma and its relationship to tumor progression. Cancer 74:1899–1903

Collado B, Gutierrez-Canas I, Rodriguez-Henche N et al (2004) Vasoactive intestinal peptide increases vascular endothelial growth factor expression and neuroendocrine differentiation in human prostate cancer LNCaP cells. Regul Pept 119:69–75

Collado B, Sanchez MG, Diaz-Laviada I et al (2005) Vasoactive intestinal peptide (VIP) induces c-fos expression in LNCaP prostate cancer cells through a mechanism that involves $Ca2+$ signalling. Implications in angiogenesis and neuroendocrine differentiation. Biochim Biophys Acta 1744:224–233

Cox ME, Deeble PD, Bissonette EA et al (2000) Activated $3',5'$-cyclic AMP-dependent protein kinase is sufficient to induce neuroendocrinelike differentiation of the LNCaP prostate tumor cell line. J Biol Chem 275:13812–13818

Deeble PD, Murphy DJ, Parsons SJ et al (2001) Interleukin-6-, cyclic AMP-mediated signalling potentiates neuroendocrine differentiation of LNCaP prostate tumor cells. Mol Cell Biol 21:8471–8482

Deftos LJ, Nakada S, Burton DW et al (1996) Immunoassay and immunohistology studies of chromogranin a as a neuroendocrine marker in patients with carcinoma of the prostate. Urology 48:58–62

Deng X, Liu H, Huang J et al (2008) Ionizing radiation induces prostate cancer neuroendocrine differentiation through interplay of CREB and ATF2: implications for disease progression. Cancer Res 68:9663–9670

di Sant'Agnese PA (1992) Neuroendocrine differentiation in carcinoma of the prostate. Diagnostic, prognostic, and therapeutic implications. Cancer 70:254–268

di Sant'Agnese PA (1998) Neuroendocrine differentiation in prostatic carcinoma: an update. Prostate 36(8):74–79
Dizeyi N, Konrad L, Bjartell A et al (2002) Localization and mRNA expression of somatostatin receptor subtypes in human prostatic tissue and prostate cancer cell lines. Urol Oncol 7:91–98
Erasmus CE, Verhagen WI, Wauters CA et al (2002) Brain metastasis from prostate small cell carcinoma: not to be neglected. Can J Neurol Sci 29:375–377
Facchini G, Caraglia M, Morabito A et al (2010) Metronomic administration of zoledronic acid and taxotere combination in castration resistant prostate cancer patients: phase I ZANTE trial. Cancer Biol Ther 10:543–548
Festuccia C, Guerra F, D'Ascenzo S (1998) *In vitro* regulation of pericellular proteolysis in prostatic tumor cells treated with bombesin. Int J Cancer 75:418–431
Fizazi K, De Bono JS, Flechon A et al (2012) Randomised phase II study of CNTO328 (CNTO 328), an anti-IL-6 monoclonal antibody, in combination with mitoxantrone/prednisone versus mitoxantrone/prednisone alone in metastatic castration-resistant prostate cancer. Eur J Cancer 48:85–93
Ghannoum JE, DeLellis RA, Shin SJ (2004) Primary carcinoid tumor of the prostate with concurrent adenocarcinoma: a case report. Int J Surg Pathol 12:167–170
Grobholz R, Bohrer MH, Siegsmund M et al (2000) Correlation between neovascularisation and neuroendocrine differentiation in prostatic carcinoma. Pathol Res Pract 196(5):277–284
Grobholz R, Griebe M, Sauer CG et al (2005) Influence of neuroendocrine tumor cells on proliferation in prostatic carcinoma. Hum Pathol 36:562–570
Guillemot F, Lo LC, Johnson JE et al (1993) Mammalian achaete-scute homolog 1 is required for the early development of olfactory and autonomic neurons. Cell 75:463–476
Hansson J, Abrahamsson PA (2003) Neuroendocrine differentiation in prostate carcinoma. Scand J Urol Nephrol 37(Suppl 212):28–36
Helpap B (2002) Morphology and therapeutic strategies for neuroendocrine tumors of the genitourinary tract. Cancer 95:1415–1420
Hofmann M, Machtens S, Stief C et al (2004) Feasibility of Ga-68-DOTABOM PET in prostate carcinoma patients [abstract]. J Nucl Med 45:449P
Huang J, di Sant''Agnese P (2002) Neuroendocrine differentiation in prostate cancer: an overview. In: Lamberts S (ed) Advances in oncology: the expanding role of octreotide. Bioscientifica Ltd, Bristol, pp 243–262
Huang J, Yao JL, Zhang L et al (2005) Differential expression of interleukin-8 and its receptors in the neuroendocrine and nonneuroendocrine compartments of prostate cancer. Am J Pathol 166:1807–1815
Huang J, Yao JL, Di Sant'agnese PA et al (2006) Immunohistochemical characterization of neuroendocrine cells in prostate cancer. Prostate 66:1399–1406
Huss WJ, Gray DR, Werdin ES et al (2004) Evidence of pluripotent human prostate stem cells in a human prostate primary xenograft model. Prostate 60:77–90
Isaacs JT (2008) Prostate stem cells and benign prostatic hyperplasia. Prostate 68:1025–1034
Isaacs JT, Coffey DS (1989) Etiology and disease process of benign prostatic hyperplasia. Prostate Suppl 2:33–50
Ishimaru H, Kageyama Y, Hayashi T et al (2002) Expression of matrix metalloproteinase-9 and bombesin/gastrinreleasing peptide in human prostate cancers and their lymph node metastases. Acta Oncol 41:289–296
Jin RJ, Wang Y, Masumori N et al (2004) NE-10 neuroendocrine cancer promotes the LNCaP xenograft growth in castrated mice. Cancer Res 64:5489–5495
Jin RJ, Lho Y, Connelly L et al (2008) The nuclear factor-kappaB pathway controls the progression of prostate cancer to androgen-independent growth. Cancer Res 68:6762–6769
Jongsma J, Oomen MH, Noordzij MA (2000) Androgen-independent growth is induced by neuropeptides in human prostate cancer cell lines. Prostate 42:34–44
Kadmon D, Thompson TC, Lynch GR et al (1991) Elevated plasma chromogranin-a concentrations in prostatic carcinoma. J Urol 146:358–361

Kamiya N, Suzuki H, Kawamura K et al (2008) Neuroendocrine differentiation in stage D2 prostate cancers. Int J Urol 15:423–428

Kawai S, Hiroshima K, Tsukamoto Y et al (2003) Small cell carcinoma of the prostate expressing prostatespecific antigen and showing syndrome of inappropriate secretion of antidiuretic hormone: an autopsy case report. Pathol Int 53:892–896

Kim J, Adam RM, Freeman MR (2002) Activation of the Erk mitogen-activated protein kinase pathway stimulates neuroendocrine differentiation in LNCaP cells independently of cell cycle withdrawal and STAT3phosphorylation. Cancer Res 62:1549–1554

Kimura N, Hoshi S, Takahaski M et al (1997) Plasma chromogranin a in prostatic carcinoma and neuro-endocrine tumors. J Urol 157:565–568

Kokubo H, Yamada Y, Nishio Y et al (2005) Immunohistochemical study of chromogranin a in stage D2 prostate cancer. Urology 66:135–140

Koutsilieris M, Mitsiades CS, Bogdanos J et al (2004) Combination of somatostatin analog, dexamethasone, and standard androgen ablation therapy in stage D3 prostate cancer patients with bone metastases. Clin Cancer Res 10:4398–4405

Lantry LE, Cappelletti E, Maddalena ME et al (2006) 177Lu-AMBA: synthesis and characterization of a selective 177Lu-labeled GRP receptor agonist for systemic radiotherapy of prostate cancer. J Nucl Med 47:1144–1152

Lee LF, Louie MC, Desai SJ et al (2004) Interleukin-8 confers androgen-independent growth and migration of LNCaP: differential effects of tyrosine kinases Src and FAK. Oncogene 23: 2197–2205

Levine L, Lucci JA, Pazdrak B et al (2003) Bombesin stimulates nuclear factor kappa B activation and expression of proangiogenic factors in prostate cancer cells. Cancer Res 63:3495–3502

Liu Y (2008) FDG PET-CT demonstration of metastatic neuroendocrine tumor of prostate. World J Surg Oncol 6:64

Liu IJ, Zafar MB, Lai YH (2001) Fluorodeoxyglucose positron emission tomography studies in diagnosis and staging of clinically organ-confined prostate cancer. Urology 57:108–115

Logothetis C, Hoosein N (1992) The inhibition of the paracrine progression of prostatic cancer as an approach to early therapy of prostatic carcinoma. J Cell Biochem Suppl 16H:128–134

Markwalder R, Reubi JC (1999) Gastrin-releasing peptide receptors in the human prostate: relation to neoplastic transformation. Cancer Res 59:1152–1159

Mazzucchelli R, Lopez-Beltran A, Scarpelli M et al (2002) Predictive factors in prostate needle biopsy. Pathologica 94:331–337

McWilliam LJ, Manson C, George NJ (1997) Neuroendocrine differentiation and prognosis in prostatic adenocarcinoma. Br J Urol 80:287–290

Meyer-Siegler K (2001) COX-2 specific inhibitor, NS-398, increases macrophage migration inhibitory factor expression and induces neuroendocrine differentiation in C4-2b prostate cancer cells. Mol Med 7:850–860

Mitsiades CS, Bogdanos J, Karamanolakis D et al (2006) Randomized controlled clinical trial of a combination of somatostatin analog and dexamethasone plus zoledronate vs. zoledronate in patients with androgen ablation-refractory prostate cancer. Anticancer Res 26:3693–3700

Mori S, Murakami-Mori K, Bonavida B (1999) Interleukin-6 induces G1 arrest through induction of p27(Kip1), a cyclin-dependent kinase inhibitor, and neuron-like morphology in LNCaP prostate tumor cells. Biochem Biophys Res Commun 257:609–614

Mori R, Xiong S, Wang Q et al (2009) Gene profiling and pathway analysis of neuroendocrine transdifferentiated prostate cancer cells. Prostate 69:12–23

Morris MJ, Akhurst T, Larson SM et al (2005) Fluorodeoxyglucose positron emission tomography as an outcome measure for castrate metastatic prostate cancer treated with antimicrotubule chemotherapy. Clin Cancer Res 11:3210–3216

Nagakawa O, Murakami K, Ogasawara M et al (1999) Effect of chromogranin a (pancreastatin) fragment on invasion of prostate cancer cells. Cancer Lett 147:207–213

Nagakawa O, Ogasawara M, Murata J et al (2001) Effect of prostatic neuropeptides on migration of prostate cancer cell lines. Int J Urol 8:65–70

Oyama N, Akino H, Suzuki Y (2001) FDG PET for evaluating the change of glucose metabolism in prostate cancer after androgen ablation. Nucl Med Commun 22:963–968

Palapattu GS, Wu C, Silvers CR et al (2009) Selective expression of CD44, a putative prostate cancer stem cell marker, in neuroendocrine tumor cells of human prostate cancer. Prostate 69:787–798

Pearse AG, Takor T (1979) Embryology of the diffuse neuroendocrine and its relationship to the common peptides. Fed Proc 38:2288–2294

Pinski J, Wang Q, Quek ML et al (2006) Genistein-induced neuroendocrine differentiation of prostate cancer cells. Prostate 66:1136–1143

Powles T, Murray I, Brock C (2007) Molecular position emission tomography and PET/CT imaging in urological malignancies. Eur Urol 51:1511–1521

Reubi JC (2003) Peptide receptors as molecular targets for cancer diagnosis and therapy. Endocr Rev 24:389–427

Reubi JC, Macke HR, Krenning EP (2005) Candidates for peptide receptor radiotherapy today and in the future. J Nucl Med 46(suppl 1):67S–75S

Rufini V, Calcagni ML, Baum RP (2006) Imaging of neuroendocrine tumors. Semin Nucl Med 36:228–247

Salido M, Vilches J, Lopez A (2000) Neuropeptides bombesin and calcitonin induce resistance to etoposide induced apoptosis in prostate cancer cell lines. Histol Histopathol 15:729–738

Salido M, Vilches J, Roomans GM (2004) Changes in elemental concentrations in LNCaP cells are associated with a protective effect of neuropeptides on etoposide-induced apoptosis. Cell Biol Int 28:397–402

Sarkar D, Singh SK, Mandal AK et al (2010) Plasma chromogranin a: clinical implications in patients with castrate resistant prostate cancer receiving docetaxel chemotherapy. Cancer Biomark 8:81–87

Sauer CG, Roemer A, Grobholz R (2006) Genetic analysis of neuroendocrine tumor cells in prostatic carcinoma. Prostate 66:227–234

Scher HI, Sawyers CL (2005) Biology of progressive, castration-resistant prostate cancer: directed therapies targeting the androgen receptor signaling axis. J Clin Oncol 23:8253–8261

Schmid KW, Helpap B, Totsch M et al (1994) Immunohisto-chemical localization of chromogranin a and B and secretogranin II in normal, hyperplastic and neoplastic prostate. Histopathology 24:233–239

Schoder H, Herrmann K, Gonen M (2005) 2-[18F]fluoro-2-deoxyglucose positron emission tomography for the detection of disease in patients with prostate-specific antigen relapse after radical prostatectomy. Clin Cancer Res 11:4761–4769

Schottelius M, Poethko T, Herz M et al (2004) First 18F-labeled tracer suitable for routine clinical imaging of sst receptor-expressing tumors using positron emission tomography. Clin Cancer Res 10:3593–3606

Schron DS, Gipson T, Mendelsohn G (1984) The histogenesis of small cell carcinoma of the prostate: an immunohistochemical study. Cancer 53:2478–2480

Sciarra A, Monti S, Gentile V et al (2003) Variation in chromogranin. A serum levels during intermittent versus continuous androgen deprivation therapy for prostate adenocarcinoma. Prostate 55:168–179

Seethalakshmi L, Mitra SP, Dobner PR et al (1997) Neurotensin receptor expression in prostate cancer cell line and growth effect of NT at physiological concentrations. Prostate 31:183–192

Segal NH, Cohen RJ, Haffejee Z et al (1994) BCL-2 proto-oncogene expression in prostate cancer and its relationship to the prostatic neuroendocrine cell. Arch Pathol Lab Med 118:616–618

Sehgal I, Thompson TC (1999) Novel regulation of type IV collagenase (matrix metalloproteinase-9 and −2) activities by transforming growth factor-beta1 in human prostate cancer cell lines. Mol Biol Cell 10:407–416

Tanaka M, Suzuki Y, Takaoka K et al (2001) Progression of prostate cancer to neuroendocrine cell tumor. Int J Urol 8:431–436

Tang Y, Wang L, Goloubeva O et al (2009) The relationship of neuroendocrine carcinomas to anti-tumor therapies in TRAMP mice. Prostate 69:1763–1773

Tarle M, Rados N (1991) Investigation on serum neurone-specific enolase in prostatic cancer diagnosis and monitoring: comparative study of a multiple tumor marker assay. Prostate 19:23–33

van Bokhoven A, Varella-Garcia M, Korch C et al (2003) Molecular characterization of human prostate carcinoma cell lines. Prostate 57:205–225

Van de Wiele C, Phonteyne P, Pauwels P et al (2008) Gastrin-releasing peptide receptor imaging in human breast carcinoma versus immunohistochemistry. J Nucl Med 49:260–264

Vilches J, Salido M, Fernandez-Segura E et al (2004) Neuropeptides, apoptosis and ion changes in prostate cancer. Methods of study and recent developments. Histol Histopathol 19:951–961

Waltregny D, Leav I, Signoretti S et al (2001) Androgen-driven prostate epithelial cell proliferation and differentiation *in vivo* involve the regulation of p27. Mol Endocrinol 15:765–782

Wang Q, Horiatis D, Pinski J (2004) Interleukin-6 inhibits the growth of prostate cancer xenografts in mice by the process of neuroendocrine differentiation. Int J Cancer 111:508–513

Weinstein MH, Partin AW, Veltri RW et al (1996) Neuroendocrine differentiation in prostate cancer: enhanced prediction of progression after radical prostatectomy. Hum Pathol 27:683–687

Wright ME, Tsai MJ, Aebersold R (2003) Androgen receptor represses the neuroendocrine transdifferentiation process in prostate cancer cells. Mol Endocrinol 17:1726–1737

Wu C, Huang J (2007) Phosphatidylinositol 3-kinase-AKT-mammalian target of rapamycin pathway is essential for neuroendocrine differentiation of prostate cancer. J Biol Chem 282:3571–3583

Wu G, Burzon DT, di Sant'Agnese PA et al (1996) Calcitonin receptor mRNA expression in the human prostate. Urology 47:376–381

Wu C, Zhang L, Bourne PA et al (2006) Protein tyrosine phosphatase PTP1B is involved in neuroendocrine differentiation of prostate cancer. Prostate 66:1125–1135

Xing N, Qian J, Bostwick D et al (2001) Neuroendocrine cells in human prostate over-express the anti-apoptosis protein survivin. Prostate 48:7–15

Yang JC, Ok JH, Busby JE et al (2009) Aberrant activation of androgen receptor in a new neuropeptide-autocrine model of androgen-insensitive prostate cancer. Cancer Res 69:151–160

Yao JL, Madeb R, Bourne P et al (2006) Small cell carcinoma of the prostate: an immunohistochemical study. Am J Surg Pathol 30:705–712

Yashi M, Nukui A, Kurokawa S et al (2003) Elevated serum progastrin-releasing peptide (31–98) level is a predictor of short response duration after hormonal therapy in metastatic prostate cancer. Prostate 56:305–312

Yuan TC, Veeramani S, Lin MF (2007) Neuroendocrine-like prostate cancer cells: neuroendocrine transdifferentiation of prostate adenocarcinoma cells. Endocr Relat Cancer 14:531–547

Zhang H, Chen J, Waldherr C et al (2004) Synthesis and evaluation of bombesin derivatives on the basis of pan-bombesin peptides labeled with indium-111, lutetium-177, and yttrium-90 for targeting bombesin receptor-expressing tumors. Cancer Res 64:6707–6715

Chapter 6
Metastatic Dissemination

Stefania Staibano

Abstract In spite of recent developments in diagnosis, staging and treatment, most patients with advanced prostate cancer will ultimately progress from androgen-sensitive to an irreversible castration-resistant disease. These androgen-independent cancers frequently give rise to widespread metastasis, dramatically reducing the median survival of patients (Tannock et al, N Engl J Med, 351(15):1502–1512, 2004) and accounting for more than 32, 000 deaths/year in USA (Jemal et al, CA Cancer J Clin, 60:277–300, 2010), which correspond to over 90 % of PC related mortality (Man, Gardner, Int J Biol Sci, 4(4):246–258, 2008).

It is a common belief that cancer metastasis result from a multi-stage nonrandom process characterized by intricate interactions between cancer cells and the host microenvironment, leading to the detachment of cancer cells from their tissue of origin, their dissemination through the bloodstream and to invasion of the target metastatic site (Patel et al, Future Oncol, 7(11):1285–1297, 2011).

Metastasis represents yet one of the most enigmatic aspects of prostate cancer pathogenesis, in which a cascade of proteolytic enzymes, inflammatory cytokines, growth factors, activated oncogenes, oxidative stress and hypoxia linked proteins and adhesion molecules, orchestrate a continuous loop that enable migrating cancer cells detached from the primary tumor bulk, to survive and proliferate in an adverse remote body microenvironment.

In this chapter, we discuss the nature and alterations of the signaling pathways involved in the development of prostate cancer metastasis, reporting the current status of knowledge on the changes occurring either in prostate cancer cells and in tumor-associated stromal tissue, with particular emphasis to the process of epithelial-mesenchymal transition ("phenotypic plasticity") and to the role of cancer stem cells in prostate cancer progression and metastasis.

S. Staibano (✉)
Department of Advanced Biomedical Sciences, Pathology Section, Faculty of Medicine and Surgery, University of Naples "Federico II", via S. Pansini, n.5, Naples, Italy
e-mail: staibano@unina.it

S. Staibano (ed.), *Prostate Cancer: Shifting from Morphology to Biology*,
DOI 10.1007/978-94-007-7149-9_6, © Springer Science+Business Media Dordrecht 2013

We will highlight, also, the emerging data concerning new therapeutic targets for treatment of metastatic prostate cancer that, while deserving further inquiry, look very promising to improve our chances to successful approach the advanced disease or, even, primarily reduce the risk of metastasis from castration-resistant prostate cancer (Vashisht, Bagler, PLoS One, 7(11):e49401, 2012).

Metastases represent the most fearful evolution of advanced/systemic prostate cancer progressed into a castration-resistance state after first-instance deprivation therapy.

Before the onset of metastasis, prostate cancer is usually characterized by a long latency period, in which genetic (Nguyen and Massague 2007; Zhao et al. 2013) and epigenetic (Rodenhiser 2009) cellular alterations lead to changes in cancer cells molecular phenotype, with the gain of both cytoskeletal motility and the ability to detach from the tissue of origin. The acquired abilities of epithelial prostate cancer cells are critically boosted by activated prostatic stromal cells, as tumor-associated macrophages (TAMs), cancer-associated fibroblasts (CAFs) and endothelial cells (Wang et al. 2013).

Besides their role in tumor-associated angiogenesis, CD31/CD34-positive endothelial cells lining microvessels decrease upon castration, increasing instead when prostate cancer progress to castration-resistance. Recently, it has shown *in vitro* that endothelial cells secrete high levels of IL-6. This cytokine down-regulates AR and activates the TGFbeta/MMP9 signaling pathway in prostate cancer cells, contributing then to their invasive and metastasizing ability (Wang et al. 2013).

TAMs produce several migration-stimulating factors, as CXCL12, IL-6 and TNF (Allavena et al. 2008). Activated CAFs mostly exhibit a myofibroblastic phenotype induced either by the direct physical contact with cancer cells and *via* the hyper-stimulation, by several tumor- and hypoxia associated growth factors, as EGF, FGF, IGF, VEGF. CAFs overproduce TGFbeta (Roodman 2004), which intervene in ECM remodeling (Allavena et al. 2008) and in the induction of epithelial-mesenchymal transition (EMT) (Yilmaz and Christofori 2009) of metastasizing cells.

Extracellular matrix proteins, facilitating either tumor growth and metastasis, continuously accumulate in tumor stroma. This is the case for versican, a large proteoglycan associated with metastasis and poor outcome of prostate cancer and several solid malignant cancers. It has been shown to regulate cancer cell adhesion, proliferation, migration, angiogenesis, invasion and metastasis mainly through physical interactions mediated by chondroitin and dermatan sulfate side chains; looking particularly attractive as a possible adjunctive therapeutic target for aggressive prostate cancers (Du et al. 2013).

Even a disturbance of the interplay between the electrical and metabolic activity of prostate cells seems to play a role in the gain of propensity to metastasize of prostate cancer. It has been, in fact, recently reported that an altered expression of connexins, which form intercellular channels involved in gap-junction-mediated intercellular coupling, might be correlated with the invasive potential of cancer cells (Czyż et al. 2012). This finding, however, deserves further investigation.

The acquired EMT ability of prostate cancer cells leading to the detachment from the bulk of primary cancer, is conditioned by the dramatic loss of adhesion proteins, as E-cadherin (Yates 2011; Lazari et al. 2013) and their regulating transcriptional inducers, as the SAM Pointed Domain ETS transcription Factor (SPDEF) (Pal et al. 2013), and by the increase of their transcriptional repressors, as the Wilms' tumor gene (WT1) (Brett et al. 2013). Recently, the altered expression of the human metastasis-associated gene 1 (MTA1) has been found strictly associated with the pAkt/E-cadherin pathway regulation and with metastatic prostate cancer (Wang et al. 2012), and, a combined testing strategy for detecting MTA1 and E-cadherin, has been proposed for selecting high-risk prostate cancer patients (Fan et al. 2012).

Before permeating blood vessels, detached tumor cells have to escape anoikis and gain survival benefits (Hu et al. 2012). The anoikis-resistance and EMT properties of prostate tumor cells are mediated by several molecular players, including members of the Notch signaling pathway, as well of the Akt survival pathway including the early-recruited focal adhesion player tallin. Tallin mediates integrin activation and induces downstream survival pathways resulting in the promotion of cancer cells progression to metastasis (Desiniotis and Kyprianou 2011).

Both Notch-related proteins and tallin appear, then, as promising candidate as either prognostic markers and therapeutic targets in metastasizing prostate cancers.

To survive in the bloodstream, prostate cancer cells activate multiple survival pathways, comprising the overexpression of several members of the anti-apoptotic Bcl-2 protein family, combined with the inactivation of the FADD death receptor pathway, or the lack of expression of pro-apoptotic effector proteins as Bax and caspases (Igney and Krammer 2002).

Moreover, circulating prostate cancer cells may activate survivin expression and undergo to autophagy, to survive in the absence of sufficient extracellular nutrients (Roca et al. 2008).

While metastasizing prostate cancer cells may optionally variously localize in several body sites, as lung and liver, they invariably hit the bone (Osanto and Van Poppel 2012) (Fig. 6.1).

In bone-metastasizing cancers, the CG-protein-coupled calcium sensing receptor (CaSR), which is primarily involved in the feedback regulation of extracellular free ionised calcium (Ca2+), may act as an oncogene, associating also with cancer progression. In prostate cancer, its altered expression seems to facilitate bone metastasis (Brennan et al. 2012).

Disseminated prostate cancer cells (DPCC) reaching the bone marrow occupy the same bone niche in which hematopoietic stem cells reside in a quiescent state (Taichman 2005). This has led to the concept that DPCCs behave as "parasites" of the hematopoietic niche.

DPCCs which evade immune attack and/or chemotherapy cytotoxicity may outlive for a variable time in bone marrow of patients after radical prostatectomy or chemotherapy (Morgan et al. 2009; Pfitzenmaier et al. 2007), in an auto-induced reversible state of growth arrest, the so-called "tumor-dormancy" (Townson and Chambers 2006; Aguirre-Ghiso 2007; Shiozawa et al. 2008b; Joyce and Pollard 2009). This underlies the troublesome unresolved phenomenon of "minimal residual

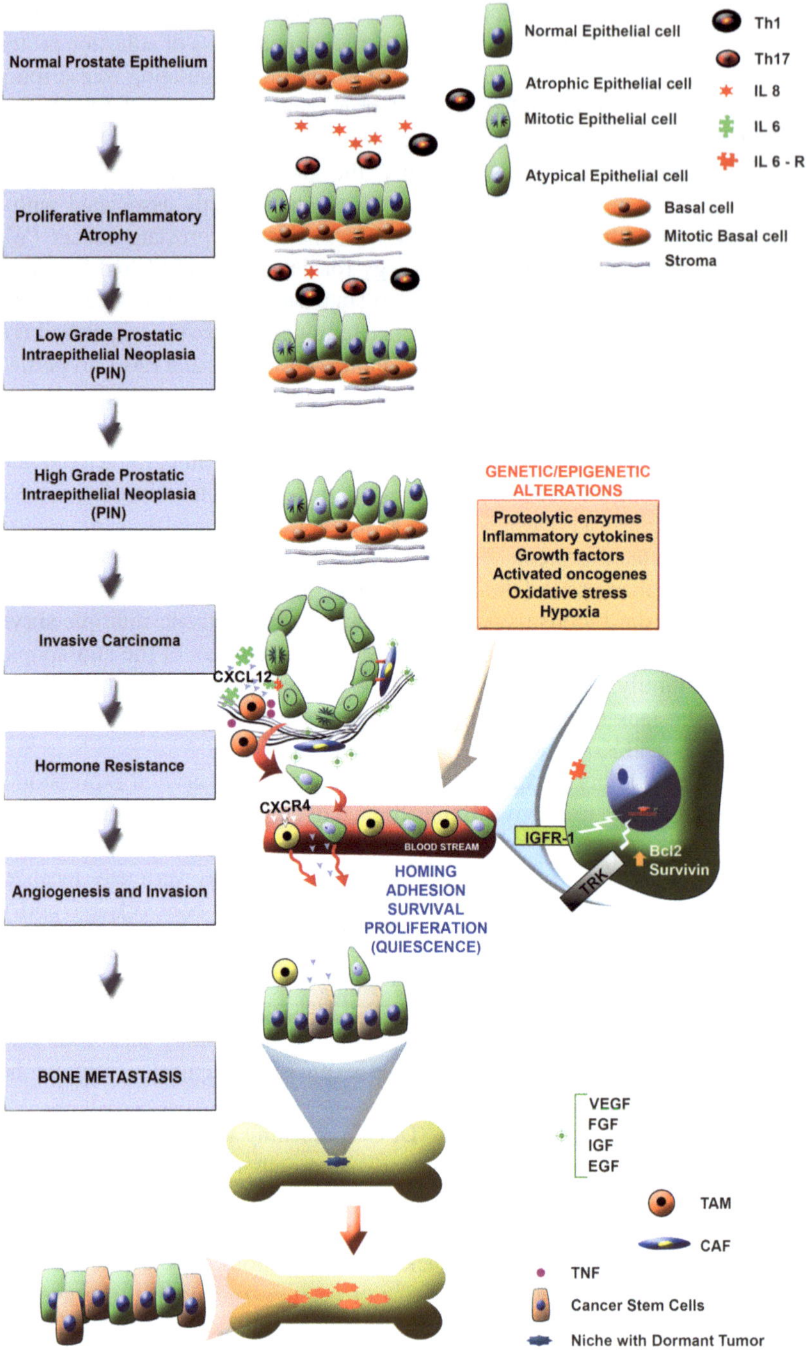
Normal Prostate Epithelium
Proliferative Inflammatory Atrophy
Low Grade Prostatic Intraepithelial Neoplasia (PIN)
High Grade Prostatic Intraepithelial Neoplasia (PIN)
Invasive Carcinoma
Hormone Resistance
Angiogenesis and Invasion
BONE METASTASIS
Normal Epithelial cell
Atrophic Epithelial cell
Mitotic Epithelial cell
Atypical Epithelial cell
Th1
Th17
IL 8
IL 6
IL 6 - R
Basal cell
Mitotic Basal cell
Stroma
GENETIC/EPIGENETIC ALTERATIONS
Proteolytic enzymes
Inflammatory cytokines
Growth factors
Activated oncogenes
Oxidative stress
Hypoxia
CXCL12
CXCR4
BLOOD STREAM
IGFR-1
TRK
Bcl2
Survivin
HOMING
ADHESION
SURVIVAL
PROLIFERATION
(QUIESCENCE)
VEGF
FGF
IGF
EGF
TAM
CAF
TNF
Cancer Stem Cells
Niche with Dormant Tumor

disease", responsible for most cases of prostate cancer recurrence and therapy failure. The overall regulation of this process is still under active investigation but it has been now accepted that it involves prostate cancer stem cells (PCSC). From the first appearance on the scenario of cancer metastasis of solid tumors, in 2003 (Al-Hajj et al. 2003), a definitive consensus about their origin and specific markers has not been reached yet.

Several putative surface markers, in fact, are shared also by normal stem cells (Patrawala et al. 2007; Collins et al. 2005) as well as by different solid tumors. This is the case for CD133/prominin-1 and CD44 (Patrawala et al. 2007) that have been found expressed in CSC of lung, breast, colon, ovarian and head and neck squamous cell carcinomas (Cui et al. 2011; Chu et al. 2009; Shi et al. 2010). As well, CD133, CD44, integrins, Sca-1, and breast cancer resistance protein (BRCP) are expressed either in PCSC and in normal prostate stem cells (Yu et al. 2012; Tang et al. 2007); Oct-3/4, beta-catenin and SMO are stemness markers expressed by most of normal and neoplastic stem cells (Patrawala et al. 2006). In addition, there is still a considerable variance among the different antibodies available for the detection of stem cells markers and this may explain, almost partially, the presence of some overlaps or discrepancies between the many existing studies on this topic.

Encouraging results indicate that ALDH1A1, a member of ALDH family of proteins involved in the intracellular production of retinoic acid, could be considered as a promising marker of stemness for prostate cancer cells (Li et al. 2010). ALDH1A1 overexpressing prostate cancer cells display, in fact, high migration and clonogenic ability *in vitro* and metastatic ability *in vivo* (van den Hoogen et al. 2010). As well, a second member of the ALDH family, ALDH7A1, seems to be involved in the bone metastasis formation (van den Hoogen et al. 2010), since its knockdown results in inhibition of experimentally induced intra-bone metastasis.

Fig. 6.1 The signaling pathways involved in the development of prostate cancer bone metastasis. Metastasis result from interactions between cancer cells and the host microenvironment that enable them to detach from the primary tumor bulk, disseminate through the bloodstream and invade of metastatic site. These steps are regulated by a cascade of proteolytic enzymes, inflammatory cytokines, growth factors, activated oncogenes, oxidative stress and hypoxia linked proteins, and adhesion molecules. Activated stromal cells, as tumor-associated macrophages (*TAMs*), cancer-associated fibroblasts (*CAFs*), and endothelial cells, favor the entire process: TAMs produce several migration-stimulating factors as CXCL12, IL-6, and TNF; CAFs intervene in ECM remodeling and in the induction of epithelial-mesenchymal transition (EMT) of metastasizing cells. Circulating cells activate multiple survival pathways, as the overexpression of several members of the anti-apoptotic Bcl2 protein family and the activation of survivin expression. Disseminated prostate cancer cells (DPCC) reaching the bone marrow occupy a bone niche and, when evade immune attack and/or chemotherapy cytotoxicity, may outlive for a variable time in an auto-induced reversible state of growth arrest, the so-called "tumor-dormancy". Dormant DPCCs give rise to bone metastatic lesions by re-entering the cell cycle and proliferating. The causative factors leading to this process are still a matter of investigation, but it seems involve prostate cancer stem cells (PCSC)

It has become clear, however, that the existence of a single reliable marker of PCSCs doesn't exist and a definite combination of markers expression may, instead, identify the metastatic profile of PCSC (Eaton et al. 2010). To support this idea, it has been shown that the co-expression of CD166 (epithelial stem cell marker) (Dalerba et al. 2007), CD151 (marker of stem-like tumor stromal cells) and the tumor rejection antigen/TRA-1-60 (Draper et al. 2002) identifies prostate cancer cells with high ability in sphere formation *in vitro* and generating, *in vivo*, tumors capable of self-renewal and differentiation, consistent with stem cells properties.

Moreover, it has been reported that the signature for stem cell markers may also vary between metastasis and primitive tumors with different Gleason grade (Castellón et al. 2012).

Traditionally, prostate CSC have been thought to derive from the basal cel layer, which express most of the known markers of stemness, as CD133, CD44, CD117, Tert, p63 (Tsujimura et al. 2002). Several findings support this hypothesis. As an example, it emerged that normal basal cells of human prostate can initiate prostate cancer in immunocompromised mice (Goldstein et al. 2010) and primary cells FACS-sorted confirmed the basal cell origin for prostate cancer (Goldstein et al. 2010; Lawson et al. 2010).

On the other side, there are line of evidence that, in several instances, PCSC could have originated from prostate luminal-cells. For instance, a genetic lineage-marking study has shown that rare prostate luminal cells express the androgen/AR-regulated transcriptional co-activator Nkx3-1 in absence of androgens (castration-resistant Nkx3-1-expressing cells, CARNs). CARNs show stem-cell properties, as they are self-renewing and reproduce prostate ducts in renal graft, and cause HGPIN and cancer following Pten deletion (Wang et al. 2009).

Besides their origin, PCSC are considered as the guest actors in the bone marrow metastasis phase (Colombel et al. 2012). They have, in fact, the necessary characteristics for survive and reproduce in the bone microenvironment.

The bone marrow niche, in turn, is critical for the progression from localized disease to distant metastases (Chung et al. 2005; Cher et al. 2006; Morrissey and Vessella 2007; Karlou et al. 2010). The niche is composed by the endothelia of sinusoids (Kiel and Morrison 2008; Doan and Chute 2012), osteoblasts, adipocytes, mesenchymal stem cells, and contains a soluble extracellular matrix rich in growth factors, cytokines (Bussard et al. 2008) and nutrients, useful for cancer cell survival. In addition, it contains adhesion molecules (Taichman 2005; Yin and Li 2006; Arai et al. 2009) as annexin II (Shiozawa et al. 2008a), which interact with tumor cells and local osteoblasts and fibroblasts to provide the framework for the stable homing of prostate cancer cells (Shiozawa et al. 2008b).

Among the several cytokines actively secreted by osteoblast, a pivotal role seems to be played by CXCL12, also known as stromal cell derived factor-1, with its receptors CXCR4 and CXCR7. These two receptors are strongly expressed by DPCC. The binding of CXCR4 and CXCR7 of prostate cancer cells with CXCL12 induces the expression of several adhesion molecules, which enhance their binding to the bone niche (Sun et al. 2005, 2007).

This finding may have relevant implication on therapy, as it has been shown that molecular antagonists of the CXC4R, as the small molecule AMD3100 and the G-CSF analog Filgrastim is able to mobilize metastasizing prostate cancer cells from the bone marrow niche.

Another protein responsible for the reversible cell-cycle arrest of DPCC is the fibroblast secreted annexin II (Anxa2) which operates with its receptor Anxa2R in a manner similar to the CXCL12/CXCR4-CXCR7 pathway (Jung et al. 2007; Shiozawa et al. 2008a).

The degree of expression of either CXCR4 and CXCR7 by prostate cancer cells has been found to correlate with a poor outcome of patients (Sun et al. 2003; Wang et al. 2008; Shiozawa et al. 2008b; Mai et al. 2000). All these considerations have rendered the targeting of bone marrow niche molecules a particularly active and attractive research field.

Several reports indicate that the alteration of multiple other signaling pathways accounting for the tumorigenic potential of PCSC may be used to control them.

For instance, targeting NF-kB with small molecule inhibitors may block sphere generation *in vitro* and tumor-initiation *in vivo*, by purified naïve stem-like human prostatic cells (Rajasekhar et al. 2011), thus supporting the reported adverse prognostic significance in terms of biochemical recurrence risk of the presence of NF-kB stained cells in positive margins of radical prostatectomy specimens (Ross et al. 2004). Similarly, the therapeutic use of WNT inhibitors has been shown to reduce the self-renewal of PCSC and improve the outcome of patients harbouring tumors co-expressing Wnt3a, nuclear beta-catenin, keratin 18, CD133 and CD44 (Bisson and Prowse 2009).

Moreover, the colonization of the skeleton by prostate cancer cells is mediated also by collagen type I, the most represented bone protein, mainly through the binding with the increased expression of integrin α(2)β(1). This integrin has been found elevated in PCa bone metastatic lesions compared to either primary tumors or their soft tissue metastases suggesting it is needed for the selective metastatization to the bone (Sottnik et al. 2013).

Dormant DPCCs give rise to bone metastatic lesions by re-entering the cell cycle and proliferating. The causative factors leading to this process are still a matter of investigation.

The striking propensity to localize to the bone is shared also by other "big killers", as lung and breast cancer (Patel et al. 2011). However, these other cancer types give rise to osteolytic (bone resorbing) bone marrow metastases, while prostate cancer can produce predominantly osteoblastic lesions (Zetter 1990; Jacobs 1983; Chappard et al. 2011), via the inhibition of osteoblast apoptosis and the increase of osteoblast proliferation and metabolism, induced by parathyroid hormone (PTH), PTH-related protein and bone morphogenic proteins BMP (Keller et al. 2001). In addition, the expression of BMP may lead to the osteoblastic differentiation of bone mesenchymal stem cells creating an autocrine and paracrine feedback loop between the prostate cancer epithelial cell and the bone microenvironment.

In contrast to the rapid progress being made in the development of anti-osteolytic therapies, the treatment of osteosclerotic MBD remains restricted to palliative radiotherapy for symptomatic solitary lesions and systemic taxane-based chemotherapy for widespread multiple lesions (Sturge et al. 2011). Thus, new therapeutic strategies focused on the complex pathology of osteoblastic bone-forming metastases of prostate cancer are urgently needed and promising results start to emerge from current preclinical studies.

The "lethal phenotype" of metastatic castrate-resistant prostate cancer depends, then, from the bi-directional action of cancer epithelial cells in the bone and host stromal response to tumor cells (Loberg et al. 2005).

Elucidating the bidirectional interactions between the cancer cell and host bone microenvironment is now an important area of prostate cancer research (Efstathiou and Logothetis 2010).

By a clinical point-of-view, these osteoblastic metastases cause bone pain, and are constituted by disorganized neo-synthetized, unstructured "woven" bone which, similarly to that observed also for osteolytic lesions, frequently give rise to painful fractures (Roudier et al. 2003, 2008; Eastham 2007).

The progressive filling of bone marrow by metastatic prostate cancer cells cause myelophtisis, leukoerythroblastic anemia (Eriksson et al. 1972; Shamdas et al. 1993), up to bone marrow failure (Spivak 1994). These phenomena are thought to be caused, at least in part, by the physical displacement out of their bone marrow niches of hematopoietic stem cells by prostate cancer cells. HPCs displaced in the bloodstream might then undergo to forced, but incomplete, differentiation into lineage-specific nonfunctional progenitors (Shiozawa et al. 2011).

Patients with bone-metastatic prostate cancer experience a significative higher risk of death for disease when compared with patients without skeletal involvement (Norgaard et al. 2010).

The rationale for this bone-forming activity could reside in its possible contribute to support availability of bone niches for the successful homing and expansion of metastasizing prostate cancer cells.

The last decade has registered significant advancement in the identification of the steps involved in the multilayered process of prostate cancer metastasis but further translational studies are needed, to shed new light on several fundamental questions:

- Do hormone receptors have a relevant role in the induction and establishment of prostate cancer metastasis?

  Mounting evidence indicates that androgen receptor (AR) signaling continues to play a critical role in the growth of advanced PC despite androgen deprivation (Zheng et al. 2013). Recent data indicate that convergence of oncogenic and hormone receptor pathways promotes the metastatic phenotype (Augello et al. 2013). However, the downstream AR target genes involved in progression of castration-resistance are largely unknown. It has been reported that cyclin D1b, a splice variant of cyclin D1 exerting a highly oncogenic function in human cancers, promote AR-mediated activation of genes associated with metastatic phenotype in tumor xenograft models of prostate cancer (Augello et al. 2013).

Moreover, Jin HJ and colleagues showed that the AR pathway induces prostate cell growth also via the induction of the synthesis of FoxA1 (Jin et al. 2013). However, this protein, which is a transcription factor essential for the prostate lineage-specific gene expression, inhibits cell motility and epithelial-to-mesenchymal transition (EMT) through AR-independent mechanism opposite to the action of AR signaling, thus behaving as an inhibitor of prostate cancer metastasis. In orthotopic mouse models, FoxA1 has been found up-regulated in localized prostate cancer and down-regulated in EMT bearing metastatic prostate cancer cells. Then, FoxA1 may be considered an AR-independent metastasis inhibitor that, following mutations, can contribute instead to prostate cancer progression.

WNT7B, as a direct AR target gene highly expressed in castration-resistant prostate cancer (CRPC), suggests that AR-regulated WNT7B signaling is critical for the growth of CRPC and development of the osteoblastic bone response characteristic of advanced PC (Zheng et al. 2013). WNT7B is necessary for the growth of PC cells and this effect is enhanced under androgen-deprived conditions; it promotes the androgen-independent growth of CRPC cells likely through the activation of protein kinase C isozymes, induces osteoblast differentiation *in vitro* through a direct cell-cell interaction, and is upregulated in human PC xenografts that cause an osteoblastic reaction when grown in bone. Contrasting data still exist about the real significance of AR reactivation in castration-resistant prostate cancer cells and its relevance for prostate cancer stem cell biology (Miki et al. 2007; Collins et al. 2005; Rajasekhar et al. 2011; Patrawala et al. 2006).

– MicroRNAs (miRs) function as either oncogenes or tumor suppressor genes in cancer (Zhu et al. 2013). Early reports suggest that in androgen-dependent prostate cancer cells, they may play a role in tumor development, progression, evolution to metastasis, response to therapy, and prognosis (Qu et al. 2013) In prostate epithelial EP156T cells, miR-182 and miR-203 have been really shown to induce MET features and growth factor independent cell growth.

On the opposite side, elevated serum levels of miR-141 have been found related with the presence of bone prostate cancer metastasis, without significant correspondence with either Gleason score of primary tumor or PSA value. By converse, miR-141 showed a positive correlation with serum alkaline phosphatase levels (Zhang et al. 2013).

However, more data are required before we reach a comprehensive knowledge about their definite roles in androgen-independent, bone metastasizing prostate cancer (Brennan et al. 2012).

The better understanding of the molecular phenotype of PCSC and DPCC could provide novel therapeutic strategies, allowing the targeting of bone metastatic prostate cancer cells, before they exit dormancy and become lethal (Patel et al. 2011).

Early profiling studies have evidenced the role of miRNA expression in prostate CSC (Liu et al. 2011), revealing that they specifically target several stem cells markers in prostate cancer. As an example, the overexpression of miR-34A leads to the decrease of CD44+ prostate cancer cells, inhibiting tumor development and metastasis, thus appearing as a promising potential new therapeutic tool for neutralize the killing potential of PSCS.

- Recently, it has been suggested that infiltrating immune cells facilitate tumor stem cell proliferation. Moreover, it has been proposed that aberrant immune cell infiltration preferentially associates with tumor capsular areas showing distinct degenerative alterations. Tumor-associated lymphocytes might cause focal disruption of prostate cancer capsule, favoring, then, tumor cell budding and metastasis (Jiang et al. 2013).

  This finding deserves further evaluation, as it may have a relevant impact on our knowledge of the prostate cancer metastasis causative events. It suggests, in fact, that the aberrant immune cell infiltration may have the same destructive impact of cancer cells on the lining capsule, offering in turn a selective proliferative advantage to prostate cancer stem cells proximal to these focal disruptions.

  Moreover, it will be also clarified if the selective tumor-associated immunoreactive infiltrate may have a causative role even for the early onset of aggressive prostate cancer at young ages, typically originating in healthy men with morphologically normal prostate (Man and Gardner 2008).

Overall, the understanding of the molecular background of prostate cancerogenesis has already changed our way to look at prostate cancer.

The growing flow of information concerning the bidirectional interactions between the epithelial cancer cells, tumor-associated stroma, and host bone microenvironment has become an impressively active area of prostate cancer research (Efstathiou and Logothetis 2010). The stromal-interacting pathways represent exciting targets for new molecular niche-directed therapies, which in the next future will guide our efforts to fight metastatic prostate cancer.

References

Aguirre-Ghiso JA (2007) Models, mechanisms and clinical evidence for cancer dormancy. Nat Rev Cancer 7(11):834–846

Al-Hajj M, Wicha MS, Benito-Hernandez A, Morrison SJ, Clarke MF (2003) Prospective identification of tumorigenic breast cancer cells. Proc Natl Acad Sci U S A 100:3983–3988

Allavena P, Sica A, Solinas G, Porta C, Mantovani A (2008) The inflammatory microenvironment in tumor progression: the role of tumor-associated macrophages. Crit Rev Oncol Hematol 66(1):1–9

Arai F, Yoshihara H, Hosokawa K, Nakamura Y, Gomei Y et al (2009) Niche regulation of hematopoietic stem cells in the endosteum. Ann N Y Acad Sci 1176:36–46

Augello MA, Burd CJ, Birbe R, McNair C, Ertel A, Magee MS, Frigo DE, Wilder-Romans K, Shilkrut M, Han S, Jernigan DL, Dean JL, Fatatis A, McDonnell DP, Visakorpi T, Feng FY, Knudsen KE (2013) Convergence of oncogenic and hormone receptor pathways promotes metastatic phenotypes. J Clin Invest 123(1):493–508

Bisson I, Prowse DM (2009) WNT signaling regulates self-renewal and differentiation of prostate cancer cells with stem cell characteristics. Cell Res 19:683–697

Brennan SC, Thiem U, Roth S, Aggarwal A, Fetahu IS, Tennakoon S, Gomes AR, Brandi ML, Bruggeman F, Mentaverri R, Riccardi D, Kallay E (2013) Calcium sensing receptor signalling in physiology and cancer. Biochim Biophys Acta 1833(7):1732–1744

Brett A, Pandey S, Fraizer G (2013) The Wilms' tumor gene (WT1) regulates E-cadherin expression and migration of prostate cancer cells. Mol Cancer 12:3

Bussard KM, Gay CV, Mastro AM (2008) The bone microenvironment in metastasis; what is special about bone? Cancer Metastasis Rev 27:41–55

Castellón EA, Valenzuela R, Lillo J, Castillo V, Contreras HR, Gallegos I, Mercado A, Huidobro C (2012) Molecular signature of cancer stem cells isolated from prostate carcinoma and expression of stem markers in different Gleason grades and metastasis. Biol Res 45(3):297–305

Chappard D, Bouvard B, Baslé MF, Legrand E, Audran M (2011) Bone metastasis: histological changes and pathophysiological mechanisms in osteolytic or osteosclerotic localizations. A review. Morphologie 95(309):65–75. Epub 2011 May 28

Cher ML, Towler DA, Rafii S et al (2006) Cancer interaction with the bone microenvironment: a workshop of the National Institutes of Health Tumor Microenvironment Study Section. Am J Pathol 168:1405–1412

Chu P, Clanton DJ, Snipas TS, Lee J, Mitchell E, Nguyen ML et al (2009) Characterization of a subpopulation of colon cancer cells with stem cell-like properties. Int J Cancer 124:1312–1321

Chung LW, Baseman A, Assikis V, Zhau HE (2005) Molecular insights into prostate cancer progression: the missing link of tumor microenvironment. J Urol 173:10–20

Collins AT, Berry PA, Hyde C, Stower MJ, Maitland NJ (2005) Prospective identification of tumorigenic prostate cancer stem cells. Cancer Res 65:10946–10951

Colombel M, Eaton CL, Hamdy F, Ricci E, van der Pluijm G, Cecchini M, Mege-Lechevallier F, Clezardin P, Thalmann G (2012) Increased expression of putative cancer stem cell markers in primary prostate cancer is associated with progression of bone metastases. Prostate 72(7): 713–720

Cui F, Wang J, Chen D, Chen YJ (2011) CD133 is a temporary marker of cancer stem cells in small cell lung cancer, but not in non-small cell lung cancer. Oncol Rep 25:701–708

Czyż J, Szpak K, Madeja Z (2012) The role of connexins in prostate cancer promotion and progression. Nat Rev Urol 9(5):274–282

Dalerba P, Dylla SJ, Park IK, Liu R, Wang X, Cho RW et al (2007) Phenotypic characterization of human colorectal cancer stem cells. Proc Natl Acad Sci U S A 104:10158–10163

Desiniotis A, Kyprianou N (2011) Significance of talin in cancer progression and metastasis. Int Rev Cell Mol Biol 289:117–147

Doan PL, Chute JP (2012) The vascular niche: home for normal and malignant hematopoietic stem cells. Leukemia 26(1):54–62

Draper JS, Pigott C, Thomson JA, Andrews PW (2002) Surface antigens of human embryonic stem cells: changes upon differentiation in culture. J Anat 200:249–258

Du WW, Yang W, Yee AJ (2013) Roles of versican in cancer biology – tumorigenesis, progression and metastasis. Histol Histopathol 28(6):701–713

Eastham JA (2007) Bone health in men receiving androgen deprivation therapy for prostate cancer. J Urol 177(1):17–24

Eaton CL, Colombel M, van der Pluijm G, Cecchini M, Wetterwald A, Lippitt J et al (2010) Evaluation of the frequency of putative prostate cancer stem cells in primary and metastatic prostate cancer. Prostate 70:875–882

Efstathiou E, Logothetis CJ (2010) A new therapy paradigm for prostate cancer founded on clinical observations. Clin Cancer Res 16:1100–1107

Eriksson S, Killander J, Wadman B (1972) Leuco-erythroblastic anaemia in prostatic cancer. Report of two cases with complete haematological remission. Scand J Haematol 9(6):648–653

Fan L, Wang H, Xia X, Rao Y, Ma X, Ma D, Wu P, Chen G (2012) Loss of E-cadherin promotes prostate cancer metastasis via upregulation of metastasis-associated gene 1 expression. Oncol Lett 4(6):1225–1233. Epub 2012 Sep 21

Goldstein AS, Huang J, Guo C, Garraway IP, Witte ON (2010) Identification of a cell of origin for human prostate cancer. Science 329:568–571

Hu YY, Zheng MH, Zhang R, Liang YM, Han H (2012) Notch signaling pathway and cancer metastasis. Adv Exp Med Biol 727:186–198

Igney FH, Krammer PH (2002) Death and anti-death: tumour resistance to apoptosis. Nat Rev Cancer 2(4):277–288. A comprehensive review of cell death pathways and survival mechanisms exploited by cancer

Jacobs SC (1983) Spread of prostatic cancer to bone. Urology 21(4):337–344

Jemal A, Siegel R, Ward E (2010) Cancer statistics. CA Cancer J Clin 60:277–300

Jiang B, Mason J, Jewett A, Liu ML, Chen W, Qian J, Ding Y, Ding S, Ni M, Zhang X, Man YG (2013) Tumor-infiltrating immune cells: triggers for tumor capsule disruption and tumor progression? Int J Med Sci 10(5):475–497

Jin HJ, Zhao JC, Ogden I, Bergan RC, Yu J (2013) Androgen receptor-independent function of FoxA1 in prostate cancer metastasis. Cancer Res 73(12):3725–3736

Joyce JA, Pollard JW (2009) Microenvironmental regulation of metastasis. Nat Rev Cancer 9(4):239–252

Jung Y, Wang J, Song J, Shiozawa Y, Havens A et al (2007) Annexin II expressed by osteoblasts and endothelial cells regulates stem cell adhesion, homing, and engraftment following transplantation. Blood 110(1):82–90

Karlou M, Tzelepi V, Efstathiou E (2010) Therapeutic targeting of the prostate cancer microenvironment. Nat Rev Urol 7:494–509

Keller ET, Zhang J, Cooper CR, Smith PC, McCauley LK et al (2001) Prostate carcinoma skeletal metastases: cross-talk between tumor and bone. Cancer Metastasis Rev 20(3–4):333–349

Kiel MJ, Morrison SJ (2008) Uncertainty in the niches that maintain haematopoietic stem cells. Nat Rev Immunol 8(4):290–301

Lawson DA, Zong Y, Memarzadeh S, Xin L, Huang J, Witte ON (2010) Basal epithelial stem cells are efficient targets for prostate cancer initiation. Proc Natl Acad Sci U S A 107:2610–2615

Lazari P, Poulias H, Gakiopoulou H, Thomopoulou GH, Barbatis C, Lazaris AC (2013) Differential immunohistochemical expression of CD44s, E-cadherin and β-catenin among hyperplastic and neoplastic lesions of the prostate gland. Urol Int 90(1):109–116. Epub 2012 Dec 5 Leukemia. 2011

Li T, Su Y, Mei Y, Leng Q, Leng B, Liu Z et al (2010) ALDH1A1 is a marker for malignant prostate stem cells and predictor of prostate cancer patients' outcome. Lab Invest 90:234–244

Liu C, Kelnar K, Liu B, Chen X, Calhoun-Davis T, Li H et al (2011) The microRNA miR-34a inhibits prostate cancer stem cells and metastasis by directly repressing CD44. Nat Med 17:211–215

Loberg RD, Logothetis CJ, Keller ET, Pienta KJ (2005) Pathogenesis and treatment of prostate cancer bone metastases: targeting the lethal phenotype. J Clin Oncol 23:8232–8241

Mai J, Waisman DM, Sloane BF (2000) Cell surface complex of cathepsin B/annexin II tetramer in malignant progression. Biochim Biophys Acta 1477(1–2):215–230

Man YG, Gardner WA (2008) Bad seeds produce bad crops: a single stage-process of prostate tumor invasion. Int J Biol Sci 4(4):246–258

Miki J, Furusato B, Li H, Gu Y, Takahashi H, Egawa S et al (2007) Identification of putative stem cell markers, CD133 and CXCR4, in hTERT-immortalized primary nonmalignant and malignant tumor-derived human prostate epithelial cell lines and in prostate cancer specimens. Cancer Res 67:3153–3161

Morgan TM, Lange PH, Porter MP, Lin DW, Ellis WJ et al (2009) Disseminated tumor cells in prostate cancer patients after radical prostatectomy and without evidence of disease predicts biochemical recurrence. Clin Cancer Res 15(2):677–683

Morrissey C, Vessella RL (2007) The role of tumor microenvironment in prostate cancer bone metastasis. J Cell Biochem 101:873–886

Nguyen DX, Massague J (2007) Genetic determinants of cancer metastasis. Nat Rev Genet 8(5):341–352

Norgaard M, Jensen AO, Jacobsen JB, Cetin K, Fryzek JP, Sorensen HT (2010) Skeletal related events, bone metastasis and survival of prostate cancer: a population based cohort study in Denmark (1999 to 2007). J Urol 184(1):162–167

Osanto S, Van Poppel H (2012) Emerging novel therapies for advanced prostate cancer. Ther Adv Urol 4(1):3–12

Pal M, Koul S, Koul HK (2013) The transcription factor sterile alpha motif (SAM) pointed domain-containing ETS transcription factor (SPDEF) is required for E-cadherin expression in prostate cancer cells. J Biol Chem 288(17):12222–12231

Patel LR, Camacho DF, Shiozawa Y, Pienta KJ, Taichman RS (2011) Mechanisms of cancer cell metastasis to the bone: a multistep process. Future Oncol 7(11):1285–1297

Patrawala L, Calhoun T, Schneider-Broussard R, Li H, Bhatia B, Tang S et al (2006) Highly purified CD44+ prostate cancer cells from xenograft human tumors are enriched in tumorigenic and metastatic progenitor cells. Oncogene 25:1696–1708

Patrawala L, Calhoun-Davis T, Schneider-Broussard R, Tang DG (2007) Hierarchical organization of prostate cancer cells in xenograft tumors: the CD44+alpha2beta1+ cell population is enriched in tumor-initiating cells. Cancer Res 67:6796–6805

Pfitzenmaier J, Ellis WJ, Hawley S, Arfman EW, Klein JR et al (2007) The detection and isolation of viable prostate-specific antigen positive epithelial cells by enrichment: a comparison to standard prostate-specific antigen reverse transcriptase polymerase chain reaction and its clinical relevance in prostate cancer. Urol Oncol 25(3):214–220

Qu Y, Li WC, Hellem MR, Rostad K, Popa M, McCormack E, Oyan AM, Kalland KH, Ke XS (2013) MiR-182 and miR-203 induce mesenchymal to epithelial transition and self-sufficiency of growth signals via repressing SNAI2 in prostate cells. Int J Cancer 133(3):544–555

Rajasekhar VK, Studer L, Gerald W, Socci ND, Scher HI (2011) Tumour-initiating stem-like cells in human prostate cancer exhibit increased NF-kappaB signalling. Nat Commun 2:162

Roca H, Varsos ZS, Mizutani K, Pienta KJ (2008) CCL2, survivin and autophagy: new links with implications in human cancer. Autophagy 4(7):969–971

Rodenhiser DI (2009) Epigenetic contributions to cancer metastasis. Clin Exp Metastasis 26(1):5–18

Roodman GD (2004) Mechanisms of bone metastasis. N Engl J Med 350(16):1655–1664

Ross JS, Kallakury BV, Sheehan CE, Fisher HA, Kaufman RP Jr, Kaur P et al (2004) Expression of nuclear factor-kappa B and I kappa B alpha proteins in prostatic adenocarcinomas: correlation of nuclear factor-kappa B immunoreactivity with disease recurrence. Clin Cancer Res 10:2466–2472

Roudier MP, True LD, Higano CS et al (2003) Phenotypic heterogeneity of end-stage prostate carcinoma metastatic to bone. Hum Pathol 34:646–653

Roudier MP, Morrissey C, True LD, Higano CS, Vessella RL et al (2008) Histopathological assessment of prostate cancer bone osteoblastic metastases. J Urol 180(3):1154–1160

Shamdas GJ, Ahmann FR, Matzner MB, Ritchie JM (1993) Leukoerythroblastic anemia in metastatic prostate cancer. Clinical and prognostic significance in patients with hormone-refractory disease. Cancer 71(11):3594–3600

Shi MF, Jiao J, Lu WG, Ye F, Ma D, Dong QG et al (2010) Identification of cancer stem cell-like cells from human epithelial ovarian carcinoma cell line. Cell Mol Life Sci 67:3915–3925

Shiozawa Y, Havens AM, Jung Y, Ziegler AM, Pedersen EA et al (2008a) Annexin II/annexin II receptor axis regulates adhesion, migration, homing, and growth of prostate cancer. J Cell Biochem 105(2):370–380

Shiozawa Y, Havens AM, Pienta KJ, Taichman RS (2008b) The bone marrow niche: habitat to hematopoietic and mesenchymal stem cells, and unwitting host to molecular parasites. Leukemia 22(5):941–950

Shiozawa Y, Pedersen EA, Havens AM, Jung Y, Mishra A et al (2011) Human prostate cancer metastases target the hematopoietic stem cell niche to establish footholds in mouse bone marrow. J Clin Invest 121(4):1298–1312

Sottnik JL, Daignault-Newton S, Zhang X, Morrissey C, Hussain MH, Keller ET, Hall CL (2013) Integrin alpha2beta 1 (α 2β 1) promotes prostate cancer skeletal metastasis. Clin Exp Metastasis 30(5):569–578

Spivak JL (1994) Cancer-related anemia: its causes and characteristics. Semin Oncol 21(2 Suppl 3):3–8

Sturge J, Caley MP, Waxman J (2011) Bone metastasis in prostate cancer: emerging therapeutic strategies. Nat Rev Clin Oncol 8(6):357–368. doi:10.1038/nrclinonc.2011.67. Epub 2011 May 10

Sun YX, Wang J, Shelburne CE, Lopatin DE, Chinnaiyan AM et al (2003) Expression of CXCR4 and CXCL12 (SDF-1) in human prostate cancers (PCa) *in vivo*. J Cell Biochem 89(3):462–473

Sun YX, Schneider A, Jung Y, Wang J, Dai J et al (2005) Skeletal localization and neutralization of the SDF-1(CXCL12)/CXCR4 axis blocks prostate cancer metastasis and growth in osseous sites *in vivo*. J Bone Miner Res 20(2):318–329

Sun YX, Fang M, Wang J, Cooper CR, Pienta KJ et al (2007) Expression and activation of alpha v beta 3 integrins by SDF-1/CXC12 increases the aggressiveness of prostate cancer cells. Prostate 67(1):61–73

Taichman RS (2005) Blood and bone: two tissues whose fates are intertwined to create the hematopoietic stem-cell niche. Blood 105(7):2631–2639

Tang DG, Patrawala L, Calhoun T, Bhatia B, Choy G, Schneider-Broussard R et al (2007) Prostate cancer stem/progenitor cells: identification, characterization, and implications. Mol Carcinog 46:1–14

Tannock IF, de Wit R, Berry WR, Horti J, Pluzanska A, Chi KN, Oudard S, Théodore C, James ND, Turesson I, Rosenthal MA, Eisenbergee MA (2004) TAX 327 investigators. Docetaxel plus prednisone or mitoxantrone plus prednisone for advanced prostate cancer. N Engl J Med 351(15):1502–1512

Townson JL, Chambers AF (2006) Dormancy of solitary metastatic cells. Cell Cycle 5(16): 1744–1750

Tsujimura A, Koikawa Y, Salm S, Takao T, Coetzee S, Moscatelli D et al (2002) Proximal location of mouse prostate epithelial stem cells: a model of prostatic homeostasis. J Cell Biol 157:1257–1265

van den Hoogen C, van der Horst G, Cheung H, Buijs JT, Lippitt JM, Guzman-Ramirez N et al (2010) High aldehyde dehydrogenase activity identifies tumor-initiating and metastasis-initiating cells in human prostate cancer. Cancer Res 70:5163–5173

Vashisht S, Bagler G (2012) An approach for the identification of targets specific to bone metastasis using cancer genes interactome and gene ontology analysis. PLoS One 7(11):e49401

Wang J, Shiozawa Y, Wang Y, Jung Y, Pienta KJ et al (2008) The role of CXCR7/RDC1 as a chemokine receptor for CXCL12/SDF-1 in prostate cancer. J Biol Chem 283(7):4283–4894

Wang X, Kruithof-de Julio M, Economides KD, Walker D, Yu H, Halili MV et al (2009) A luminal epithelial stem cell that is a cell of origin for prostate cancer. Nature 461:495–500

Wang H, Fan L, Wei J, Weng Y, Zhou L, Shi Y, Zhou W, Ma D, Wang C (2012) Akt mediates metastasis-associated gene 1 (MTA1) regulating the expression of E-cadherin and promoting the invasiveness of prostate cancer cells. PLoS One 7(12):e46888. Epub 2012 Dec 5

Wang X, Lee SO, Xia S, Jiang Q, Luo J, Li L, Yeh S, Chang C (2013) Endothelial cells enhance prostate cancer metastasis via IL-6->Androgen Receptor->TGF-β->MMP-9 signals. Mol Cancer Ther 12(6):1026–1037

Yates C (2011) Prostate tumor cell plasticity: a consequence of the microenvironment. Adv Exp Med Biol 720:81–90

Yilmaz M, Christofori G (2009) EMT, the cytoskeleton, and cancer cell invasion. Cancer Metastasis Rev 28(1–2):15–33

Yin T, Li L (2006) The stem cell niches in bone. J Clin Invest 116(5):1195–1201

Yu C, Yao Z, Jiang Y, Keller ET (2012) Prostate cancer stem cell biology. Minerva Urol Nefrol 64(1):19–33

Zetter BR (1990) The cellular basis of site-specific tumor metastasis. N Engl J Med 322(9): 605–612

Zhang HL, Qin XJ, Cao DL, Zhu Y, Yao XD, Zhang SL, Dai B, Ye DW (2013) An elevated serum miR-141 level in patients with bone-metastatic prostate cancer is correlated with more bone lesions. Asian J Androl 15(2):231–235. Epub 2013 Feb 4

Zhao J, Wu XY, Ling XH, Lin ZY, Fu X, Deng YH, He HC, Zhong W (2013) Analysis of genetic aberrations on chromosomal region 8q21-24 identifies E2F5 as an oncogene with copy number gain in prostate cancer. Med Oncol 30(1):465

Zheng D, Decker KF, Zhou T, Chen J, Qi Z, Jacobs K, Weilbaecher KN, Corey E, Long F, Jia L (2013) Role of WNT7B-induced non-canonical pathway in advanced prostate cancer. Mol Cancer Res 11(5):482–493

Zhu KC, Lu JJ, Xu XL, Sun JM (2013) MicroRNAs in androgen-dependent PCa. Front Biosci 18:748–755

Chapter 7
Resistance to Castration – Resistance to Drugs

Stefania Staibano

Abstract Up to 70 % of newly diagnosed patients with advanced prostate cancer (PCa) will progress to castration-resistant prostate cancer (CRPC) and, in most cases (from 50 to 70 %), will develop hematogenous bone metastasis. Once PCa cells spread to the skeleton, cancer-related death becomes inevitable, with a death burden of more than 28,000 cases in 2012, in the United States (Semenas et al, Curr Drug Target, 13(10):1308–1323, 2012).

To date, therapeutic regimens are unable to revert this fatal progression (Semenas et al, Curr Drug Target, 13(10):1308–1323, 2012).

Thus, PCa bone metastatic prostate cancer still represents a major clinical challenge.

Prostate cancer biology is tightly linked to AR, which regulates epithelial proliferation and suppresses apoptosis both in normal and in cancer prostate tissue, and is involved in the progression of the disease toward a castration-resistant state (Hodgson et al, World J Urol, 30(3):279–285, 2012). Our knowledge of the molecular mechanisms, responsible for the acquired resistance to ADT in prostate cancer, has exponentially progressed during the last years. For instance, we have recently learnt that it may be associated with the occurrence of AR splicing variants (Hu et al. 2011).

Surgical castration has shown to induce regression of advanced disease 40-years before the cloning of androgen receptor (AR) (Huggins et al, Arch Surg, 43:209–223, 1941; Lubahn et al, Science, 240:327–330, 1988).

Since then, hormonal therapy was held over as the main available therapeutic option for aggressive prostate cancers. In the last decade, however, chemotherapy was introduced to targeting the epithelium of metastatic, hormone-resistant prostate

S. Staibano (✉)
Department of Advanced Biomedical Sciences, Pathology Section, Faculty of Medicine and Surgery, University of Naples "Federico II", via S. Pansini, n.5, Naples, Italy
e-mail: staibano@unina.it

S. Staibano (ed.), *Prostate Cancer: Shifting from Morphology to Biology*,
DOI 10.1007/978-94-007-7149-9_7, © Springer Science+Business Media Dordrecht 2013

cancer (Pinto et al, Tumour Biol, 33(2):421–426, 2012; Hodgson et al, World J Urol, 30(3):279–285, 2012). The cytotoxic conventional drug Docetaxel was approved by the Food and Drug Administration in 2004, and still represents the standard first-line treatment for patients with castration-resistant prostate cancer (CRPC) (Sartor et al, Oncologist, 16(11):1487–1497, 2011). It produces sensible palliative effects on bone-metastasis-related symptoms, but prolongs only modestly the survival of patients (Hodgson et al, World J Urol, 30(3):279–285, 2012; Tannock et al, N Engl J Med, 351:1502–1512, 2004; Petrylak et al, N Engl J Med, 351:1513–1520, 2004). Docetaxel acts mainly by inducing apoptosis of target epithelial cells. The common intrinsic defects of mCRPC in apoptosis pathways, such as BCL-2 overexpression and/or phosphatase and tensin homolog (PTEN) loss (Mathew, Dipaola, J Urol, 178:S36–S41, 2007; Galsky, Vogelzang, Ann Oncol, 21:2135–2144, 2010), may constitute the rationale of the unsatisfactory rate of cure attributable to this drug (Srigley et al, Histopathology, 60(1):153–165, 2012). In recent years, similar effects on survival have been demonstrated also for several other chemotherapeutic agents, such as mitoxantrone, etoposide, cisplatinum, vinblastine–estramustine and taclitaxel.

Following progression after treatment with docetaxel, new cabazitaxel (XRP6258)-prednisone treatment regimens have led to a significantly longer overall survival, and other novel agents are currently being evaluated, including the cell-based immunotherapy sipuleucel-T, the androgen biosynthesis inhibitors abiraterone acetate and MDV3100, the chemotherapic Cabazitaxel, as well as the radionuclide alpharadin/Radium 223 (bone microenvironment targeting agents) (Sartor et al, Oncologist, 16(11):1487–1497, 2011; Liu et al, Front Endocrinol (Lausanne), 3:72, 2012; Antonarakis, Armstrong, Prostate Cancer Prostatic Dis, 14(3):206–218, 2011). To date, they seem to offer a survival advantage to patients, and look promising to improve the prognosis of metastatic CRPC.

However, the real clinical benefit of these systemic therapies remains still transient, probably due also to the well-known clonal heterogeneity of advanced prostate cancers, and the overall survival of patients that holds frustratingly steady.

The high cost of these therapies and the increasing complexity of clinical decision making, further underscore the need to multiply the efforts to develop more potent chemotherapy agents and/or novel AR/inhibitors agents that may better overcome resistance mechanisms to existing therapies (Liu et al, Front Endocrinol (Lausanne), 2012; Hodgson et al, World J Urol, 30(3):279–285, 2012; Armstrong, George, Urol Oncol, 26:430–437, 2008; Schrijvers et al, Adv Ther, 27:285–296, 2010).

Several recently developed drug candidates, directed against the metastatic cancer microenvironments or niches, show promising results in this direction (Hodgson et al, World J Urol, 30(3):279–285, 2012).

The efficacy of the standard-of-care therapeutic intervention directed to mCRPC will be greatly improved by our increasing understanding of molecular mechanisms of the acquired resistance to ADT and chemotherapy, which is expected to provide valuable insights also to new unfailing biomarkers of resistance, therapeutic response and disease progression of prostate cancer, allowing us to personalize the

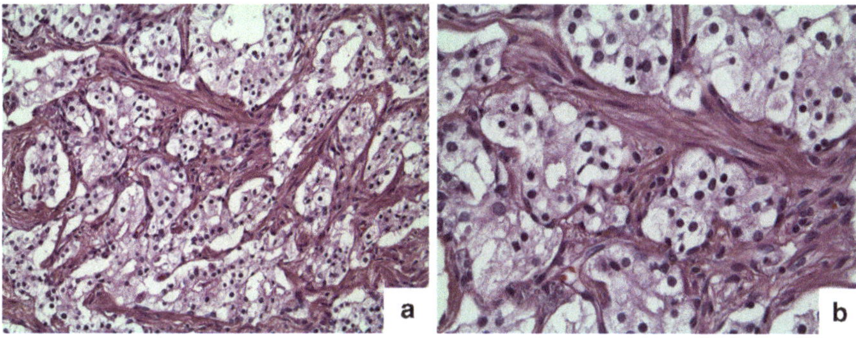

Fig. 7.1 Therapy modifications. (**a**) and (**b**) Two examples of therapy-modified neoplastic cells characterized by nuclear pyknosis, hyperchromasia, cytoplasmic clearing and vacuolation. Chronic inflammation usually flanks epithelial tumor changes

therapy for the single patients with mCRPC (Liu et al, Front Endocrinol (Lausanne), 3:72, 2012; Antonarakis and Armstrong, Prostate Cancer Prostatic Dis, 14(3):206–218, 2011).

The knowledge of the molecular mechanisms underpinning prostate cancer progression is changing dramatically our therapeutic approach to its advanced, metastasizing phase, opening up the chance to design and develop novel agents targeting the multiple pathways responsible for the lethal cancer phenotype, in a more efficient and safer manner (Corcoran and Gleave, Histopathology, 60(1): 216–231, 2012).

During the last decade, the landscape of treatment of prostate cancer has registered dramatic changes, due to the progressive advances in molecular biology. However, the acquired resistance to AR- and/or chemotherapy, so far, represents the unresolved cause of treatment failure in metastatic castration-resistant prostate cancers (Sun et al. 2012; El-Amm and Aragon-Ching 2013).

From the early seminal studies of Huggins and Hodges (1941) demonstrating the androgen-dependent nature of prostate cancer, maximal androgen blockade therapy still constitutes the cornerstone for the initial treatment for advanced disease (el-Rayes and Hussain 2002; Beekman and Hussain 2008) and, its fall-out on clinical response of patients, endlessly continues to be a matter of study (Srigley et al. 2012).

The effects androgen deprivation therapy (ADT) has on prostate tissue are strictly related to its duration, and become strikingly evident after 3 months of treatment. They are constituted by the overall decrease in number of the epithelial cancer cells, either clustered in small, atrophic glands, or present as thin cords/individual tumor cells. Therapy-modified neoplastic cells are, in fact, characterized by nuclear pyknosis, hyperchromasia, cytoplasmic clearing and vacuolation. Chronic inflammation usually flanks epithelial tumor changes (Humphrey 2003; Petraki and Sfikas 2007; Têtu 2008; Evans et al. 2011; Têtu et al. 1991; Bullock et al. 2002; Vallancourt et al. 1996; Civantos et al. 1995; Armas et al. 1994) (Fig. 7.1).

The rationale for ADT is that the competitive binding of the androgen receptors causes the block of testosterone-driven proliferation of prostate cancer cells, leading to apoptosis and clinical remission for about 18–36 months (Pienta and Bradley 2006; Beekman and Hussain 2008).

FDA-approved AR inhibitors for ADT lead to the achievement of castrate levels of circulating testosterone, corresponding to at least <50 ng/dL (Morote et al. 2007). Gonadal suppression can also be further supported by medical therapy targeting the hypothalamic-pituitary-testicular axis, such as the synthetic gonadotropin-releasing hormone (GnRH) agonists, which bind to GnRH pituitary receptors. This binding results in an initially increased luteinizing hormone (LH) and follicle stimulating hormone (FSH) surge, causing a transient elevation of circulating testosterone, but leads, from one to two weeks after the onset of therapy, to GnRH receptors down-regulation, decreased LH and FSH production, followed by the final drop of testosterone to castrate level.

Several approaches to ADT have been proposed, since its introduction in the clinics. They range from surgical to pharmacological gonadal suppression, from monotherapy to combined ADT (Tannock et al. 2004), from early to delayed, from intermittent to continuous, and from primary gonadal suppression to peripheral blockade (Beekman and Hussain 2008) both at high and low doses (Chodak et al. 1995; Scher et al. 1997; Tyrrell et al. 2005).

Randomized phase III trials actually indicate that primary gonadal suppression, by continuous androgen deprivation therapy dosing, seems reliable as the standard for treating advanced prostate cancers (Langenhuijsen et al. 2013).

To date, besides the substantial absence of well-designed randomized trials analyzing the overall survival of patients, the most widely spread treatment, in this scenario, is the hormonal therapy with LH-RH analogues considering that the attractive alternative hormonal treatments (intermittent treatment, antiandrogen monotherapy, or antiandrogen plus 5 alpha reductase inhibitors) should be still evaluated with caution due to the short time experience (Antolin et al. 2012).

However, to date, patients receiving androgen deprivation therapy develop castration-resistant prostate cancer (CRPC) recurrences and, a surprisingly frequent finding shows that, even with stable castrate levels of serum testosterone, prostate cancer bone metastases continue to rely on androgen signaling for their growth (Scher and Sawyers 2005; Montgomery et al. 2008).

Starting from the hypothesis that progression to castration resistance is a function of permanent androgen withdrawal, it has been postulated then that intermittent regimens of androgen deprivation therapy, exposing prostate cancer cells periodically to androgens, may help maintain their androgen sensitive/dependent state. This approach is attractive, as it may minimize adverse side-effects of long-standing androgen deprivation.

Once again, clinical evidences frequently disregard the expected results. This could be due to the finding that AR-dependent signaling almost always occurs in CRPC, but it shows a substantial functional heterogeneity in tumor tissue (Antonarakis and Armstrong 2011).

Potential pathogenetic mechanisms could be represented by intratumoral AR amplification, further (second) mutations in the AR gene, that allow activation; low androgen levels or other endogenous steroids; truncated or alternatively spliced AR transcripts; constitutively activated AR; changes in levels of AR cofactors involved in ligand-independent activation of AR signaling; increased expression of enzymes involved in androgen synthesis; androgen synthesis by CYP17-independent pathways (and genetic changes in the *CYP17A* gene preventing its inhibition by the CYP17 inhibitors abiraterone and Orteronel (TAK-700)) (Antonarakis and Armstrong 2011); intracellular conversion of adrenal androgens to testosterone and dihydrotestosterone (Mostaghel et al. 2009); vicious loops mediated by cytokines and growth factors (Scher and Sawyers 2005; Debes and Tindall 2004; Feldman and Feldman 2001; Mohler 2008; Taplin 2008).

In other words, all these postulated processes allow prostate cancer cells metabolism to shift from endocrine "physiological" sources of androgens (testes and adrenal glands) to paracrine, autocrine, and intracrine aberrant, intratumoral sources.

Even considering these eveniences, tumors may still respond to the agents that block AR signaling within the tumor microenvironment. Second-generation AR-antagonists are available, including MDV3100.

MDV3100 is an oral non-steroidal AR antagonist with a binding affinity for the AR, which is five times greater than that of bicalutamide and shows a strong activity as AR-antagonist in castration-resistant tumors, even in the setting of overexpressed or constitutively activated AR. In addition, it does not exhibit any measurable agonistic activity, and reduces the AR translocation from the cytoplasm into the nucleus, with resultant tumoricidal activity as opposed to the cytostatic activity that first-generation anti-androgens (Antonarakis and Armstrong 2011).

Some additional androgen receptor (AR)-directed therapies with higher receptor affinity and specificity are currently under evaluation in clinical trials, and hold promises, in the near future, to improve the outcome of patients with advanced prostate cancer.

Prostate cancers with lower AR activity, or those exposed to prolonged periods of androgen suppression, may show up-regulation of other oncogenic pathways, including Src kinase (Park et al. 2008), clusterin, epithelial to mesenchymal transition pathways, PI3K, c-MET and others (Antonarakis and Armstrong 2011).

This has fueled the preclinical and clinical exploration of myriad molecular targets comprising alternative oncogenic pathways, targeting angiogenesis, tumor microenvironment, cell growth and proliferation, apoptosis, cell nutrition, DNA repair and epigenetic regulation (Hodgson et al. 2012).

So, we now hold a growing number of epithelial, stromal and epithelial-stromal targeting therapeutics. Specific biomarkers permit quantization and localization of therapy-induced effects within each compartment. For example, PSA levels reflect modulation of cancer epithelial cells, bone-specific alkaline phosphatase (BAP) levels reflect modulation of osteoblast activity, and urinary N-telopeptide (uNTx) levels reflect modulation of osteoclast activity (Cook et al. 2006).

The first therapy regimens with mitoxantrone and prednisone, was established in 1996 (NCCN 2011; Tannock et al. 2004). From 2004, the standard chemotherapy for the first-line treatment of patients with metastatic CRPC considers the use of docetaxel and prednisone (Petrylak et al. 2004; Tannock et al. 2004; Pinto et al. 2012).

At the time of writing we have great expectations concerningthe results of a randomized phase III trial comprising 1,500 patients with progressing mCRPC. The trial was performed comparing standard-schedule docetaxel and prednisone with and without the multi-targeted kinase inhibitor dasatinib, a molecule active against Src signaling. Src is a non-receptor tyrosine kinase that, in prostate cancer cells, is associated with testosterone-mediated cell proliferation; its overexpression is considered an important mediator of the transition to androgen-independent growth (Lee et al. 2001, 2004).

Src phosphorylation induces, in addition, the expression of pro-angiogenic factors, including VEGF, which, in turn, can activate Src also in endothelial cells, mediating the increases in vessels permeability and tumor-associated neo-angiogenesis (Park et al. 2007; Araujo and Logothetis 2010; Agarwal et al. 2012).

Among the therapeutic agents specifically targeting tumor microenvironment, several have shown target effects, but none has demonstrated yet any beneficial impact on disease progression or overall patients survival (Saad et al. 2002; Mathew et al. 2007; Carducci et al. 2007), when used as monotherapy in patients with mCRPC. This was the case of zoledronic acid (osteoclast inhibitor), imatinib (multitarget tyrosine kinase inhibitor), and atrasentan (selective endothelin, a receptor antagonist, which inhibits osteoblast proliferation).

The use of the osteoclast suppressor humanized monoclonal IgG2 antibody, Denosumab, significantly extended bone metastasis free survival by 4.3 months compared with placebo, delaying the onset of radiation to the bone. Denosumab is directed against RANKL (Schwarz and Ritchlin 2007), the receptor activator of NF-κB ligand, which is produced by bone marrow stromal cells and osteoblasts and stimulates osteoclasts differentiation, activation, and survival, RANKL is overexpressed by bone metastatic prostate cancer epithelial cells. The rationale of this therapeutic approach resides in the concept that, although prostate cancer bone metastases are osteoblastic, the development of these lesions involves an osteolytic response mediated by osteoclasts. However, the overall survival of patients was similar in both arms of the trial (Fizazi et al. 2011).

Endothelin antagonists represent another promising new class of stromal-targeting agents (Pinto et al. 2012). They are directed against the Endothelin-1 (ET-1) signalling peptide, which is involved in prostate cancer progression. ET-1 binds to and activates the ET_A receptor, which is overexpressed on prostate cancer cells and osteoblasts surface. ET-1 is overexpressed in prostate cancer cells as compared with benign tissue, and ET_A activation in tumors, has been shown to promote tumor cell proliferation and invasion, pro-angiogenic factors secretion, and apoptosis resistance (Nelson et al. 1996, 2003). In osteoblasts, ET_A activation promotes proliferation, survival, invasion, secretion of pro-angiogenic factors, resistance to apoptosis, and generation of osteoblastic metastatic disease (Guise et al. 2003). In preclinical models, several newly-generated inhibitors of ET-1 signalling have shown to inhibit tumor cell proliferation and invasion, as well as the metastases development (Growcott 2009).

Nevertheless, the real improvement of patients survival, as well as the eventual correct protocol regimen, are still to be clarified (Pinto et al. 2012).

Thus, at present, stromal-targeting agents are commonly used in combination with epithelial-targeting chemotherapics.

Among the multiple self-protective molecular mechanisms acting in mCRPC, it has recently emerged the role of clusterin (Antonarakis and Armstrong 2011). Clusterin (CLU) is a stress-induced chaperone protein overexpressed in prostate tumors treated with androgen ablation or chemotherapy (Mita et al. 2009; de Bono et al. 2010; Zoubeidi et al. 2010). It is considered of importance in the cytoprotective defense from radio- and chemotherapy of the mCRPC (Tiligata et al. 2002).

Overexpression of CLU in prostate cancer is linked to the emergence of the treatment-resistant phenotype. In animal models, it is consistently up-regulated in castration-resistant regrowth (Miyake et al. 2000). In human tissue, the expression of CLU increases in hormone-naïve prostate cancer with increasing Gleason grade, and is up-regulated within weeks of androgen withdrawal (July et al. 2002). CLU binds to (and stabilizes) a wide variety of client proteins, and promotes cell survival and transformation through multiple mechanisms, including activation of the extracellular signal-related kinase (ERK) and Akt pathways, inhibition of ER stress, suppression of Bax activity, and release of nuclear factor kappaB (NF-κB) inhibition (Zoubeidi et al. 2010).

Expression of CLU is up-regulated by a number of different mechanisms, including stress-activated transcription factors (e.g. heat-shock factor-1), in response to endoplasmic reticulum (ER) stress, and as a downstream response to cytokines and insulin-like growth factor-1 receptor (Zellweger et al. 2001).

In prostate cancer cell lines, inhibition of clusterin resulted associated with a greater susceptibility to cytotoxic agents and radiation (Gleave et al. 2001; Zellweger et al. 2002). Therapeutic approaches with second-generation antisense anti-clusterin oligonucleotides (custirsen, OGX-011) (Chi et al. 2008), have produced the increase of sensitivity to androgen deprivation as well as chemotherapy in prostate cancer cell lines and xenograft models (Beer et al. 2004; Berthold et al. 2005).

Evaluation of OGX-011 in prostate cancer continues in a large phase III trial (SYNERGY) that is currently accruing. An anticipated 800 men with mCRPC will be randomized to treatment with standard-schedule docetaxel and prednisone, with or without OGX-011 640 mg by weekly intravenous infusion, until disease progression, unacceptable toxicity, or the completion of ten cycles. The primary endpoint is overall survival, and study completion is expected in early 2014 (Fig. 7.2).

Very interestingly, chemotherapy with genotoxic chemotherapy (mitoxantrone and docetaxel) has been shown to generate a response in micro-environmental stromal cells, promoting prostate cancer cell growth and resistance to subsequent cycles of treatment. This stromal response is due to DNA damage in fibroblasts

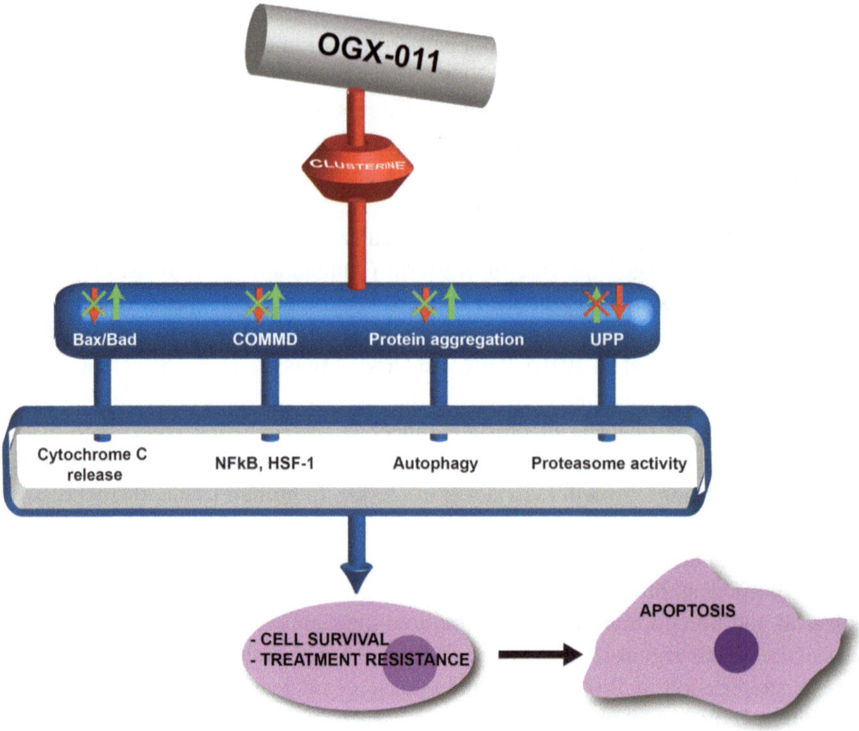

Fig. 7.2 Novel therapies designed to target non-ARmediated pathways: chaperone proteins. Clusterin, a chaperone protein, in prostate cancer cell lines, if overexpressed, results in androgen-independent growth, while clusterin gene silencing induces apoptosis and reduction in growth. Its expression is upregulated in patients with prostate cancer who have received androgen-deprivation therapy. Custirsen (OGX-011) is an antisense inhibitor of clusterin that acts suppressing clusterin expression in tumor tissue, when administered to patients with localized prostate cancer

and smooth muscle cells leading to a "30-fold" overproduction of WNT16B, which is secreted (and interacts) with adjacent prostate cancer cells, facilitating their proliferation, invasion, and therapy resistance (Sun et al. 2012) (Fig. 7.3).

7.1 What Considerations Can Be Drawn Basing on This Information?

Without any doubt, recent advances in the knowledge of prostate tumor biology, have lead us to no longer consider prostate cancer as a disease arising only from abnormally proliferating epithelial cells, but rather as the result of intricate interactions between prostate cancer epithelial and stromal cells. This has produced remarkable achievements in the development of therapy, particularly for metastatic castrate-resistant prostate cancer.

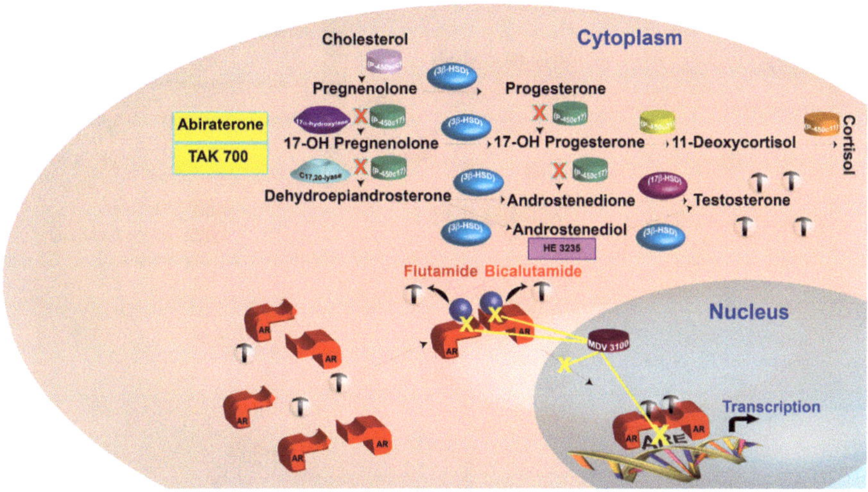

Fig. 7.3 Therapies designed to target AR-mediated pathways. Persistent AR activation is an important mediator of disease progression in CRPC and mechanisms involved include AR gene amplification or overexpression; AR gene mutation; enhanced AR signal transduction mediated via coactivators; and endocrine or autocrine activation of the AR, for example, by adrenal androgens or intratumoral production of dihydrotestosterone (DHT). AR-directed approaches include drugs that antagonize the AR or that reduce of androgen precursors. Among AR antagonists are included bicalutamide and flutamide, that inhibit the binding between AR and testosterone, and MDV3100, which exert its function by inhibiting interaction between testosterone and AR and between AR-testosterone complex and the ARE-sequences on DNA. Therapies that decrease androgen production are also being developed. Abiraterone acetate is a selective and irreversible inhibitor of cytochrome P450 (CYP450)c17, involved in androgen synthesis. TAK-700 is a novel CYP450c17 inhibitor similar to abiraterone

The goal of the next-coming therapy, then, will be to disrupt the crosstalk between epithelial and cancer cells and their microenvironment, through the use of new drugs targeting multiple signaling pathways, from androgen receptor signaling, to kinase receptor signaling, and immune surveillance. At last, we now hold a paramount variety of targets that can be manipulated to overcome AR- and chemo-resistance, and different strategies emerge for inhibiting their function.

The increasing knowledge of the crystal structures of the ligand-sites, or specific domains, of each protein active in the induction and maintenance of prostate cancer aggressiveness and therapy resistance, allows the progression of design and synthesis of novel inhibitors. By converse, in other instances the inhibition of metastasizing ability of cancer cells may be reached through the targeted disruption of protein transcription with antisense oligonucleotides.

In addition, new therapies may also consider the charming chance of a different delivery of the "old" drugs (Batist 2007; Gabizon et al. 2003; Ewer et al. 2004; Koukourakis et al. 2000). For instance, therapeutic nanoparticles targeted against the prostate-specific membrane antigen (PSMA) protein, which is expressed on the surface of prostate tumor cells that accumulate in tumors while bypassing healthy cells, have shown promising results in ongoing clinical trials.

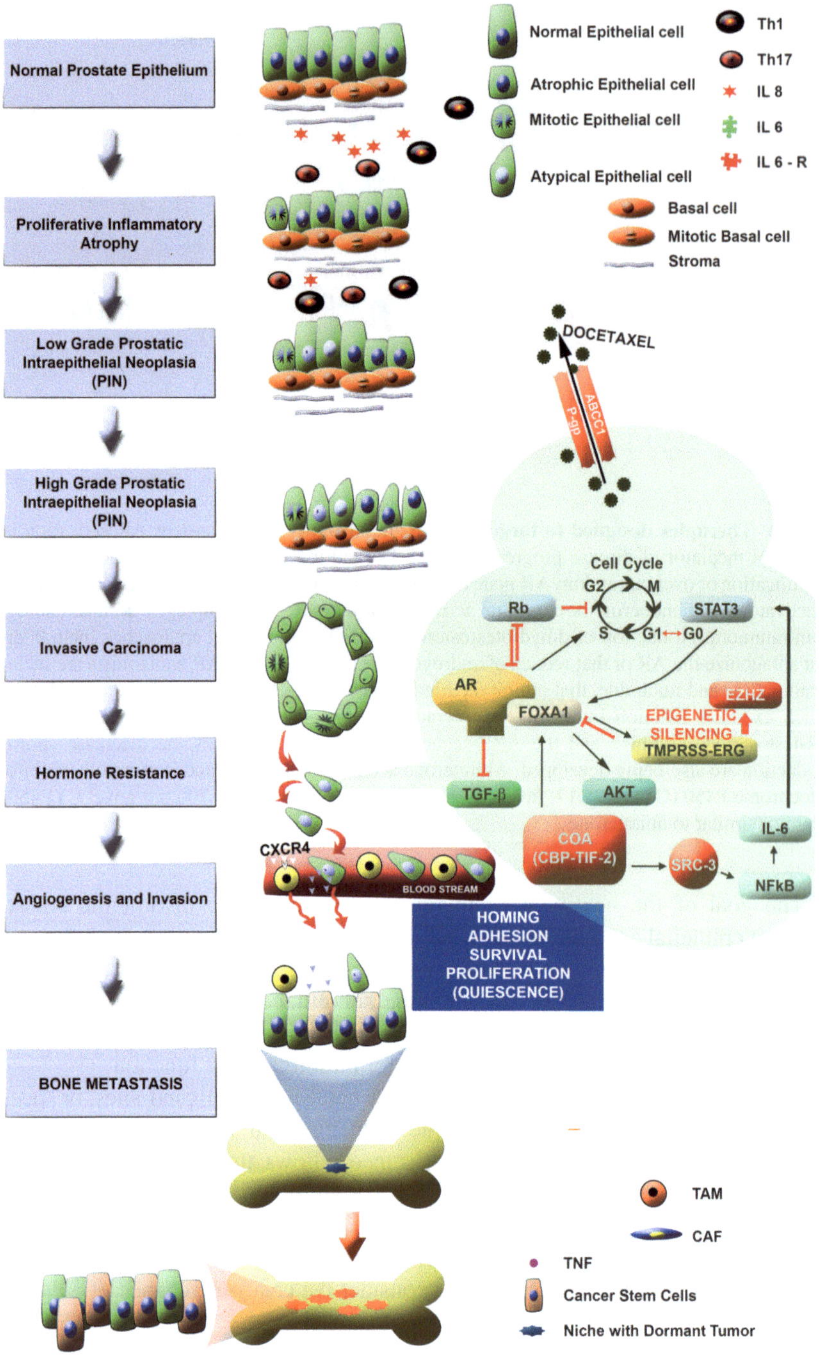
Normal Prostate Epithelium
Proliferative Inflammatory Atrophy
Low Grade Prostatic Intraepithelial Neoplasia (PIN)
High Grade Prostatic Intraepithelial Neoplasia (PIN)
Invasive Carcinoma
Hormone Resistance
Angiogenesis and Invasion
BONE METASTASIS
Normal Epithelial cell
Atrophic Epithelial cell
Mitotic Epithelial cell
Atypical Epithelial cell
Th1
Th17
IL 8
IL 6
IL 6 - R
Basal cell
Mitotic Basal cell
Stroma
DOCETAXEL
ABCC1
P-gp
Cell Cycle
G2
M
S
G1
G0
Rb
STAT3
AR
FOXA1
EZHZ
EPIGENETIC SILENCING
TMPRSS-ERG
TGF-β
AKT
COA (CBP-TIF-2)
SRC-3
IL-6
NFkB
CXCR4
BLOOD STREAM
HOMING
ADHESION
SURVIVAL
PROLIFERATION
(QUIESCENCE)
TAM
CAF
TNF
Cancer Stem Cells
Niche with Dormant Tumor

Nanoparticles homing docetaxel have been designed at Massachusetts Institute of Technology (MIT) and Brigham and Women's Hospital in Boston, and the early results have shown their selective accumulation at tumor sites, without side-effects, producing tumors shrunk even at lower drug doses than those usually administered (Corcoran and Gleave 2012).

There is great expectation on the successful therapeutic effect of an impressive number of new drugs in the near future, which could be safer and more manageable than old cytotoxic agents, and then could be used also at the earliest phases of prostate cancer, before it becomes lethal.

However, despite impressive preclinical activity, to date, most targeted therapies have failed in early clinical development (Corcoran and Gleave 2012). Pathway redundancy, target mutation, difficulty with drug delivery, toxicity, or overestimation of a target's importance in preclinical models, could all be responsible for this. Since international guidelines are obviously lacking, the gamble is, therefore, either to identify the more reliable targets to be translated into clinically useful drugs, and to design a rational approach to optimal treatment sequencing or even a combination therapy with these drugs (El-Amm and Aragon-Ching 2013).

Notwithstanding the several new therapies that have been shown to extend survival of mCRPC patients, but none of these approaches are curative, and annual mortality rates, from prostate cancer in the Western Countries, remain unacceptably high (Antonarakis and Armstrong 2011).

Fig. 7.4 Alterations in AR function in PCa and the development of CRPC. Following androgen ablation, increased AR levels and a dramatic shift in AR function is observed ultimately leading to the development of CRPC. AR binding sites (AR cistrome) are frequently marked by specific chromatin modifications introduced by pioneer factors as FOXA1, prior to hormone treatment. FOXA1 ablation causes massive reprogramming of the AR cistrome and consequentially its function with a survival and growth advantage. Androgens promote proliferation through signals that modulate critical regulators of the cell cycle. For the most part, the AR regulates the cell cycle through induction of signals that regulate G1-S phase transition through the promotion of G1 cyclin-dependant kinases (CDKs) and inhibition of the retinoblastoma (Rb) tumor suppressor gene. AR coregulators, such as SRC3, stimulate or decrease AR activity in a promoter specific manner. Elevated levels of SRC-3 are expressed in primary tumors. To activate AR transcriptional activity, SRC-3 coactivates AP1 that positively regulates Akt levels, leading to increase in proliferation and reduced apoptosis. Moreover, SRC-3 stimulates cellular motility by activating focal adhesion kinase signaling and invasion by activating AR-dependent expression of matrix metalloproteinases 2 and 13. IL-6 potentiate AR function and increased levels are associated with androgen independent growth, resistance to chemotherapeutic drugs and neuroendocrine differentiation. Continuous activation of the NF-kB pathway inhibits prostate regression following castration, maintains nuclear AR and sustains epithelial proliferation. TGF-b and AR signaling cooperate to maintain the differentiated state of the stroma in the benign prostate. In malignant epithelial cells, AR suppresses TGF-b receptor II (TbR-II) transcription and reduces TGF-b1 driven apoptosis, suggesting that AR action in malignant epithelial cells provides a growth advantage by suppressing TGF mediated pathways. CTC have the same TMPSS2-ERG fusion status as primary tumor and its expression is significantly increased by AR signaling. Overexpression of the oncogenic transcription factor ERG causes expression of epigenetic factors such as the methyltransferase EZH2 that epigenetically silences differentiating factors and tumor suppressors

The imperative goal for advanced prostate cancer therapy will be to successfully hit the specific driver mutations responsible for AR- and/or drug-resistance of advanced, metastasizing prostate cancer (Gerlinger et al. 2012). Many questions of interest have still to be properly addressed. The first refers to the optimal treatment of heterogeneous tumors harbouring different levels of the target mutation, which deserves to set up new personalized treatment algorithms based on the results of the genetic profiling of patients (Dora Dias-Santagata et al. 2010). In addition, we have to be aware that AR inhibition/chemotherapeutic drugs, as well as radiation therapy, may induce new advantageous mutations in prostate cancer cells, which after an initial positive clinical response, may increase again their survival and resistance (Semenas et al. 2012).

The future direction of prostate cancer care, then, will rely not only on our ability to detect and hit the molecular patterns responsible for the AR/chemotherapy-resistant phenotype of advanced, metastasizing cancers, but also on the chance to really personalize and potentially change therapy when resistance eventually recurs (Fig. 7.4).

References

Agarwal N, Sonpavde G, Sternberg CN (2012) Novel molecular targets for the therapy of castration-resistant prostate cancer. Eur Urol 61(5):950–960. Epub 2011 Dec 22

Antolín AR, Ojeda JM, Otero JR, Rodríguez AC, Castellano D, Esteban MD, Sicilia LD, González RD (2012) Hormonal treatment in biochemical recurrence after radical prostatectomy. Arch Esp Urol 65(1):111–121. Review. Spanish

Antonarakis ES, Armstrong AJ (2011) Emerging therapeutic approaches in the management of metastatic castration-resistant prostate cancer. Prostate Cancer Prostatic Dis 14(3):206–218. Epub 2011 May 17. Review

Araujo J, Logothetis C (2010) Dasatinib: a potent SRC inhibitor in clinical development for the treatment of solid tumors. Cancer Treat Rev 36:492–500

Armas OA, Aprikian AG, Melamed J et al (1994) Clinical and pathological effects of neoadjuvant total androgen ablation therapy on clinically localized prostatic adenocarcinoma. Am J Surg Pathol 18:979–991

Armstrong AJ, George DJ (2008) New drug development in metastatic prostate cancer. Urol Oncol 26:430–437

Batist G (2007) Cardiac safety of liposomal anthracyclines. Cardiovasc Toxicol 7:72–74

Beekman KW, Hussain M (2008) Hormonal approaches in prostate cancer: application in the contemporary prostate cancer patient. Urol Oncol 26(4):415–419

Beer TM, Garzotto M, Henner WD, Eilers KM, Wersinger EM (2004) Multiple cycles of intermittent chemotherapy in metastatic androgen-independent prostate cancer. Br J Cancer 91:1425–1427

Berthold DR, Sternberg CN, Tannock IF (2005) Management of advanced prostate cancer after first-line chemotherapy. J Clin Oncol 23:8247–8252

Bradley DA, Hussain M (2008) Promising novel cytotoxic agents and combinations in metastatic prostate cancer. Cancer J 14(1):15–19. Review

Bullock MJ, Srigley JR, Klotz LH et al (2002) Pathologic effects of neoadjuvant cyproterone acetate on nonneoplastic prostate, prostatic intraepithelial neoplasia, and adenocarcinoma. A detailed analysis of radical prostatectomy specimens from a randomized trial. Am J Surg Pathol 26:1400–1413

Carducci MA, Saad F, Abrahamsson PA et al (2007) A phase 3 randomized controlled trial of the efficacy and safety of atrasentan in men with metastatic hormone-refractory prostate cancer. Cancer 110:1959–1966

Chi KN, Zoubeidi A, Gleave ME (2008) Custirsen (OGX-011): a second-generation antisense inhibitor of clusterin for the treatment of cancer. Expert Opin Investig Drug 17:1955–1962

Chodak G, Sharifi R, Kasimis B, Block NL, Macramalla E, Kennealey GT (1995) Single-agent therapy with bicalutamide: a comparison with medical or surgical castration in the treatment of advanced prostate carcinoma. Urology 46(6):849–855

Civantos F, Marcial MA, Banks ER et al (1995) Pathology of androgen deprivation therapy in prostate carcinoma: a comparative study of 173 patients. Cancer 75:1634–1641

Cook RJ, Coleman R, Brown J et al (2006) Markers of bone metabolism and survival in men with hormone-refractory metastatic prostate cancer. Clin Cancer Res 12:3361–3367

Corcoran NM, Gleave ME (2012) Targeted therapy in prostate cancer. Histopathology 60(1): 216–231

de Bono JS, Oudard S, Ozguroglu M et al (2010) Prednisone plus cabazitaxel or mitoxantrone for metastatic castration-resistant prostate cancer progressing after docetaxel treatment: a randomised open-label trial. Lancet 376:1147–1154

Debes JD, Tindall DJ (2004) Mechanisms of androgen-refractory prostate cancer. N Engl J Med 351(15):1488–1490

Dias-Santagata D, Akhavanfardy S, David SS, Vernovsky K, Kuhlmann G, Boisvert SL, Stubbs H, McDermott U, Settleman J, Kwak EL, Clark JW, Isakoff SJ, Sequist LV, Engelman JA, Lynch TJ, Haber DA, Louis DN, Ellisen LW, Borger DR, John A (2010) Iafrate Rapid targeted mutational analysis of human tumours: a clinical platform to guide personalized cancer medicine. EMBO Mol Med 2:146–158

El-Amm J, Aragon-Ching JB (2013) The changing landscape in the treatment of metastatic castration-resistant prostate cancer. Ther Adv Med Oncol 5(1):25–40

el-Rayes BF, Hussain MH (2002) Hormonal therapy for prostate cancer: past, present and future. Expert Rev Anticancer Ther 2(1):37–47

Evans AJ, Ryan P, van der Kwast T (2011) Treatment effects in the prostate including those associated with traditional and emerging therapies. Adv Anat Pathol 18:281–293

Ewer MS, Martin FJ, Henderson C, Shapiro CL, Benjamin RS, Gabizon AA (2004) Cardiac safety of liposomal anthracyclines. Semin Oncol 31:161–181

Feldman BJ, Feldman D (2001) The development of androgen-independent prostate cancer. Nat Rev Cancer 1(1):34–45. Review

Fizazi K, Carducci M, Smith M et al (2011) Denosumab versus zoledronic acid for treatment of bone metastases in men with castration-resistant prostate cancer: a randomised, double-blind study. Lancet 377:813–882

Gabizon A, Shmeeda H, Barenholz Y (2003) Pharmacokinetics of pegylated liposomal Doxorubicin: review of animal and human studies. Clin Pharmacokinet 42:419–436

Galsky MD, Vogelzang NJ (2010) Docetaxel-based combination therapy for castration-resistant prostate cancer. Ann Oncol 21:2135–2144

Gerlinger M, Rowan AJ, Horswell S, Larkin J, Endesfelder D, Gronroos E, Martinez P, Matthews N, Stewart A, Tarpey P, Varela I, Phillimore B, Begum S, McDonald NQ, Butler A, Jones D, Raine K, Latimer C, Santos CR, Nohadani M, Eklund AC, Spencer-Dene B, Clark G, Pickering L, Stamp G, Gore M, Szallasi Z, Downward J, Futreal PA, Swanton C (2012) Intratumor heterogeneity and branched evolution revealed by multiregion sequencing. N Engl J Med 366(10):883–892. Erratum in: N Engl J Med. 2012 Sep 6;367(10):976

Gleave ME, Miyake H, Zellweger T et al (2001) Use of antisense oligonucleotides targeting the antiapoptotic gene, clusterin/testosterone-repressed prostate message 2, to enhance androgen sensitivity and chemosensitivity in prostate cancer. Urology 58:39–49

Growcott JW (2009) Preclinical anticancer activity of the specific endothelin A receptor antagonist ZD4054. Anticancer Drug 20:83–88

Guise TA, Yin JJ, Mohammad KS (2003) Role of endothelin-1 in osteoblastic bone metastases. Cancer 97:779–784

Hodgson MC, Bowden WA, Agoulnik IU (2012) Androgen receptor footprint on the way to prostate cancer progression. World J Urol 30(3):279–285. Epub 2011 Sep 17. Review

Hu R, Isaacs WB, Luo J (2011) A snapshot of the expression signature of androgen receptor splicing variants and their distinctive transcriptional activities. Prostate. 71(15):1656–1667

Huggins C, Hodges CV (1941) Studies on prostatic cancer: (1) the effect of estrogen and of androgen injection on serum phosphates in metastatic carcinoma of the prostate. Cancer Res 1:293–297

Huggins C, Stevens RE, Hodges CV (1941) Studies on prostatic cancer: (II) the effects of castration on advanced carcinoma of the prostate gland. Arch Surg 43:209–223

Humphrey PA (2003) Prostate pathology. American Society for Clinical Pathology, Chicago, pp 456–476

July LV, Akbari M, Zellweger T, Jones EC, Goldenberg SL, Gleave ME (2002) Clusterin expression is significantly enhanced in prostate cancer cells following androgen withdrawal therapy. Prostate 50:179–188

Koukourakis MI, Koukouraki S, Giatromanolaki A, Kakolyris S, Georgoulias V, Velidaki A, Archimandritis S, Karkavitsas NN (2000) High intratumoral accumulation of stealth liposomal doxorubicin in sarcomas–rationale for combination with radiotherapy. Acta Oncol 39:207–211

Langenhuijsen JF, Badhauser D, Schaaf B, Kiemeney LA, Witjes JA, Mulders PF (2013) Continuous vs. intermittent androgen deprivation therapy for metastatic prostate cancer. Urol Oncol. 31(5):549–556

Lee LF, Guan J, Qiu Y, Kung HJ (2001) Neuropeptide-induced androgen independence in prostate cancer cells: roles of nonreceptor tyrosine kinases EtkBmx, Src, and focal adhesion kinase. Mol Cell Biol 21:8385–8397

Lee LF, Louie MC, Desai SJ et al (2004) Interleukin-8 confers androgenindependent growth and migration of LNCaP: differential effects of tyrosine kinases Src and FAK. Oncogene 23: 2197–2205

Liu Y, Hegde P, Zhang F, Hampton G, Jia S (2012) Prostate cancer – a biomarker perspective. Front Endocrinol (Lausanne) 3:72

Lubahn DB, Joseph DR, Sullivan PM, Willard HF, French FS, Wilson EM (1988) Cloning of human androgen receptor complementary DNA and localization to the X chromosome. Science 240:327–330

Mathew P, Dipaola R (2007) Taxane refractory prostate cancer. J Urol 178:S36–S41

Mathew P, Thall PF, Bucana CD et al (2007) Platelet-derived growth factor receptor inhibition and chemotherapy for castration-resistant prostate cancer with bone metastases. Clin Cancer Res 13:5816–5824

Mita AC, Denis LJ, Rowinsky EK et al (2009) Phase I and pharmacokinetic study of XRP6258 (RPR 116258A), a novel taxane, administered as a 1-hour infusion every 3 weeks in patients with advanced solid tumors. Clin Cancer Res 15:723–730

Miyake H, Nelson C, Rennie PS, Gleave ME (2000) Testosterone-repressed prostate message-2 is an antiapoptotic gene involved in progression to androgen independence in prostate cancer. Cancer Res 60:170–176

Mohler JL (2008) A role for the androgen-receptor in clinically localized and advanced prostate cancer. Best Pract Res Clin Endocrinol Metab 22(2):357–372. Review

Montgomery RB, Mostaghel EA, Vessella R et al (2008) Maintenance of intratumoral androgens in metastatic prostate cancer: a mechanism for castration-resistant tumor growth. Cancer Res 68:4447–4454

Morote J, Orsola A, Planas J, Trilla E, Raventós CX, Cecchini L, Catalán R (2007) Redefining clinically significant castration levels in patients with prostate cancer receiving continuous androgen deprivation therapy. J Urol 178(4 Pt 1):1290–1295

Mostaghel EA, Montgomery B, Nelson PS (2009) Castration-resistant prostate cancer: targeting androgen metabolic pathways in recurrent disease. Urol Oncol 27(3):251–257

NCCN (National Comprehensive Cancer Network) (2011) Guidelines for Patientes. Proste cancer, Version 1

Nelson JB, Chan-Tack K, Hedican SP et al (1996) Endothelin-1 production and decreased endothelin B receptor expression in advanced prostate cancer. Cancer Res 56:663–668
Nelson J, Bagnato A, Battistini B, Nisen P (2003) The endothelin axis: emerging role in cancer. Nat Rev Cancer 3:110–116
Park SI, Shah AN, Zhang J, Gallick GE (2007) Regulation of angiogenesis and vascular permeability by Src family kinases: opportunities for therapeutic treatment of solid tumors. Expert Opin Ther Target 11:1207–1217
Park SI, Zhang J, Phillips KA et al (2008) Targeting SRC family kinases inhibits growth and lymph node metastases of prostate cancer in an orthotopic nude mouse model. Cancer Res 68: 3323–3333
Petraki CD, Sfikas CP (2007) Histopathological changes induced by therapies in the benign prostate and prostate adenocarcinoma. Histol Histopathol 1:107–118
Petrylak DP, Tangen CM, Hussain MH et al (2004) Docetaxel and estramustine compared with mitoxantrone and prednisone for advanced refractory prostate cancer. N Engl J Med 351: 1513–1520
Pienta KJ, Bradley D (2006) Mechanisms underlying the development of androgen-independent prostate cancer. Clin Cancer Res 12(6):1665–1671
Pinto A, Merino M, Zamora P, Redondo A, Castelo B, Espinosa E (2012) Targeting the endothelin axis in prostate carcinoma. Tumour Biol 33(2):421–426. Epub 2011 Dec 29
Saad F, Gleason DM, Murray R et al (2002) A randomized, placebo-controlled trial of zoledronic acid in patients with hormone-refractory metastatic prostate carcinoma. J Natl Cancer Inst 94:1458–1468
Sartor O, Michels RM, Massard C, de Bono JS (2011) Novel therapeutic strategies for metastatic prostate cancer in the post-docetaxel setting. Oncologist 16(11):1487–1497. Epub 2011 Nov 2.Review
Scher HI, Sawyers CL (2005) Biology of progressive, castration-resistant prostate cancer: directed therapies targeting the androgen-receptor signaling axis. J Clin Oncol 23:8253–8261
Scher HI, Liebertz C, Kelly WK, Mazumdar M, Brett C, Schwartz L, Kolvenbag G, Shapiro L, Schwartz M (1997) Bicalutamide for advanced prostate cancer: the natural versus treated history of disease. J Clin Oncol 15(8):2928–2938
Schrijvers D, Van Erps P, Cortvriend J (2010) Castration-refractory prostate cancer: new drugs in the pipeline. Adv Ther 27:285–296
Schwarz EM, Ritchlin CT (2007) Clinical development of anti-RANKL therapy. Arthritis Res Ther 9(suppl 1):S7
Semenas J, Allegrucci C, Boorjian SA, Mongan NP, Persson JL (2012) Overcoming drug resistance and treating advanced prostate cancer. Curr Drug Target 13(10):1308–1323. Review
Srigley JR, Delahunt B, Evans AJ (2012) Therapy-associated effects in the prostate gland. Histopathology 60(1):153–165
Sun Y, Campisi J, Higano C, Beer TM, Porter P, Coleman I, True L, Nelson PS (2012) Treatment-induced damage to the tumor microenvironment promotes prostate cancer therapy resistance through WNT16B. Nat Med 18(9):1359–1368
Tannock IF, de Wit R, Berry WR, Horti J, Pluzanska A, Chi KN, Oudard S, Théodore C, James ND, Turesson I, Rosenthal MA, Eisenberger MA (2004) TAX 327 Investigators. Docetaxel plus prednisone or mitoxantrone plus prednisone for advanced prostate cancer. N Engl J Med 351(15):1502–1512
Taplin ME (2008) Androgen receptor: role and novel therapeutic prospects in prostate cancer. Expert Rev Anticancer Ther 8(9):1495–1508. Review
Têtu B (2008) Morphological changes induced by androgen blockade in normal prostate and prostatic carcinoma. Best Pract Res Clin Endocrinol Metab 22:271–283
Têtu B, Srigley JR, Boivin JC et al (1991) Effect of combination endocrine therapy (LHRH agonist and flutamide) on normal prostate and prostatic adenocarcinoma. A histopathologic and immunohistochemical study. Am J Surg Pathol 15:111–120
Tiligada E, Miligkos V, Delitheos A (2002) Cross-talk between cellular stress, cell cycle and anticancer agents: mechanistic aspects. Curr Med Chem Anticancer Agent 2:553–566

Tyrrell CJ, Payne H, See WA, McLeod DG, Wirth MP, Iversen P, Armstrong J, Morris C (2005) Casodex' Early Prostate Cancer Trialists Group. Bicalutamide ('Casodex') 150 mg as adjuvant to radiotherapy in patients with localised or locally advanced prostate cancer: results from the randomised Early Prostate Cancer Programme. Radiother Oncol 76(1):4–10

Vallancourt L, Têtu B, Fradet Y et al (1996) Effect of neoadjuvant endocrine therapy (combined androgen blockade) on normal prostate and prostatic carcinoma: a randomized study. Am J Surg Pathol 20:86–93

Zellweger T, Miyake H, July LV, Akbari M, Kiyama S, Gleave ME (2001) Chemosensitization of human renal cell cancer using antisense oligonucleotides targeting the antiapoptotic gene clusterin. Neoplasia 3:360–367

Zellweger T, Chi K, Miyake H et al (2002) Enhanced radiation sensitivity in prostate cancer by inhibition of the cell survival protein clusterin. Clin Cancer Res 8:3276–3284

Zoubeidi A, Chi K, Gleave M (2010) Targeting the cytoprotective chaperone, clusterin, for treatment of advanced cancer. Clin Cancer Res 16:1088–1093

Chapter 8
Crossroads of Signaling Pathways

Stefania Staibano

Abstract As studies on PC progression continue to uncover a growing number of crosstalks and co-occurrences of mutations and epigenetic alterations, new drugs are getting approved bringing significant changes in the treatment paradigm of these tumors.

This chapter recapitulates the best known examples of molecular interactions potentially targetable to achieve these therapeutic evolutionary changes, to allow a better control of PC which, in 2012 alone, has still killed more than 28,000 men, in USA (Siegel et al, CA Cancer J Clin, 62:10–29, 2012; El-Amm, Aragon-Ching, Ther Adv Med Oncol, 5(1):25–40, 2013).

8.1 Background and Aims

Prostate cancer (PC) is a multifocal disease, composed by several independent tumor foci that may show different degrees of molecular alterations. The heterogeneous nature of PC which, at a molecular level, derives from the crosstalk of multiple signal transductions variously acting in promoting growth, survival and therapy-resistance of PC cells (Pittoni et al. 2011).

The development of new therapeutic strategies, particularly focused toward castration-resistant prostate cancer (CRPC), relies on a better understanding of the resistance pathways selectively adopted from prostate cancer cells (El-Amm and Aragon-Ching 2013).

Recently, it has been shown that androgens residual from androgen-deprivation therapy may indirectly favor cancer growth, with a progressive increase of the PSA levels, *via* the over-expression of many HIF-1 dependent, hypoxia-inducible

S. Staibano (✉)
Department of Advanced Biomedical Sciences, Pathology Section, Faculty of Medicine and Surgery, University of Naples "Federico II", via S. Pansini, n.5, Naples, Italy
e-mail: staibano@unina.it

S. Staibano (ed.), *Prostate Cancer: Shifting from Morphology to Biology*,
DOI 10.1007/978-94-007-7149-9_8, © Springer Science+Business Media Dordrecht 2013

genes. The interplay between hypoxia and AR, further cross-talks with several oxidative stress mediators, cytokines, growth factors, DNA-repair pathways, and epigenetic regulators, in a cooperative effort to ensure the survival of neoplastic cells in a highly adverse metabolic (and environmental) background. These interacting signaling mechanisms, indeed, may either potentiate or counteract each other, leading alternatively to cell death or adaptation and radio-chemoresistance. It thus becomes apparent that resistance to therapy can be overcome only through a proper therapeutic manipulation of the right factor(s) that, in turn, will influence the others, triggering PC cells death (Marignol et al. 2008).

As an example, it has been surprisingly shown that cell death can be induced in castration-resistant tumors "still" *via* AR, which can modulate apoptosis and autophagy if targeted in conjunction with PKA pathway members (Attar et al. 2009).

As well, considering that AR is largely expressed in tumor microenvironmental stromal cells, drugs targeting AR signaling in PC cells give rise to a therapeutic favourable effect also on the stromal compartments, as AR is largely expressed in tumor microenvironmental stromal cells (Mantalaris et al. 2001).

Thus, a favourable response to AR targeting will encompass both reductions in serum PSA and bone-specific, osteoblast derived, alkaline phosphatase.

Overall, mounting evidence suggest that the cell fate, in response to therapeutic attack, depends on a plethora of variable factors, ranging from metabolic stress, functional status of cells, the interaction level between the stress-response pathways, paracrine mediators produced by tumor microenvironment, and the epigenetic interactions on DNA damage response and DNA repair (Murr 2010).

During these last years, the efforts of the scientific community have been focused on the correct interpretation of the *complementary pathways* which could kill radio- and chemoresistant cancer cells.

The end-point of such a program will require carefully designed clinical trials, a rigorous patient selection and retrospective analyses of clinical, pathological, and follow-up data.

In the rapidly evolving field of prostate cancer therapy, new drugs are being used; as well, new insights indicating possible different rational approaches to the treatment sequencing with "old" drugs are being proposed. The search for novel biomarkers, useful for the individualized prediction of treatment response and outcome of PC patients, is actively on. This is a particularly complex investigational field (considering that the malignant phenotype of prostate cancer cells results from a highly variable combination of functional, genetic and epigenetic defects in cell cycle metabolism, checkpoint control and DNA-repair pathways, working together to render PC a lethal disease). The real challenge lies not only in detecting all these alterations, but also in defining the multiple layers of their reciprocal intersection (Sarwar and Persson 2011).

We are thus requested to critically review our knowledge about the role of the plethora of molecular guests acting on the scenario of prostate cancer progression, which comprises both epithelial and stromal cells, both contributing to tumor heterogeneity and growth dynamics (Cho-Chung 1989, 1990; Camps et al. 1990; Cunha et al. 1996).

That prostate cancer development and growth is dependent on androgens and can be suppressed by androgen ablation monotherapy is an old concept (Zhu and Kyprianou 2008).

The appearance of androgen-independent prostate tumor growth, leading to cancer recurrence and highly metastatic disease, is a well-known phenomenon as well (Wang et al. 2007).

During the entire life-span of prostate cancer, the androgen axis actively cross-talks with a plethora of growth factors, driving the shift of prostate cancer cell toward survival and invasion advantage. Androgenic control of growth and differentiation is tightly regulated in both stromal and epithelial cells (Sar et al. 1990).

This explain why the successful treatment of PC with drugs targeting (AR) signaling (defined as *Epithelial-Stromal Targeting Agents*), leads to reduction in either serum PSA and bone-specific alkaline phosphatase (Mantalaris et al. 2001; Niu et al. 2010).

A poor clinical outcome for prostate cancer patients has been associated, instead, with low-AR levels in the stromal microenvironment (Henshall et al. 2001), and this finding has been proposed as one of the mechanisms involved in the emergence of androgen-independent cancer (Dayyani et al. 2011). The propulsive effect of androgens on prostate epithelial cell proliferation and survival are indirectly regulated by paracrine mediators produced by stromal cells, such as insulin-like growth factor (IGF), fibroblast growth factor (FGF), epidermal growth factor (EGF), (Cunha and Donjacour 1989), vascular endothelial growth factor (VEGF) and transforming growth factor-β (TGF-β) (Byrne et al. 1996).

The epidermal growth factor-1 (EGF) and its receptor (EGFR), (Russell et al. 1998) are frequently up-regulated in advanced stages of PC (Di Lorenzo et al. 2002) Targeting EGFR with monoclonal antibodies or with tyrosine kinase inhibitors, suppresses growth and invasion of androgen-dependent and -independent prostate cancer cells *in vitro*, leading to the conclusion that the multi-crossed signals between EGF/EGFR and androgen signaling is crucial for the acquisition and the maintenance of androgen sensitivity (Bonaccorsi et al. 2004; Festuccia et al. 2005; Leotoing et al. 2007).

Both AR and EGF can activate MAPK and, in a 'functional-symmetry', the EGF-activated MAPK/extracellular signaling-regulated kinase kinase-1 (MEKK1) cascade. This allows EGF to interfere with AR function, modulating the androgen response and blocking androgen-dependent transcription in differentiated cells. MAPK extracellular kinase (MEK) inhibition reverses the EGF-mediated AR down-regulation in differentiated cells (Peterziel et al. 1999).

The alteration of this EGF–AR interplay is an important contributor to prostate tumor progression. The modulation of AR signaling activity by ERBB2 (HER-2/neu), a lead member of the EGFR family of receptor tyrosine kinases, has been found correlated with prostate cancer progression to cell growth of androgen-independent metastatic disease (Heinlein and Chang 2004), *in vitro* and *in vivo* (Craft et al. 1999; Yeh et al. 1999; Mellinghoff et al. 2004; Liu et al. 2005).

Similar positive feedbacks with AR activity in prostate cancer cells have been described also for several other growth factors (Orio et al. 2002). Evidence supports a strict interaction between AR and the IGF signaling. The high IGF1 signaling in prostate cancer cells (HepG2 and LNCaP cells) (Wu et al. 2007) likely depends upon AR up-regulation of IGF1 receptor expression and/or, alternatively, upon the modulation of IGF-binding proteins (IGFBPs), which, in turn, are up-regulated by either androgens and IGF1 in androgen-responsive human fibroblasts (Yoshizawa and Ogikubo 2006) IGF1 enhances AR transactivation under low/absent androgen levels (Culig et al. 1994; Orio et al. 2002) and promotes prostate tumor cell proliferation (Burfeind et al. 1996). According to several reports, high IGF1 levels in the serum can be considered a marker of an increased risk of prostate cancer (Pollak et al. 1998; Wolk et al. 1998).

Even more data concern the cross-talks between AR and Transforming growth factor-β (TGFβ). This ubiquitous cytokine, for instance, contributes to the regulation of proliferation, growth and differentiation of prostate stromal and epithelial cells.

Cofilin and prohibitin, two novel signaling effectors of TGFB1, that serve as potential intracellular effectors of its apoptotic action, were identified in human prostate cancer cells (Zhu et al. 2006). Androgens can inhibit TGFB1-induced apoptosis in prostate cancer cells (Chipuk et al. 2002) via the AR-associated protein 55 (ARA55/Hic-5; LIM protein superfamily). Overexpression of ARA55 inhibits TGFB-mediated up-regulation of SMAD transcriptional activity in rat prostate epithelial cells, as well as human prostate cells, via an interaction between ARA55 and SMAD3 (Wang et al. 2005). Cancer cells become refractory to the growth inhibitory activity of TGFB due to the loss (or mutation) of transmembrane receptors or intracellular TGFB signaling effectors during tumor initiation (Akhurst and Derynck 2001). In advanced prostate cancer and in PC bone metastasis, TGFB is over-expressed and TGFB1 ligand overexpression results in pro-oncogenic rather than growth suppressive effect (Coffey et al. 1986; Roberts et al. 1986; Derynck and Zhang 2003; Zhu and Kyprianou 2005).

The androgenic-mediated TGFB enhancement seems to play a role on the epithelial-mesenchymal transition (EMT) in metastasizing cancers (Zavadil and Bottinger 2005), with a further interplay with E-cadherin. The effects of TGFB1 expression on stromal cell proliferation and differentiation, depend on the specific stromal cell type, microenvironment, and interactions with other growth factor (Sporn and Roberts 1992). AR and TGFB1 levels significantly correlate in the stromal component of prostatic intraepithelial neoplasia (Cardillo et al. 2000). Very interestingly, TGFB1 triggers the AR translocation from nucleus to cytoplasm in prostate stromal cells underlying to myodifferentiation (Gerdes et al. 1998, 2004), while androgens enhance TGFB1-mediated proliferation of prostatic smooth muscle cells (Salm et al. 2000).

Prostate cancer progression toward androgen-independent disease has been linked also to changes in the expression of several members of the FGF family, characterized by a broad spectrum of functions on cell differentiation, migration, and angiogenesis (Ornitz and Itoh 2001). FGF2 can stimulate also the proliferation of prostate stromal cells, in a synergistic fashion with DHT (Niu et al. 2001).

The synthesis of FGF2 and FGF7 in prostate epithelial cells seems to be mainly regulated by estrogen receptors (ER), whereas ER act in coordination with AR to mediate the synthesis of these growth factors in stromal cells. Androgen depletion rapidly reduces stromal IGF1 expression, after castration, favoring PC cells apoptosis (Ohlson et al. 2007). AR can otherwise directly influence the expression of FGF1, FGF2, FGF8, and FGF10 in either prostate tumor epithelial cells and stromal cells (Saric and Shain 1998; Nakano et al. 1999; Rosini et al. 2002), while a paracrine secretion of FGF10 exert a positive feedback on AR, up-regulating its expression (Memarzadeh et al. 2007). In response to FGFs, AR potentiates FGF-induced survival of prostate cancer cells, possibly through BCL-2 induction, allowing the escape of selected clones from androgenic control (Rosini et al. 2002; Gonzalez-Herrera et al. 2006).

Among the cross-actions between AR and the *multifunctional growth factor signaling pathways*, the interplay between the cellular responses to androgen and hypoxia is emerging as a further key phenomenon in the developing of androgen-independent, metastasizing prostate cancer cell clones (Marignol et al. 2008). In prostate cells, androgens and hypoxia share several regulatory molecular mechanisms: both androgens and hypoxia, in fact, due to the presence of a hypoxia-responsive region in the human PSA promoter, can induce in fact PSA expression (Horii et al. 2007).

It has been then hypothesized that hypoxia, mainly through the hypoxia-inducible factor (HIF1A), may facilitate PC progression through the cross-talk with AR. To further support this idea, it has been recently reported that residual androgens following androgen deprivation induce the expression of hypoxia-inducible genes and stimulate cancer re-growth (Marignol et al. 2008).

This is of particular interest if we consider that, as for most solid cancers, hypoxia is a common feature of prostate tumors. Then, targeting hypoxia looks as a very appealing complementary strategy for the management of aggressive prostate cancers.

The 'hypoxia-response' signaling system up-regulates the expression of a a wide spectrum of effectors that increase the ability of tumor cells to turn poor oxygenation into survival advantage (Anastasiadis et al. 2003) and radio- and chemoresistance (Zhou et al. 2006).

In response to the decrease in the micro-environmental oxygen (Ellis et al. 2009) HIF-1α regulates gene expression of several genes involved in multiple physiological responses, such as erythropoiesis and glycolysis (short term solutions) and angiogenesis (long term solution) (Semenza 1998), via the expression of VEGF (Delongchamps et al. 2006).

VEGF, is "the" angiogenic cytokine, driving endothelial cell proliferation and migration, and vessel assembly (Fong et al. 1995). The expression of HIF1, AR and VEGF expression are tightly correlated (Boddy et al. 2005; Banham et al. 2007). So, AR regulates angiogenesis in androgen-sensitive PC through the HIF1-induced VEGF increase (Boddy et al. 2005). Following androgen-deprivation, the intracellular reactive oxygen species induce, instead, the direct up-regulation of VEGF-C, which favor AR transactivation mediated by the AR co-activator BAG-1L

(Rinaldo et al. 2007). Clonal selection for cells with higher expression of HIF-1a and/or apoptotic resistance pathways contributes to determine cell specific responses to hypoxia (Zhou et al. 2006).

HIF-1α over-expression/hyperfunction may be induced by genetic loss of expression/function of pVHL (Ivan and Kaelin 2001, p 53; An et al. 1998) and/or PTEN (Zundel et al. 2000) leading to the activation of PI3K/AKT/mTOR pathway (Zundel et al. 2000; Stiehl et al. 2002) which, in turn, plays a well-known role in proliferation, survival and metastatic ability of hormone independent prostate cancers, as demonstrated by the correlation between elevated phosphorylated AKT and high Gleason grade of PC (Yuan et al. 2007). HIF-1α may be induced, in addition, by several cytokines and growth factors including insulin (Zelzer et al. 1998), insulin like growth factors (Feldser et al. 1999), P42/44 mitogen activated kinase (MAPK) (Richard et al. 1999).

Furthermore, alternative mechanisms to those mentioned above have been identified which include the tumor microenvironment (Weidemann and Johnson 2008) and mutations within the ODD domain of HIF-1α (Mabjeesh and Amir 2007). This eloquently explain why targeting only one member of the hypoxia-related angiogenetic pathway is insufficient to permanently inhibit tumor angiogenesis and why tumor cells, treated with a mono-drug therapy, develop resistance by selection of 'hypoxia resistant' cells or by activating alternate angiogenic pathways (Kerbel and Folkman 2002).

To further complicate this scenario, under severe hypoxia, radio-/chemo-resistance and clonal selection may develop as a response of opposing signals delivered by survival and death pathways that allow selection of cells that have a growth advantage either genetically or epigenetically determined (Zhou et al. 2006).

Post-translational epigenetic modifications, including acetylation mediated by histone acetyltransferases (HATS) and deacetylation by histone deacetylases (HDACs), have been shown to be critical to HIF-1α signaling (Ellis et al. 2009).

HIF-1α signaling up-regulates the activity of HDACs. Then, HDAC inhibitors are emerging as a new class of anti-angiogenetic cancer therapeutics.

However, the anti-angiogenic properties of HDACI have been associated also with the alteration of numerous pro- and anti-angiogenic genes (Liu et al. 2006) other than HIF-1α and VEGF. They encompass FGF, *angiopoietin*, tunica intima endothelial kinase 2 (*TIE2*), endothelial nitric oxide synthase (*eNOS*) (Qian et al. 2004; Rossig et al. 2002, p 53), pVHL, thrombospondin 1 (Kim et al. 2001; Kwon et al. 2002; Sasakawa et al. 2003; Kang et al. 2008, p 21)$^{WAF1/CIP1}$, and survivin (Qian et al. 2004). Even for HDACI, monotherapy has shown limited responses in the clinic, but it seems very promising, as a part of combination therapies. Pre-clinical and clinical studies indicate that HDACI have positive effects on the expression of pro- and anti-angiogenic genes, suggesting their useful role in reinforcing the actions of anti-VEGF therapies.

As it has been largely shown in the past decade, the intricate molecular cross-talks underlying the malignant phenotype and the emergence of androgen-independent prostate tumors encompass the expression and functional defects in HR, single-strand break- (DNA-ssb) repair, MMR and base-excision repair (BER).

Prostate cancer cell lines (Chen et al. 2003; Yeh et al. 2001) have been found to be defective in mismatch–repair (MMR), and up to 23 % of prostate cancers display a high level of microsatellite instability associated with mutations in MMR genes and deficient MMR protein expression (Norris et al. 2007; Prtilo et al. 2005; Sun et al. 2006).

This, in turn, may lead to high mutation rates among microsatellites, ending in a mutator phenotype.

As well, DNA polymorphisms in BER- or DNA-ssb repair associated Xrcc1, Ogg1 and DNA polymerase-b genes have been associated with higher risk for prostate cancer (Chen et al. 2003; Rybicki et al. 2004; Xu et al. 2002).

Most of the molecular therapies targeting the key control pathways involved in prostate cancerogenesis and progression may indirectly influence the DNA-dsb repair activity of neoplastic cells. This has been observed for therapies inhibiting EGFR, IGFR, HDAC and proteasome pathways (Ma et al. 2003; Chinnaiyan et al. 2005; Dittmann et al. 2005a, b) have documented that the radiosensitization mediated by inhibiting EGFR can be related with the altered DNA-PKcs expression, function and cytoplasmic sequestration; as well, an increased DNA-dsbs (Rochester et al. 2005) has been shown to be induced by the inhibition of IGF-1R, via the altered ATM activation, and could be used in combination with radiotherapy (Choudhury et al. 2006) in hypoxic cancers. By converse, the RAS-mediated tumor cell radioresistance could be linked to the augmented DNA-dsb repair (Chang et al. 2005) induced by the use of farnesyl transferase inhibitors (FTIs) via the increase of Ku80 expression.

An additional crosstalk involves DNA-repair genes in human cancers. It concerns the occurrence of silencing DNA repair genes such as *MLH1* and *O*-6-methylguanine-DNA methyltransferase (*MGMT*) leading to microsatellite instability and a failure to repair DNA lesions (Jones and Baylin 2002). This phenomenon is still a matter of investigation in PC.

Another intriguing example of molecular co-sharing in PC is represented by the c-kit receptor (Pittoni et al. 2011). c-Kit receptor is normally expressed by prostate stem cells, that apparently require the c-Kit signaling for prostate regeneration (Leong et al. 2008) in humans, after hormone ablation, c-Kit expression may be observed in a considerable percentage of high-risk prostate cancer cells (Di Lorenzo et al. 2004) and in 10–30 % of this subset of prostate cancers NE differentiation occurs.

In normal prostate, a resident stromal population of mast-cells (MCs), also express the c-kit receptor (Leong et al. 2008).

It was initially thought that MCs can promote tumor growth of WD adenocarcinoma synthetizing MMP-9. MMP-9 has been indicated to correlate with progression of prostate tumor in humans (Castellano et al. 2008). As an ECM-degrading enzyme, it facilitates cell migration and invasion of tumor cells, allowing also the cleavage and activation of growth factors concurrently acting in tumor progression. In addition, peritumoral MCs were shown to stimulate prostate tumor growth in rats by providing proangiogenic factors (Johansson et al. 2010).

For these reasons, it was hypothesized that targeting MCs would be considered useful to counteract the growth of prostate cancers.

However, it was demonstrated, both in mouse and in humans, that poorly-differentiated prostate tumors with features of EMT show an autocrine production of MMP-9 and are devoid of infiltrating MCs.

This implies that MC inactivation would be ineffective when used in therapy for advanced and poorly-differentiated PC, and lead to the intriguing consideration that MC may contribute to the maintenance of prostate stem cell homeostasis and counteract NE tumor formation, serving as "natural decoys" that sequester stimulating growth factors, thus limiting c-Kit signaling in prostate cancer stem cells (Pittoni et al. 2011).

This hypothesis strongly discourages the idea of MC inhibition in PC. Otherwise, the therapeutic use of c-Kit tyrosine kinase inhibitors, such as imatinib, would instead offer the advantage of targeting both adenocarcinoma-promoting MCs (stroma targeting) and NE tumor variants (tumor targeting (Pittoni et al. 2010).

Overall, prostate cancerogenesis emerges as an extremely complex field, involving genetic and epigenetic alterations with multiple layers of merging.

We are still far away to use molecular classifications to unequivocally define different prognostic subcategories of prostate cancer. Key questions remain to be answered before the full range of mutations and genetic alterations will be elucidated.

Nevertheless, it is likely that, as for most of human cancers, also in PC the genes with a high incidence of mutation frequently participate to the evenience of abnormal epigenetic events, and this co-occurrence may be related, for instance, to the abnormal expansion of neoplastic stem cell population (Coussens and Werb 2002; Meng and Riordan 2006) which, in turn, may further select the addiction of oncogenic gene mutations, which drive PC cells to invasion, metastasis, and resistance to therapy.

Our knowledge of the intricated cross-links between genetic and epigenetic events occurring in PC has registered exciting progresses during the last decade.

The role of the three-dimensional texture and regulation of chromatin function in PC has been partially uncovered, and this has led to hypothesize the therapeutic use of drugs and small molecules such as HDAC inhibitors or DNA methylating and demethylating agents, acting as epigenetic modulators, as alternative or complementary tools for fighting aggressive PC (Murr 2010). The emerging data confirm that prostate carcinogenesis bases upon a definite group of interconnecting key signaling pathways.

Large scales of studies and carefully designed clinical trials will be required to validate novel effective therapeutic strategies for the treatment of PC.

The availability of next-generation sequencing will provide us with a broad genotyping platform which contribute to further refine our notions, shortening the time occurring to set-up multi-faceted molecular strategies tailored against the multiple molecular alterations responsible for the killing ability of advanced, androgen-independent, prostate cancers (Dias-Santagata et al. 2010) (Fig. 8.1).

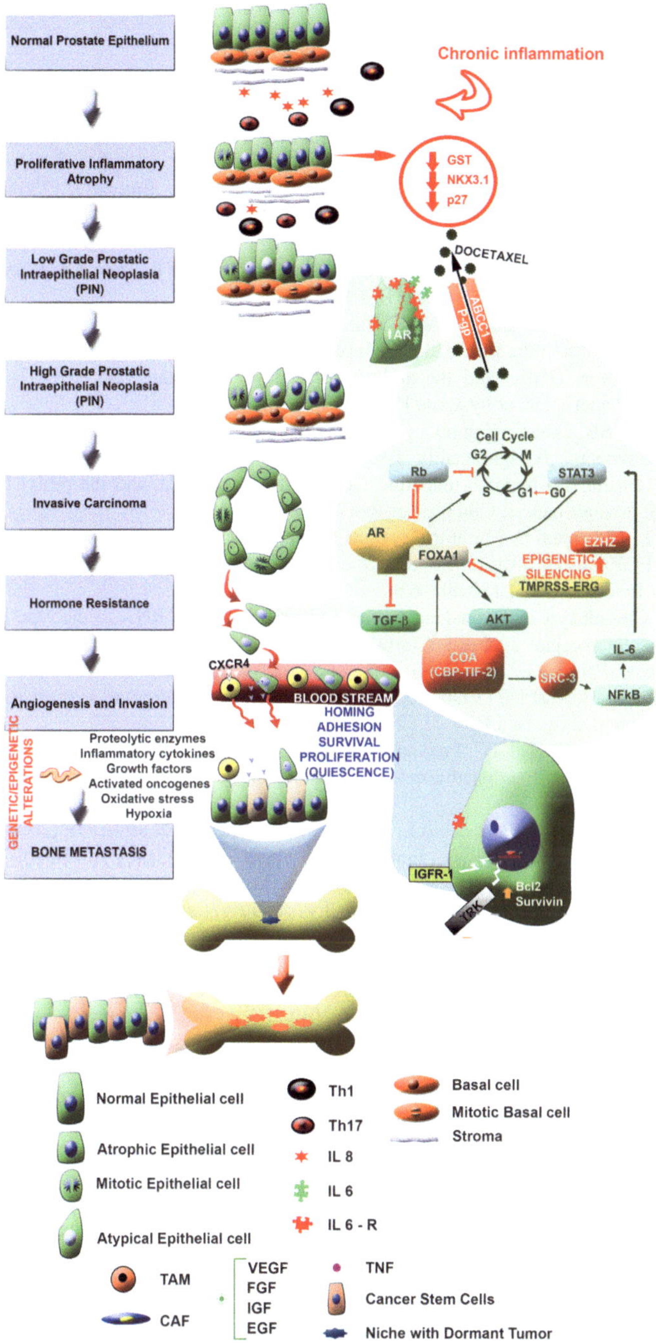

Fig. 8.1 Crossroads of molecular pathways involved in PCa. In this picture the most frequent intersections between the major molecular pathways promoting growth, survival, invasiveness, metastasis and therapy-resistance of PC cells, potentially targetable for therapeutic strategies are summarized

References

Akhurst RJ, Derynck R (2001) TGF-beta signaling in cancer – a double-edged sword. Trends Cell Biol 11:S44–S51

An WG et al (1998) Stabilization of wild-type p53 by hypoxia-inducible factor 1alpha. Nature 392(6674):405–408

Anastasiadis AG, Bemis DL, Stisser BC, Salomon L, Ghafar MA, Buttyan R (2003) Tumor cell hypoxia and the hypoxia-response signaling system as a target for prostate cancer therapy. Curr Drug Target 4:191–196

Attar RM, Takimoto CH, Gottardis MM (2009) Castration-resistant prostate cancer: locking up the molecular escape routes. Clin Cancer Res 15:3251–3255

Banham AH, Boddy J, Launchbury R, Han C, Turley H, Malone PR, Harris AL, Fox SB (2007) Expression of the forkhead transcription factor FOXP1 is associated both with hypoxia inducible factors (HIFs) and the androgen receptor in prostate cancer but is not directly regulated by androgens or hypoxia. Prostate 67:1091–1098

Boddy JL, Fox SB, Han C, Campo L, Turley H, Kanga S, Malone PR, Harris AL (2005) The androgen receptor is significantly associated with vascular endothelial growth factor and hypoxia sensing via hypoxia-inducible factors HIF-1a, HIF-2a, and the prolyl hydroxylases in human prostate cancer. Clin Cancer Res 11:7658–7663

Bonaccorsi L, Marchiani S, Muratori M, Forti G, Baldi E (2004) Gefitinib ('IRESSA', ZD1839) inhibits EGFinduced invasion in prostate cancer cells by suppressing PI3 K/AKT activation. J Cancer Res Clin Oncol 130:604–614

Burfeind P, Chernicky CL, Rininsland F, Ilan J (1996) Antisense RNA to the type I insulin-like growth factor receptor suppresses tumor growth and prevents invasion by rat prostate cancer cells *in vivo*. PNAS 93:7263–7268

Byrne RL, Leung H, Neal DE (1996) Peptide growth factors in the prostate as mediators of stromal epithelial interaction. Br J Urol 77:627–633

Camps JL, Chang SM, Hsu TC, Freeman MR, Hong SJ, Zhau HE, von Eschenbach AC, Chung LW (1990) Fibroblastmediated acceleration of human epithelial tumor growth *in vivo*. PNAS 87:75–79

Cardillo MR, Petrangeli E, Perracchio L, Salvatori L, Ravenna L, Di Silverio F (2000) Transforming growth factor-beta expression in prostate neoplasia. Anal Quant Cytol Histol 22:1–10

Castellano G, Malaponte G, Mazzarino MC, Figini M, Marchese F, Gangemi P et al (2008) Activation of the osteopontin/matrix metalloproteinase-9 pathway correlates with prostate cancer progression. Clin Cancer Res 14:7470–7480

Chang IY, Youn CK, Kim HB et al (2005) Oncogenic H-Ras upregulates expression of Ku80 to protect cells from gamma-ray irradiation in NIH3T3 cells. Cancer Res 65:6811–6819

Chen L, Elahi A, Pow-Sang J, Lazarus P, Park J (2003) Association between polymorphism of human oxoguanine glycosylase 1 and risk of prostate cancer. J Urol 170:2471–2474

Chinnaiyan P, Vallabhaneni G, Armstrong E, Huang SM, Harari PM (2005) Modulation of radiation response by histone deacetylase inhibition. Int J Radiat Oncol Biol Phys 62:223–229

Chipuk JE, Cornelius SC, Pultz NJ, Jorgensen JS, Bonham MJ, Kim SJ, Danielpour D (2002) The androgen receptor represses transforming growth factor-beta signaling through interaction with Smad3. J Biol Chem 277:1240–1248

Cho-Chung YS (1989) Site-selective 8-chloro- cyclic adenosine 3′,5′-monophosphate as a biologic modulator of cancer: restoration of normal control mechanisms. J Natl Cancer Inst 81:982–987

Cho-Chung YS (1990) Role of cyclic AMP receptor proteins in growth, differentiation, and suppression of malignancy: new approaches to therapy. Cancer Res 50:7093–7100

Choudhury A, Cuddihy A, Bristow RG (2006) Radiation and new molecular agents part I: targeting ATM-ATR checkpoints, DNA repair, and the proteasome. Semin Radiat Oncol 16:51–58

Coffey RJ Jr, Shipley GD, Moses HL (1986) Production of transforming growth factors by human colon cancer lines. Cancer Res 46:1164–1169

Coussens LM, Werb Z (2002) Inflammation and cancer. Nature 420:860–867

Craft N, Shostak Y, Carey M, Sawyers CL (1999) A mechanism for hormone-independent prostate cancer through modulation of androgen receptor signaling by the HER-2/neu tyrosine kinase. Nat Med 5:280–285
Culig Z, Hobisch A, Cronauer MV, Radmayr C, Trapman J, Hittmair A, Bartsch G, Klocker H (1994) Androgen receptor activation in prostatic tumor cell lines by insulinlike growth factor-I, keratinocyte growth factor, and epidermal growth factor. Cancer Res 54:5474–5478
Cunha GR, Donjacour AA (1989) Mesenchymal–epithelial interactions in the growth and development of the prostate. Cancer Treat Res 46:159–175
Cunha GR, Hayward SW, Dahiya R, Foster BA (1996) Smooth muscle–epithelial interactions in normal and neoplastic prostatic development. Acta Anat 155:63–72
Dayyani F, Gallick GE, Logothetis CJ, Corn PG (2011) Novel therapies for metastatic castrate-resistant prostate cancer. J Natl Cancer Inst 103(22):1665–1675
Delongchamps NB, Peyromaure M, Dinh-Xuan AT (2006) Role of vascular endothelial growth factor in prostate cancer. Urology 68:244–248
Derynck R, Zhang YE (2003) Smad-dependent and Smadindependent pathways in TGF-beta family signalling. Nature 425:577–584
Di Lorenzo G, Tortora G, D'Armiento FP, De Rosa G, Staibano S, Autorino R, D'Armiento M, De Laurentiis M, De Placido S, Catalano G et al (2002) Expression of epidermal growth factor receptor correlates with disease relapse and progression to androgen-independence in human prostate cancer. Clin Cancer Res 8:3438–3444
Di Lorenzo G, Autorino R, D'Armiento FP, Mignogna C, De Laurentiis M, De Sio M et al (2004) Expression of proto-oncogene c-kit in high risk prostate cancer. Eur J Surg Oncol 30:987–992
Dias-Santagata D, Akhavanfard S, David SS, Vernovsky K, Kuhlmann G, Boisvert SL, Stubbs H, McDermott U, Settleman J, Kwak EL, Clark JW, Isakoff SJ, Sequist LV, Engelman JA, Lynch TJ, Haber DA, Louis DN, Ellisen LW, Borger DR, Iafrate AJ (2010) Rapid targeted mutational analysis of human tumours: a clinical platform to guide personalized cancer medicine. EMBO Mol Med 2(5):146–158
Dittmann K, Mayer C, Fehrenbacher B et al (2005a) Radiationinduced epidermal growth factor receptor nuclear import is linked to activation of DNA-dependent protein kinase. J Biol Chem 280:31182–31189
Dittmann K, Mayer C, Rodemann HP (2005b) Inhibition of radiationinduced EGFR nuclear import by C225 (Cetuximab) suppresses DNA-PK activity. Radiother Oncol 76:157–161
El-Amm J, Aragon-Ching JB (2013) The changing landscape in the treatment of metastatic castration resistant prostate cancer. Ther Adv Med Oncol 5(1):25–40
Ellis L, Hammers H, Pili R (2009) Targeting tumor angiogenesis with histone deacetylase inhibitors. Cancer Lett 280(2):145–153. Review
Feldser D et al (1999) Reciprocal positive regulation of hypoxia-inducible factor 1alpha and insulin-like growth factor 2. Cancer Res 59(16):3915–3918
Festuccia C, Muzi P, Millimaggi D, Biordi L, Gravina GL, Speca S, Angelucci A, Dolo V, Vicentini C, Bologna M (2005) Molecular aspects of gefitinib antiproliferative and pro-apoptotic effects in PTEN-positive and PTENnegative prostate cancer cell lines. Endocr-Relat Cancer 12:983–998
Fong GH, Rossant J, Gertsenstein M, Breitman ML (1995) Role of the Flt-1 receptor tyrosine kinase in regulating the assembly of vascular endothelium. Nature 376:66–70
Gerdes MJ, Dang TD, Larsen M, Rowley DR (1998) Transforming growth factor-beta1 induces nuclear to cytoplasmic distribution of androgen receptor and inhibits androgen response in prostate smooth muscle cells. Endocrinology 139:3569–3577
Gerdes MJ, Larsen M, Dang TD, Ressler SJ, Tuxhorn JA, Rowley DR (2004) Regulation of rat prostate stromal cell myodifferentiation by androgen and TGF-beta1. Prostate 58:299–307
Gonzalez-Herrera IG, Prado-Lourenco L, Pileur F, Conte C, Morin A, Cabon F, Prats H, Vagner S, Bayard F, Audigier S et al (2006) Testosterone regulates FGF-2 expression during testis maturation by an IRES-dependent translational mechanism. FASEB J 20:476–478
Heinlein CA, Chang C (2004) Androgen receptor in prostate cancer. Endocr Rev 25:276–308

Henshall SM, Quinn DI, Lee CS, Head DR, Golovsky D, Brenner PC, Delprado W, Stricker PD, Grygiel JJ, Sutherland RL (2001) Altered expression of androgen receptor in the malignant epithelium and adjacent stroma is associated with early relapse in prostate cancer. Cancer Res 61:423–427

Horii K, Suzuki Y, Kondo Y, Akimoto M, Nishimura T, Yamabe Y et al (2007) Androgen-dependent gene expression of prostatespecific antigen is enhanced synergistically by hypoxia in human prostate cancer cells. Mol Cancer Res 5:383–391

Ivan M, Kaelin WG Jr (2001) The von Hippel–Lindau tumor suppressor protein. Curr Opin Genet Dev 11(1):27–34

Johansson A, Rudolfsson S, Hammarsten P, Halin S, Pietras K, Jones J et al (2010) Mast cells are novel independent prognostic markers in prostate cancer and represent a target for therapy. Am J Pathol 177:1031–1041

Jones PA, Baylin SB (2002) The fundamental role of epigenetic events in cancer. Nat Rev Genet 3:415–428

Kang JH et al (2008) CCAAT box is required for the induction of human thrombospondin-1 gene by trichostatin A. J Cell Biochem 104(4):1192–1203

Kerbel R, Folkman J (2002) Clinical translation of angiogenesis inhibitors. Nat Rev Cancer 2(10):727–739

Kim MS et al (2001) Histone deacetylases induce angiogenesis by negative regulation of tumor suppressor genes. Nat Med 7(4):437–443

Kwon HJ et al (2002) Histone deacetylase inhibitor FK228 inhibits tumor angiogenesis. Int J Cancer 97(3):290–296

Leong KG, Wang BE, Johnson L, Gao WQ (2008) Generation of a prostate from a single adult stem cell. Nature 456:804–808

Leotoing L, Manin M, Monte D, Baron S, Communal Y, Lours C, Veyssiere G, Morel L, Beaudoin C (2007) Crosstalk between androgen receptor and epidermal growth factor receptor-signalling pathways: a molecular switch for epithelial cell differentiation. J Mol Endocrinol 39:151–162

Liu Y, Majumder S, McCall W, Sartor CI, Mohler JL, Gregory CW, Earp HS, Whang YE (2005) Inhibition of HER-2/neu kinase impairs androgen receptor recruitment to the androgen responsive enhancer. Cancer Res 65:3404–3409

Liu T et al (2006) Histone deacetylase inhibitors: multifunctional anticancer agents. Cancer Treat Rev 32(3):157–165

Ma BB, Bristow RG, Kim J, Siu LL (2003) Combined-modality treatment of solid tumors using radiotherapy and molecular targeted agents. J Clin Oncol 21:2760–2776

Mabjeesh NJ, Amir S (2007) Hypoxia-inducible factor (HIF) in human tumorigenesis. Histol Histopathol 22(5):559–572

Mantalaris A, Panoskaltsis N, Sakai Y et al (2001) Localization of androgen receptor expression in human bone marrow. J Pathol 193:361–366

Marignol L, Coffey M, Lawler M, Hollywood D (2008) Hypoxia in prostate cancer: a powerful shield against tumour destruction? Cancer Treat Rev 34(4):313–327

Mellinghoff IK, Vivanco I, Kwon A, Tran C, Wongvipat J, Sawyers CL (2004) HER2/neu kinase-dependent modulation of androgen receptor function through effects on DNA binding and stability. Cancer Cell 6:517–527

Memarzadeh S, Xin L, Mulholland DJ, Mansukhani A, Wu H, Teitell MA, Witte ON (2007) Enhanced paracrine FGF10 expression promotes formation of multifocal prostate adenocarcinoma and an increase in epithelial androgen receptor. Cancer Cell 12:572–585

Meng X, Riordan NH (2006) Cancer is a functional repair tissue. Med Hypotheses 66:486–490

Murr R (2010) Interplay between different epigenetic modifications and mechanisms. Adv Genet 70:101–141

Nakano K, Fukabori Y, Itoh N, Lu W, Kan M, McKeehan WL, Yamanaka H (1999) Androgen-stimulated human prostate epithelial growth mediated by stromal-derived fibroblast growth factor-10. Endocr J 46:405–413

Niu Y, Xu Y, Zhang J, Bai J, Yang H, Ma T (2001) Proliferation and differentiation of prostatic stromal cells. BJU Int 87:386–393

Niu Y, Chang TM, Yeh S, Ma WL, Wang YZ, Chang C (2010) Differential androgen receptor signals in different cells explain why androgen-deprivation therapy of prostate cancer fails. Oncogene 29:3593–3604

Norris AM, Woodruff RD, D'Agostino RB Jr, Clodfelter JE, Scarpinato KD (2007) Elevated levels of the mismatch repair protein PMS2 are associated with prostate cancer. Prostate 67:214–225

Ohlson N, Bergh A, Stattin P, Wikstrom P (2007) Castrationinduced epithelial cell death in human prostate tissue is related to locally reduced IGF-1 levels. Prostate 67:32–40

Orio F Jr, Terouanne B, Georget V, Lumbroso S, Avances C, Siatka C, Sultan C (2002) Potential action of IGF-1 and EGF on androgen receptor nuclear transfer and transactivation in normal and cancer human prostate cell lines. Mol Cell Endocrinol 198:105–114

Ornitz DM, Itoh N (2001) Fibroblast growth factors. Genome Biol 2(3):REVIEWS3005. Epub 2001 Mar 9. Review. PubMed PMID: 11276432; PubMed Central PMCID: PMC138918

Peterziel H, Mink S, Schonert A, Becker M, Klocker H, Cato AC (1999) Rapid signalling by androgen receptor in prostate cancer cells. Oncogene 18:6322–6329

Pittoni P, Piconese S, Tripodo C, Colombo MP (2010) Tumor-intrinsic and -extrinsic roles of c-Kit: mast cells as the primary off-target of tyrosine kinase inhibitors. Oncogene 30:757–769

Pittoni P, Tripodo C, Piconese S, Mauri G, Parenza M, Rigoni A, Sangaletti S, Colombo MP (2011) Mast cell targeting hampers prostate adenocarcinoma development but promotes the occurrence of highly malignant neuroendocrine cancers. Cancer Res 71(18):5987–5997

Pollak M, Beamer W, Zhang JC (1998) Insulin-like growth factors and prostate cancer. Cancer Metastasis Rev 17:383–390

Prtilo A, Leach FS, Markwalder R et al (2005) Tissue microarray analysis of hMSH2 expression predicts outcome in men with prostate cancer. J Urol 174:1814–1818, discussion p. 18

Qian DZ et al (2004) The histone deacetylase inhibitor NVP-LAQ824 inhibits angiogenesis and has a greater antitumor effect in combination with the vascular endothelial growth factor receptor tyrosine kinase inhibitor PTK787/ZK222584. Cancer Res 64(18):6626–6634

Richard DE et al (1999) P42/p44 mitogen-activated protein kinases phosphorylate hypoxia-inducible factor 1alpha (HIF-1alpha) and enhance the transcriptional activity of HIF-1. J Biol Chem 274(46):32631–32637

Rinaldo F, Li J, Wang E, Muders M, Datta K (2007) RalA regulates vascular endothelial growth factor-C (VEGF-C) synthesis in prostate cancer cells during androgen ablation. Oncogene 26:1731–1738

Roberts AB, Sporn MB, Assoian RK, Smith JM, Roche NS, Wakefield LM, Heine UI, Liotta LA, Falanga V, Kehrl JH et al (1986) Transforming growth factor type beta: rapid induction offibrosis and angiogenesis in vivo and stimulation of collagen formation in vitro. PNAS 83:4167–4171

Rochester MA, Riedemann J, Hellawell GO, Brewster SF, Macaulay VM (2005) Silencing of the IGF1R gene enhances sensitivity to DNA-damaging agents in both PTEN wild-type and mutant human prostate cancer. Cancer Gene Ther 12:90–100

Rosini P, Bonaccorsi L, Baldi E, Chiasserini C, Forti G, De Chiara G, Lucibello M, Mongiat M, Iozzo RV, Garaci E et al (2002) Androgen receptor expression induces FGF2, FGF-binding protein production, and FGF2 release in prostate carcinoma cells: role of FGF2 in growth, survival, and androgen receptor down-modulation. Prostate 53:310–321

Rossig L et al (2002) Inhibitors of histone deacetylation downregulate the expression of endothelial nitric oxide synthase and compromise endothelial cell function in vasorelaxation and angiogenesis. Circ Res 91(9):837–844

Russell PJ, Bennett S, Stricker P (1998) Growth factor involvement in progression of prostate cancer. Clin Chem 44:705–723

Rybicki BA, Conti DV, Moreira A, Cicek M, Casey G, Witte JS (2004) DNA repair gene XRCC1 and XPD polymorphisms and risk of prostate cancer. Cancer Epidemiol Biomark Prev 13:23–29

Salm SN, Koikawa Y, Ogilvie V, Tsujimura A, Coetzee S, Moscatelli D, Moore E, Lepor H, Shapiro E, Sun TT et al (2000) Generation of active TGF-beta by prostatic cell cocultures using novel basal and luminal prostatic epithelial cell lines. J Cell Physiol 184:70–79
Sar M, Lubahn DB, French FS, Wilson EM (1990) Immunohistochemical localization of the androgen receptor in rat and human tissues. Endocrinology 127:3180–3186
Saric T, Shain SA (1998) Androgen regulation of prostate cancer cell FGF-1, FGF-2, and FGF-8: preferential downregulation of FGF-2 transcripts. Growth Fact 16:69–87
Sarwar M, Persson JL (2011) The protein kinase a (PKA) intracellular pathway and androgen receptor: a novel mechanism underlying the castration-resistant and metastatic prostate cancer. J Cancer Sci Ther S5:003
Sasakawa Y et al (2003) Antitumor efficacy of FK228, a novel histone deacetylase inhibitor, depends on the effect on expression of angiogenesis factors. Biochem Pharmacol 66(6):897–906
Semenza GL (1998) Hypoxia-inducible factor 1: master regulator of O2 homeostasis. Curr Opin Genet Dev 8(5):588–594
Siegel R, Naishadham D, Jemal A (2012) Cancer statistics, 2012. CA Cancer J Clin 62(10):29
Sporn MB, Roberts AB (1992) Transforming growth factorbeta: recent progress and new challenges. J Cell Biol 119:1017–1021
Stiehl DP et al (2002) Normoxic induction of the hypoxia-inducible factor 1alpha by insulin and interleukin-1beta involves the phosphatidylinositol 3-kinase pathway. FEBS Lett 512(1–3):157–162
Sun X, Chen C, Vessella RL, Dong JT (2006) Microsatellite instability and mismatch repair target gene mutations in cell lines and xenografts of prostate cancer. Prostate 66:660–666
Wang H, Song K, Sponseller TL, Danielpour D (2005) Novel function of androgen receptor-associated protein 55/Hic-5 as a negative regulator of Smad3 signaling. J Biol Chem 280:5154–5162
Wang X, Yin L, Rao P, Stein R, Harsch KM, Lee Z, Heston WD (2007) Targeted treatment of prostate cancer. J Cell Biochem 102:571–579
Weidemann A, Johnson RS (2008) Biology of HIF-1alpha. Cell Death Differ 15(4):621–627
Wolk A, Mantzoros CS, Andersson SO, Bergstrom R, Signorello LB, Lagiou P, Adami HO, Trichopoulos D (1998) Insulin-like growth factor 1 and prostate cancer risk: a population-based, case–control study. J Natl Cancer Inst 90:911–915
Wu Y, Zhao W, Zhao J, Pan J, Wu Q, Zhang Y, Bauman WA, Cardozo CP (2007) Identification of androgen response elements in the insulin-like growth factor I upstream promoter. Endocrinology 148:2984–2993
Xu J, Zheng SL, Turner A et al (2002) Associations between hOGG1 sequence variants and prostate cancer susceptibility. Cancer Res 62:2253–2257
Yeh S, Lin HK, Kang HY, Thin TH, Lin MF, Chang C (1999) From HER2/Neu signal cascade to androgen receptor and its coactivators: a novel pathway by induction of androgen target genes through MAP kinase in prostate cancer cells. PNAS 96:5458–5463
Yeh CC, Lee C, Dahiya R (2001) DNA mismatch repair enzyme activity and gene expression in prostate cancer. Biochem Biophys Res Commun 285:409–413
Yoshizawa A, Ogikubo S (2006) IGF binding protein-5 synthesis is regulated by testosterone through transcriptional mechanisms in androgen responsive cells. Endocr J 53:811–818
Yuan TC, Veeramani S, Lin MF (2007) Neuroendocrine-like prostate cancer cells: neuroendocrine transdifferentiation of prostate adenocarcinoma cells. Endocr Relat Cancer 14:531–547
Zavadil J, Bottinger EP (2005) TGF-beta and epithelial-to-mesenchymal transitions. Oncogene 24:5764–5774
Zelzer E et al (1998) Insulin induces transcription of target genes through the hypoxia-inducible factor HIF-1alpha/ARNT. EMBO J 17(17):5085–5094
Zhou J, Schmid T, Schnitzer S, Brüne B (2006) Tumor hypoxia and cancer progression. Cancer Lett 237(1):10–21
Zhu B, Kyprianou N (2005) Transforming growth factor beta and prostate cancer. Cancer Treat Res 126:157–173

Zhu ML, Kyprianou N (2008) Androgen receptor and growth factor signaling cross-talk in prostate cancer cells. Endocr Relat Cancer 15(4):841–849
Zhu B, Fukada K, Zhu H, Kyprianou N (2006) Prohibitin and cofilin are intracellular effectors of transforming growth factor beta signaling in human prostate cancer cells. Cancer Res 66:8640–8647
Zundel W et al (2000) Loss of PTEN facilitates HIF-1-mediated gene expression. Genes Dev 14(4):391–396

Part III
Genetic and Epigenetic Events in Prostate Cancer

Chapter 9
Gene Polymorphisms

Maria Siano, MariaLuisa Vecchione, and Gennaro Ilardi

Abstract Over the last decade, the explosive advances in sequencing and genotyping technologies fueled by major financial investments in basic science, have evidenced that hundreds of genes harboring variations contribute to human cancers and that genetic variability may influence patients' responses to post-surgical treatments (Hamburg and Collins, N Engl J Med 363(4):301–304, 2010).

Several studies have reported the association between one or multiple single nucleotide polymorphisms (SNPs) in multiple pathways linked to prostate cancer onset and progression (Sfanos and De Marzo, Histopathology (1):199–215, 2012).

For instance, a SNP (GG genotype) in the promoter region of alpha-1-antichymotrypsin (ACT), an acute-phase protein up-regulated in inflammatory conditions, has been reported as linked to the increased risk of prostate cancer (Licastro et al., Anticancer Res 28:395–399, 2008). Moreover, a correlation between circulating levels of PSA and the ACT GG genotype was reported in younger prostate cancer patients too. ACT is of particular interest because is bound, in men, to most of circulating PSA.

As well, the association between increased risk for prostate cancer or aggressive prostate cancer and the IL10–1082GG variant of IL-10 has been recently reported (Zabaleta et al., Carcinogenesis 29:573–578, 2008; Zabaleta et al., Carcinogenesis 30:1358–1362, 2009).

In addition, SNPs between multiple different cytokines have been investigated as a potential source of increased risk of prostate cancer (Zabaleta et al., Carcinogenesis 29:573–578, 2008; Zabaleta et al., Carcinogenesis 30:1358–1362, 2009; Kwon et al., Cancer Epidemiol Biomarkers Prev 20:923–933, 2011).

M. Siano (✉) • M. Vecchione • G. Ilardi
Department of Advanced Biomedical Sciences, Pathology Section, Faculty of Medicine and Surgery, University of Naples "Federico II", via S. Pansini, n.5, Naples, Italy
e-mail: dysian@tin.it; marialuisa.vecchione@unina.it; gennaro.ilardi@gmail.com

S. Staibano (ed.), *Prostate Cancer: Shifting from Morphology to Biology*,
DOI 10.1007/978-94-007-7149-9_9,

143 SNPs in 16 inflammation-related genes [CXC ligand 12 (CXCL12), IL-4, IL-6, IL-6ST, prostaglandin-endoperoxide synthase 2 (PTGS2), signal transducer and activator of transcription 3 (STAT3), TNF, protein kinase B (AKT1), CXCR4, IL-6R, IL-8, IL-10, nuclear factor kappa B (NFκB), phosphatidylinositol 3-kinase (PIK3)R1, PTGS1 and vascular endothelial growth factor (VEGF)] have been examined in a case–control study of African American versus Caucasian men (Kwon et al., Cancer Epidemiol Biomarkers Prev 20:923–933, 2011).

SNPs in IL-4, IL-6ST, PTGS2 and STAT3 resulted independently associated with prostate cancer susceptibility, while SNPs in AKT1, PIK3R1 and STAT3 were associated with aggressive prostate cancer. Overall, men carrying multiple 'high-risk' alleles have been found at an elevated risk for prostate cancer development. These studies strongly support the importance of inflammatory pathways in conferring prostate cancer risk.

A multitude of emerging findings support the increasing efforts of the scientific community to look at the molecular background of prostate cancer as the key brick to develop new therapies targeted to the patient, for its "molecularly unique" tumor.

Obviously, the ultimate goal of this exciting line of research will be to develop genetic tests readily and safely transferable to clinical use for the diagnosis and prediction of patients' responses to therapy.

The way to reach this objective is sprinkled by many obstacles. We are still unaware of the real clinical significance of several genetic markers and the available data concerning the side-effects of the existing gene-based therapies are far away to be conclusive.

Nevertheless, the endlessly mounting reports on this topic are very encouraging, for their promising relevance in prostate cancer patients therapeutic management.

Prostate cancer (PCa) is the most commonly diagnosed cancer in men in the United States, with estimated 241,740 new cases and 28,170 deaths, representing the second leading cause of cancer deaths (Siegel et al. 2012). In the last decades, its incidence has been increasing (AIHW – Australian Institute of Health and Welfare and AACR-Australasian Association of Cancer Registries 2007; Parkin et al. 2003). The use of prostate specific antigen (PSA) for screening and the availability of extended pattern biopsy techniques have allowed us to diagnose the majority of tumors in an early stage, being of low grade, limited volume and clinically localized (Cooperberg et al. 2003). However, etiology remains still poorly understood. Several risk factors have been reported but the only well-established are advanced age, family history, with a risk of disease in first-degree relatives of cases approximately two fold greater that in the general population (Carter et al. 2006; Edwards and Eeles 2004; Johns and Houlston 2003), and racial/ethnic background (Crawford 2006). The reason for the large variation among different ethnic groups, with the highest reported rates in african-americans and the lowest in Chinese (Ritchey et al. 2005), is until unknown, but inherited susceptibility, environmental exposure, lifestyle and hormonal levels have been involved (Mononen et al. 2006; Chang et al. 2011).

Actually, prognostic prevision for patients with PCa is based upon clinical and pathological measures, such as PSA levels, Gleason's grade and tumor stage, that are incorporated in a variety of nomograms helping to predict a poor prognosis and the potential for recurrence and/or death for disease (Shariat et al. 2008; Cheng et al. 2010). However, despite these prognostic measures, most men diagnosed with PCa undergo radical prostatectomy or radiotherapy, with important effects on the quality of life (Cheng et al. 2010). In fact, while the PCa incidence continues to increase, due to the ability to diagnose the tumor in a very early stage, the death rate is unvariated (Ries et al. 2000), indicating that indolent form of the neoplasia are increasingly diagnosed and treated (Gallagher et al. 2010), highlighting the importance of sparing men from overtreatment. The aggressiveness of prostate tumors varies substantially, with some tumors remaining indolent and others becoming life-threatening, suggesting that individual variation contributes to tumor aggressiveness. Therefore, it is mandatory to identify specific additional markers, to integrate into prediction models, that could allow to distinguish tumors that will likely recur, progress and be life-threatening from those that may have not impact on morbidity or mortality, once treated (Cheng et al. 2010).

To date, prognostic models in localized cancers do not incorporate genetic susceptibility (Lughezzani et al. 2010), ignoring that it has been shown to be correlated not only with cancer risk, but also with disease aggressiveness (Cornu et al. 2011; Burmester et al. 2004). For example, germline mutations correlate not only with PC risk, but also with its clinical outcome (Gallagher et al. 2010). It seems to be the case for BRCA mutations, which have resulted associated with higher grade of prostate cancer and worse clinical outcome (Kote-Jarai et al. 2008; Kirchhoff et al. 2004).

Moreover, genome-wide association studies (GWAS), linkage analyses and fine-mapping techniques have shown that several single nucleotide polymorphism (SNPs) could predict PCa risk and, in few studies, aggressiveness (Fitzgerald et al. 2009; Helfand et al. 2010), providing useful information at the time of diagnosis or during initial disease.

However, Pca have a multicausal etiology with multiple risk variants (genetic and environmental) acting synergistically to influence cancer-susceptibility. Therefore, single-gene approaches could be of limited value in predicting risk and a comprehensive approach is required, such as a method that integrate multiple polymorphism involved in the same pathway. Combining multiple SNPs, could amplify the predictive power (Beuten et al. 2009).

9.1 Single Nucleotide Polymorphism in Prostate Cancer

A single nucleotide polymorphism (SNP) is defined as an inherited mutation present in more than 1 % of the population. These markers can be analyzed from any normal tissue.

Association studies and genome-wide association studies in multiple case families, attempting to identify the heritable component of prostate cancer basing on candidate gene, have suggested numerous prostate cancer susceptibility genes and loci. However, the incapability of reproducing results obtained by linkage and associations studies, suggests that PCa is genetically complex and involves multiple common low-penetrance genes in its predisposition (Schaid 2004).

GWAS, allowing to genotype hundreds of thousands of SNPs simultaneously and, therefore, to screen the complete genome for common genetic variants associated with a disease, represents a successfull method in identifying genomic low-risk susceptibility regions (Chung et al. 2010; Aly et al. 2011). The first susceptibility region identified was at chromosome 8q24 (Gudmundsson et al. 2007; Haiman et al. 2007; Zheng et al. 2007; Ghoussaini et al. 2008; Yeager et al. 2008, 2009; Al Olama et al. 2009), but many more cancer susceptibility variants have been (and many more will be) discovered.

Moreover, several studies have explored the capability of well-known prostate cancer risk variants in discerning between less aggressive and more aggressive disease (Aly et al. 2011). Overall, results are contrasting; in fact, while some studies report a strong association for some of these variants with more aggressive tumor, others do not. For example, two large replication studies, from PRACTICAL (Kote-Jarai et al. 2008) and from Fitzgerald and co. (2009), evaluated the association between a series of genetic variants and aggressiveness in PCa, but no correlation was found between any of the evaluated variant and disease behaviour. On the contrary, Bao et al. (2010) in a study evaluating 20 SNPs, previously identified associated to prostate cancer, discovered that they might have a cumulative association with disease aggressiveness.

To date, several pathways and chromosomal loci have been involved in tumor development and/or progression in prostate cancer and corresponding SNPs have been identified.

9.2 Androgen Pathway

Androgens are essential in the development, growth and secretory functions of the prostate gland (Beuten et al. 2009). Free testosterone is converted into dihydrotestosterone (DHT) that binds to the androgen receptor (AR) so forming a complex that, in association with cofactor and transcription factors, induces androgen-regulated expression of genes involved in cellular proliferation and differentiation. Risk factors can include both highly penetrant susceptibility genes and low-penetrant genes (with higher frequencies in the population). In prostate cancer cases, due to their hormone dependence, genetic alterations involving low-penetrant and moderate-penetrant alleles in genes involved in prostatic functional pathways, such as androgen hormones and their metabolizing enzymes, are the most likely natural candidates for conferring susceptibility to this cancer (Beuten et al. 2009; Mononen and Schleutker 2009). Polymorphism and variants of these genes are distributed differently across the population groups and, since variation in their DNA sequence could alter protein function and result variability in disease risk, they reach an increasing interest (Coughlin and Hall 2002). However, many studies evaluating the role of genes involved in androgen pathway have reported conflicting results, attributable to population differences and biological, technical or statistical factors (Beuten et al. 2009). Moreover, considering that the endogenous factors affecting the functional genome may be different, it is important to define the polymorphic spectrum of genes implicated in cancer causation in different populations. For several reported associations it is still uncertain whether the variants have a causal role in PCa. The majority of prostate cancer cases are unlikely to be explained by allelic variability at a single locus of major susceptibility genes while genetic polymorphism seems to be more important (Coughlin and Hall 2002). Moreover, prostate carcinogenesis is a multistep process involving a multifactorial interplay between genetic and environmental factors; as a result, the effect of individual SNPs is unlikely to be substantial and may be of limited value in predicting risk. A comprehensive approach is required, such as the pathway based multigenic method integrating multiple polymorphism that interact in the same pathway. Combining multiple low-penetrant to modestly penetrant SNPs may amplify predictive power (Beuten et al. 2009).

9.2.1 *CYP Subfamily 11A1 Enzyme*

This enzyme is located on 15q23-q24 and catalyze the first step in the androgen synthesis pathway. Polymorphism in a repeated sequence at the 5′ UTR has been associated with an increased risk of metastatic disease and high grade PCa (Kumazawa et al. 2004), but other studies have failed to confirm these results (Douglas et al. 2005; Cunningham et al. 2007).

9.2.2 *Androgen Receptor*

Androgen receptor is coded by the AR gene on Xq11.2-q12. Its transactivation domain has two highly polymorphic trinucleotide repeat segments, including CAG and GGN in exon 1. A positive correlation has been reported between GGN length and AR transcriptional activity, while an inverse correlation has been demonstrated for CAG. Longer CAG repeat length has also been described to reduce the expression of AR mRNA and protein and inhibit androgen receptor transactivation in cultured prostate epithelial cells, suggesting that a decreased number of this repeat may render prostate tissue more vulnerable to long-term androgen stimulation, leading to increased proliferative activity, which in turn may increase the rate of somatic mutations. However, several groups have tried to establish the role of the variation in CAG repeat length in regard to PCa risk (Linja and Visakorpi 2004), with conflicting results. In a meta-analysis of 19 studies, an increased risk was observed in patients with shorter CAG repeats (Zeegers et al. 2004) while Lindstrom et al. (2006), in a study in a Swedish population, obtained the opposite result. Moreover, recent observations support an etiological role of genetic interactions between polymorphism that act together with CAG repeats. The distribution of GGN length polymorphism was, instead, noted to be significantly different among racial groups, suggesting a possible association between these repeat microsatellites and PCa development. Several studies substained that risk of PCa was higher in man with a short GGN repeat, but no association was observed in more recent studies (Mononen and Schleutker 2009). A germline mutation (R726L) is located in the ligand binding domain of AR, causing altered binding and transactivation properties. This mutation was reported to increase the PCa risk but, in a recent study, this association did not persist (Mononen and Schleutker 2009).

9.2.3 Steroid 5alpha-Reductase Type II

This enzyme is encoded by a gene located on chromosome 2p23 and converts testosterone into dihydrotestosterone. The polymorphism A49T has been shown to increase the activity of the enzyme substantially *in vitro* (Makridakis et al. 1999), but have been reported data *in vivo* substaining that men with this polymorphism show lower concentration of a metabolite of DHT than men with homozygous WT allele. In detail, a significant association of this polymorphism and PCa risk for clinically advanced disease has been reported for African-American carriers, while the variant is absent in low-risk population (Hsing et al. 2001). However, a meta-analysis estimated that in European descent, the fraction of PCa attributable to A49T is less than 1 % (Ntais et al. 2003).

Another variant, V89L, is associated with decreased 5alpha-reductase activity *in vitro* and *in vivo*. It is mostly common in Chinese and Japanese and in men in Greenland, who have an extremely low incidence of PCa (Hsing et al. 2001).

The polymorphic TA dinucleotide repeat is located on the 3′-UTR region and the functional consequences are due to the instability of mRNA transcripts, but no significative association with PCa risk was found (Mononen and Schleutker 2009).

9.2.4 Luteinizing Hormone (LH)

LH is an alpha-beta heterodimer in which the alpha unit is common to glycoprotein hormones and the subunit beta, coded by the gene LHB on 9q13 and determines biological specificity. The variant LHB-V, carrying two mutations (W8R and I15T), has higher bioactivity and shorter half-life than WT protein. No such association has been demonstrated with PCa risk (Mononen and Schleutker 2009).

9.2.5 CYP Subfamily 17A Enzyme

CYP17A1, located on chromosome 10q24.3, encodes for the enzyme P450c17alpha that regulates the activity of two enzymes at key points in the testosterone biosynthesis. A single base pair change (−34T>C) has been described that create an additional binding site for the transcription factor Sp-1 which may cause an increased transcription of the enzyme and enhanced steroid hormone production. Contradictory results have been reported concerning the potential role in PCa cases but, in a meta-analysis, no overall effect of this variant have been observed (Ntais et al. 2003).

9.2.6 HSD3B Family

B1 and B2 genes are located on 1p13.1 and express three enzymes involved in androgen biosynthesis and in the conversion of active DHT into inactive metabolites in steroid target tissues. The N367T variant in HSD3B1 has been found to increase the PCa risk in Caucasian men but has no effect on clinical behavior. A higher risk for PCa, moreover, has been found when a second variant on 3′-UTR of HSD3B2 (rs1819698) was present (Chang et al. 2002).

9.2.7 3Alpha-Hydroxysteroid Dehydrogenase

These enzymes regulate the transactivation of the AR and play a critical role in the hepatic clearance of steroid hormones, protecting against excess circulating steroid hormone. AKR1C3 (HSD17B5) is a gene that encodes for an enzyme that catalyzes the conversion of androstenedione to testosterone and of DHT to androstenediol; it is not expressed in epithelial cells but over expressed in prostatic carcinomas. The studies regarding the role of two polymorphism (Q5H and P180S) in PCa etiology have not found any association (Mononen et al. 2006; Berndt et al. 2007). Another polymorphism (E77G) was found in Caucasian but not in Asian individuals and was associated with lower serum testosterone (Jakobsson et al. 2007).

9.2.8 HSD17B Family

This gene family encodes for 17betaHSDs which play a pivotal role in the production of steroid hormones. The isoform 1 polymorphisms have shown no association with PCa risk or tumor stage. HSD17B2 gene (16q24) is expressed in normal prostate, BPH and PCa, with the highest levels in BPH; it is involved in conversion of active androgens in less active forms, protecting prostate from excessive sex hormone action. An association with the loss of heterozygosity at chromosomal region 16q24.1-16q24.2, that includes the HSD17B2 gene, has been correlated with aggressive forms of PCa, but not with cancer risk (Cunningham et al. 2007). Instead, a missense substitution G289S in the HSD17B3 gene, that is expressed exclusively in testis, seemed to confer a significant increase in PCa risk in a single study but, this result, has not been confirmed in others study (Margiotti et al. 2002).

9.2.9 CYP Subfamily 19A Enzyme

The CYP19A1 gene encodes for the aromatase. The mutation T201M showed a borderline significant increase in PCa in unselected cases and association with clinically less significant form in stratified analysis (Mononen et al. 2006). Polymorphism R264C, instead, showed a tendency in increased risk of high grade carcinoma but, this feature, has not been confirmed in others studies (Modugno et al. 2001; Suzuki et al. 2003). Overall, these two polymorphism showed no significant changes in protein activity or level compared to WT forms. Polymorphism in (TTTA) repeat showed, in several studies, no association with pathological grade or stage, patient age and preoperative PSA levels and contrasting results regarding to association with PCa risk (Mononen and Schleutker 2009).

9.2.10 CYP3A Locus

The CYP3A locus consists of four genes including the isoforms 3A4, 3A5, 3A7 and 3A43, located on chromosome 7q21.1.

CYP3A4 has a role in testosterone deactivation. The CYP3A4*1B polymorphism has been reported to disrupt a transcriptional regulatory element in the 5′ regulatory region. This alteration shows ethnic/geographic differences, reflecting the highest frequence of PCa in African-American men. Contradictory reports exist regarding the association with clinical stage and grade (Mononen and Schleutker 2009).

CYP3A5 catalyzes 6beta-hydroxylation of testosterone. The CYP3A5*1 form is more common in African-Americans than in Caucasian or Asian men. Some studies suggest that CYP3A4*1B is in linkage disequilibrium with CYP3A5*1, with association with a risk of more aggressive PCa (Rebbeck et al. 1998; Paris et al. 1999).

CYP3A43 is the most recent member of the family. Its variant, CYP3A43*3, has been shown to be a risk factor in familial cases. Stone et al. (2005) observed that African-American homozygous for this variant are at increased risk for PCa compared to those homozygous for others variants.

9.2.11 *UGT2B15*

It is an enzyme, located in the luminal cells, that catalyze conjugation of steroids by glucuronidation and the resulting products are water soluble and more easily excreted from the body. Its gene is located on chromosome 14q13-q21.1. A single base pair change results in an aminoacid change at residue 85 from aspartic to tyrosine (D85Y) originating an Y85 isoform more efficient and, thus, a lower androgen exposure. For this reason, the D85 allele has been correlated to increased PCa risk and aggressiveness (Mononen and Schleutker 2009).

9.2.12 *Sex Hormone Binding Globulin*

This protein regulates levels of free plasma androgens and mediate the androgen and estrogen signaling at the cell membrane. Its gene is located on 17p13-p12. Two variants, D356N and -67G>A have been studied. Only patients heterozygous for D356N showed a slight association with PCa risk, with greater risk in patient <65 years aged compared to WT carriers, while -67G>A variant did not show clear association with PCa risk (Berndt et al. 2007).

9.2.13 *Sulfotransferase 2A1 (SULT2A1)*

Sulfotransferase 2A1 is a protein expressed in the liver, adrenal cortex and small intestine, member of the SULT family that includes enzymes involved in regulation of activity of adrenal androgens, via their inactivation by sulfation, protecting against the mitogenic effects of androgens. Three gene alterations, A63P, L227E and A261T, have been described for this gene, present only in African-American patients, but they have not been associated with PCa (Mononen and Schleutker 2009).

9.3 Repair Genes

Genomic stability and integrity are essential for the maintenance of an accurate DNA replication. In fact, endogenous processes and exogenous factors that cause DNA disruptions and, consequently, gene rearrangements, translocations, amplifications and deletions may participate in cancer development (Ritchey et al. 2005). DNA repair mechanisms protect human genome from damages caused by endogenous and exogenous agents (Hirata et al. 2007). Several DNA repair pathways, involved in repairing different type of DNA damage, exist: base excision repair (BER) removes simple base modifications (such as single-strand breaks, oxidative DNA damage and alkylation); nucleotide excision repair (NER) removes larger lesions; alkyltransferase reverse DNA damage by transferring alkyl groups from damaged DNA; homologous recombination repair pathway repair double-stranded DNA breaks (Ritchey et al. 2005). Only few studies have investigated the role of polymorphisms in DNA-repair genes in prostate carcinogenesis. Ritchey et al. (2005) investigated SNPs in four genes (XRCC1, XRCC3, MGMT and XPD) in a Chinese population and their combined effects with a number of risk factors previously reported. They found that variants of XRCC1 (−399) and of MGMT (−84) are independently associated with prostate cancer risk, while no associations were observed for XRCC3 (−241), XPD (−751) or MGMT (−143) variants. Hirata et al. (2007), examined the association between XRCC1(Arg194Trp, Arg399Gln), XPC Lys939Gln and XRCC1G6721T polymorphisms in a series of prostate cancer and their relation to smoking status. They demonstrated that XRCC1Arg194Trp variants are more frequent in PCa patients compared to normal controls and, for the first time, reported inverse correlation between XPC variants and PCa risk. Moreover, in this study, the combined effect on variants of these two genes with an increase in PCa risk was evaluated (Hirata et al. 2007).

Mismatch repair (MMR) genes activity may be associated with cancer; mutations in these genes (MLH1, MSH2, MSH3, MSH6, PMS1 and PMS2) can lead to microsatellites instability (MSI) and inability in repair DNA damage during DNA replication. Some studies have correlated the reduction or loss of MMR protein expression with PCa, either in tissue that in cell lines (Boyer et al. 1995; Leach et al. 2000; Yeh et al. 2001; Chen et al. 2001, 2003; Norris et al. 2007) and those regarding the association between MLH1 polymorphism and PCa risk have reached conflicting results (Burmester et al. 2004; Fredriksson et al. 2006). However, in a recent study, Landeberg et al. have described a polymorphism in MLH1 associated with a modest increase in overall PCa risk, a more aggressive disease and risk of recurrence (Langeberg et al. 2010).

9.4 Cell Cycle Control

Alterations in cell cycle regulation play a central role in malignancies and contributes to an increased risk of metastatic disease; in fact the lack of cell's ability to respond appropriately to DNA damage and potential carcinogenic events (Kibel and Isaacs 2000; Tomlins et al. 2007) causes accumulation of genetic defects, which could promote the tumor growth and the development of a more aggressive phenotype (Elledge 1996). Polymorphism in genes of cell cycle regulators have been already described for several malignancies (such as breast, colorectal carcinoma, bladder cancer, head and neck, lung and prostate cancer); in particular, they have been associated with carcinoma development polymorphisms in TP53, CCND1, CDKN1A, CDKN1B, CDKN2A and MDM2 genes (Kibel et al. 2008). In summary, signals affecting cell cycle progression converge on two main pathways, the Arf/MDM2/p53/p21 and the INK4a/cdk4/pRb/E2F axis (Dianat et al. 2009). Several studies have investigated the role of SNPs of cell cycle control genes in determining risk of cancer and/or cancer prognosis in prostate. Two recent studies have investigated the role of polymorphisms in TP53, CCND1, CDKN1A, CDKN1B, CDKN2A, MDM2, p21, PTEN, GNAS1 and Bcl-2 in prostate cancer and has emerged MDM2 as the most promising allele associated to PCa, especially with advanced stage, in androgen-independent disease and in patients younger at diagnosis, while specific GNAS1 and Bcl-2 polymorphisms have been associated with biochemical recurrence (Kibel et al. 2008; Hirata et al. 2009).

9.5 Angiogenesis

In general, angiogenesis is required for the growth of cancers from microscopic to clinically relevant, in cancerogenesis, in the development of tumor characteristics and in clinical outcome (Jacobs et al. 2008). Like all others solid tumors, the progressive growth and metastasis of prostate cancer are angiogenesis-dependent. Tumor angiogenesis is related to an "angiogenic switch" that origins from an excess of pro-angiogenic molecules over anti-angiogenic factors produced from tumor and stroma cells (Sfar et al. 2009). Vascular endothelial growth factor (VEGF) plays a pivotal role in angiogenesis (Nicholson and Theodorescu 2004), as a specific endothelial mitogen and increasing endothelial cells permeability, migration and chemotaxis, in an autocrine and/or paracrine manner (Ferrara and Davis-Smyth 1997). Hypoxia inducible factor 1 (HIF1A) is a transcriptional factor that induces transcription of VEGF; its activity is inhibited by the hypoxia inducible factor 1 alpha subunit inhibitor (HIF1AN) (Stolze et al. 2004). Epidermal growth factor (EGF) activates several pro-oncogenic intracellular pathways, inducing proliferation, differentiation and tumorigenesis in epithelial cells (Dianat et al. 2009); moreover it has been described to increase VEGF expression in prostate cancer cell lines (Ravindranath et al. 2001). Nitric oxide is an important factor involved in both carcinogenesis mediated by angiogenesis and tumorigenesis inhibition through induction of cell death. Three isoforms of enzymes for its synthesis exist: the neuronal (NOS1), the inducible (NOS2) and the endothelial (NOS3). The inducible isoform expression is regulated by a series of transcription factors, proinflammatory cytokines (that increase its expression) and anti-inflammatory cytokines (that downregulate its expression) (Dianat et al. 2009). Matrix metalloproteinase 2 (MMP2) and 9 (MMP9) degrade basement membranes and extracellular matrix, required for new vessel formation and tumor invasion (Yoon et al. 2003; Stetler-Stevenson et al. 1993; Murphy and Gravilovic 1999). The production of MMP9 and other proteases by prostate cancer cells and stromal cells facilitates the degradation of ECM, resulting in tumor invasion and subsequent metastasis. Moreover, it has been demonstrated that MMP9 triggers VEGF release from extracellular stores, facilitating the angiogenic switch. Thrombospondin-1 (TSP1) acts as antiangiogenic factor by multiple mechanism, such as inhibition of endothelial cells proliferation and migration, induction of endothelial cells apoptosis, inhibition of MMP3-dependent activation of pro-MMP9 and interaction with VEGF (Sfar et al. 2009). Several studies have been conducted to investigate the role of SNPs in angiogenesis-related genes. For VEGF, a significantly increased risk of prostate cancer associated with the VEGF-634 (GC+CC) polymorphism has emerged, and the C allele seems to be associated with an aggressive phenotype, defined by the histological grade; similarly, the -460T allele has been associated to an increased risk of cancer, without statistically significant differences in grade or stage and between patients responders to hormonal therapy and hormone-refractory patients (Lin et al. 2003); however, contrasting results have been reported. The CC and TC genotypes of the polymorphism were associated with significantly higher rates of PSA recurrence after RP higher than the TT genotypes. On the contrary, patients with metastatic disease presenting the TT genotype had significantly worse survival compared to the CC and TC genotypes (Langsenlehner et al. 2008). Regarding the NOS3 gene polymorphism, two sites at intron 4 and exon 7 have been investigated and the a-allele has resulted associated with prostate cancer risk and more frequent in patients with high-grade tumor, indicating its role in more advanced disease. The Glu-Asp298 polymorphism at exon 7 did not show significative association with PCa risk, but patients were younger than patients with no polymorphism and a difference was observed between patients with localized or advanced disease (Medeiros et al. 2002). In a study conducted by Lee et al., investigating four SNPs in NOS2 gene locus (−2892T/C, +14 C/T, +88T/G and +524G/A) and five for NOS3 gene locus (−762c/t, -43c/t, -26a/g, -63g/t, -62g/t), all SNPs, except NOS2-2892T/C, were significantly associated with cancer risk, with a certain relation, for some variants, with ethnic groups (Lee et al. 2009). For EGF, the functional polymorphism at +61G/A resulted associated with cancer risk, with an higher age-adjusted risk of metastatic disease and with higher Gleason's grade tumor, without significative differences in response or resistance to hormone-deprivation therapy (Teixeira et al. 2008). One HIF1A SNP(P582S) was associated with advanced and overall prostate cancer but contrasting results have been also reported. Three SNPs in MMP2 were also associated with advanced and overall prostate cancer, while the homozygous variant of the P41A SNP in HIF1AN has been associated with a statistically significant reduction in risk of advanced prostate cancer (Jacobs et al. 2008). In a study from Sfar et al., although they have previously reported that the SNP 8831A/G in TSP1 gene was not significantly associated with prostate cancer risk, a significant impact in prostate cancer risk has been observed by combining the high-risk genotypes of VEGF (1154G/A and VEGF-634G/c) and TSP1 SNPs (Sfar et al. 2009).

9.6 Chromosomal Loci

Several GWAS have been conducted to detect chromosomal loci as genetic risk factors of prostate cancer. A recent meta-analysis (Liu et al. 2011) of 21 GWASs was made to improve the power to detect PCa risk loci. 37 SNPs have been reported in more than one study impacting PCa susceptibility which have significance to public health and, among these, 9 were significantly associated with PCa. Moreover, 14 were independent risk loci for PCa.

8q24 is the most frequently involved chromosomal region in prostate tumors and the rs1447295 is its most reported SNP. Genes closest to 8q24 are FAM84B, a breast cancer membrane-associated protein, and the oncogene c-MYC. In PCa c-MYC is overexpressed while its reduction inhibits tumor growth. It has been hypothesized that the risk variant in 8q24 modifies c-MYC regulation, predisposing to genomic instability and has been associated with aggressive tumors, hormone independence and poor prognosis.

The SNP rs10486567 on chromosome 7 occurs in the second intron of the JAZF zinc finger 1 gene (JAZF1) that encodes for a transcriptional repressor of NR2C2 (highly expressed in prostate tissue, interacts with androgen receptor to repress its target gene expression). In a study from Gallagher et al. (2010), this SNP resulted associated with biochemical recurrence and castration-resistant metastases.

On chromosome 10 two SNPs (rs10993994 and rs7920517) have been identified in the region containing the microsemino-protein beta gene (MSMB) that encodes PSP94, a member of the immunoglobulin binding factor family synthesized by epithelial cells in prostate and secreted in seminal plasma. Loss of PSP94 is associated with recurrence after prostatectomy, while MSMB is silenced by EZH2 in androgen-insensitive PCa. Therefore, variants in MSMB gene may represent a genetic risk marker for PCa.

A SNP (rs2735839) is located between the KLK2 and KLK3 genes, that have been reported to influence the PCa risk. SNPs in the promoter region of KLK3 and it has been associated with PSA concentrations and, in some cases, with risk of PCa and with stage of disease. rs4430796 and rs7501939 SNPs are located on 17q12 in the first and second intron of HNF1B gene, which encodes for a transcription factor.

rs4962416 falls in the fifth intron of the CTBP2 gene that encodes for a transcriptional corepressor activated under metabolic stress, highly expressed in prostate tissue. Its expression is associated with decreased PTEN expression and activation of the posphatidylinositol3-kinase pathway.

The SNPs rs5945619 and rs5945572 are highly correlated. The former is on Xp between NUDT10 and NUDT11 genes, while the latter downstream of the NUDT11 gene, that encodes isoform of diphosphoinositol polyphosphate phosphohydrolase that determines the rate of phosphorylation in DNA repair, stress responses and apoptosis. These genes showed an association with PSA concentration and with increased PCa risk.

rs9623117 and rs7291619 are on 22q13 and are in linkage disequilibrium. They occurs within the TNRC6B gene that encodes a protein that mediate miRNA-guided mRNA cleavage. Its expression is suppressed in hormone-refractory metastatic PCa compared to prostate carcinoma and genetic variation and may alter the levels of mRNA species under its control and, therefore, contribute to carcinogenesis.

9.7 Cell Adhesion Genes

The E-cadherin gene (CDH1) is located on 16q22.1; its aberrant expression has been associated with malignant transformation of prostatic epithelium, metastatic potential and poor prognosis. The -160C/A polymorphism (A-allele) in gene promoter causes a reduction in transcriptional activity. Several studies have investigated the role of the A-allele in predisposition to, origin and progression of prostate cancer, with contrasting results. In fact, some Authors report that the A-allele frequency is higher in patients with prostate cancer than in control subjects, its association with a higher stage and that its frequency is higher among European American (Verhage et al. 2002; Kamoto et al. 2005; Bonilla et al. 2006), while others have described contrary results. However, a recent meta-analysis concluded that the A-allele carriers have an increased risk of prostate cancers, in both European and Asian populations (Qiu et al. 2009).

CTNND2 (delta-cadherin) is a junction-associated protein, located on 5p15. It favors cellular spreading by disrupting the E-cadherin-based adherens junction; its expression is up-regulated in the majority of prostatic adenocarcinomas and is correlated with higher Gleason scores (Lu et al. 2005).

Intercellular adhesion molecules (iCAMs) are a group of protein involved in cell adhesion and signaling and play an important role in the development of several cancers, such as prostate cancer. In fact, their altered expression may lead to tissue architectural destruction and distant metastasis. Two gene polymorphism (−9A/C and K469E) have been described to be associated with PCa risk in men with a positive family history (Chen et al. 2006).

9.8 Vitamin D Pathway Genes

Several studies have examinated whether polymorphism in the vitamin D receptor (VDR) gene are related to prostate cancer risk, but with conflicting results. Some lines of evidence suggest that vitamin D can influence PCa risk (Coughlin and Hall 2002). In prostate cancer, vitamin D exerts an anticancer action as well as a reduction of metastatic potential, inducing increases in apoptosis, inhibition of cell cycle progression and interaction with the insulin-like growth factor axis. The antiproliferative effects of activated vitamin D are mediated trough a pathway involving VDRs; therefore, polymorphism in VDR gene could affect the binding of biological active vitamin D and modulate its antiproliferative effects. Loci of more extensively studied SNPs are Cdx2, Fokl, Bmsl, Apal, Taql and the poly-A microsatellite but no significative associations between these polymorphism and tumor characteristics was found.

9.9 THE FAS/FASL Pathway

Apoptosis, by modulating the elimination of unwanted or dangerous cells, plays an essential role in cellular homeostasis. Deregulation of this mechanism, with the consequent acquisition of resistance to apoptotic stimuli and alterations in the apoptotic pathway, may lead to development of cancer (Lima et al. 2008). The FAS/FASL system is one of the major pathways that regulate apoptosis. FAS is a cell-surface receptor which interacts with its ligand FASL, both members of the TNF superfamily. Decreased expression of FAS and/or increased expression of FASL reduce the tumor-cell apoptosis, favoring malignant progression. Polymorphism in the FAS or FASL gene have been associated with a high risk of several types of cancer, as for bladder (Li et al. 2006a), cervical cancer (Lai et al. 2003, 2005; Ueda et al. 2006), cutaneous melanoma (Li et al. 2006b), esophageal and head and neck squamous cell carcinoma (Sun et al. 2004; Zhang et al. 2006), lung and nasopharyngeal cancer (Bel Hadj Jrad et al. 2006; Wang et al. 2003; Zhang et al. 2005). Lima et al., in 2008, have, for the first time, investigated the role of polymorphisms in the FAS/FASL pathway in prostate carcinoma, with particular attention to -670A/G polymorphism. From this study emerged that this isoform may influence the PCa development, in fact, individuals carrying AG and GG genotypes present statistically significant protection for extracapsular invasion. The Authors explained these results considering that two form of FAS exist and derived from alternative splicing of the same gene: a soluble form with anti-apoptotic effect and a membranc-bound form, with pro-apoptotic properties. In previous studies, overexpression of the soluble form had been correlated to more aggressive form of PCa. In this study, Lima indicate that the reduction of Fas caused by G allele reduces the sFAS expression levels, preventing the antiapoptotic effect of this protein. However, others studies are needed to evaluate the role of polymorphism in this pathway in the PCa prognosis.

9.10 Translational Relevance

Information on prostate cancer risk has a relevant impact on clinical and public health. To date, only family history can help in identifying men with a genetic predisposition to prostate cancer. The recent technological advances, however, allow to investigate thousands of single nucleotide polymorphisms (SNPs) across the genome, to search for specific genetic markers associated with risk of developing this disease. Furthermore, several linkage and genome-wide association studies have identified several polymorphisms that influence cancer risk (Salinas et al. 2009), leading to hypothesize the development of genetic tests to predict risk. To date, none of the SNP associations reported has resulted useful in this kind of prevision and SNP genotypes alone are of limited value for predicting risk of developing prostate cancer or more aggressive disease. Therefore, further studies are necessary to understand if these genetic variants may be of real utility in risk stratification and in communicating risk-based information to individuals interested in early detection and prostate cancer prevention (Salinas et al. 2009). However, though modest, the potential of risk estimation for "single SNP" is mandatory combine various SNP variants for use in a clinical setting and for the development of a clinical tool (nomogram) that could help to integrate the information into practice (Nam et al. 2009). The identification of the effects of SNPs on gene expression and on the pathogenesis of patients, could assist in diagnosis, prognosis and tailored patient management (Huppi and Chandramouli 2004).

Therefore, men so identified to be at higher risk of prostate cancer may choose to begin PSA based screening at an earlier age and take preventive measures, such as an accurate diet and lifestyle or chemoprevention. For example, finasteride is a chemopreventive agent for prostate cancer which has been described to reduce prostate cancer risk by 25 % (Thompson et al. 2003). The preventive effect might be stronger among men resulted at higher risk basing upon a polygenic model (Wray et al. 2008). In fact, even if the effect is the same for men with high or low risk for prostate cancer, the net gain would be larger for men at higher risk who may choose to adhere earlier to a chemoprevention regimen. In any case, the potential clinical utility of this approach has to be tested in a clinical trial. However, the major limitation emerged from studies is that genetic prediction models do not distinguish aggressive from non-aggressive cancer, and therefore may exasperate the potential problem of over-diagnosis and over-treatment of prostate cancer. Nonetheless, chemoprevention could decrease the number of men developing any prostate cancer. Once more, the benefits and risks of this type of risk prediction model needs to be further evaluated.

References

AIHW (Australian Institute of Health and Welfare), AACR (Australasian Association of Cancer Registries) (2007) Cancer in Australia: an overview, 2006. Cancer series no. 37. Cat. no. CAN 32. AIHW, Canberra

Al Olama AA, Kote-Jarai Z, Giles GG, Guy M, Morrison J, Severi G et al (2009) Multiple loci on 8q24 associated with prostate cancer susceptibility. Nat Genet 41:1058–1060

Aly M, Wiklund F, Grönberg H (2011) Early detection of prostate cancer with emphasis on genetic markers. Acta Oncol 50(Suppl 1):18–23

Bao BY, Pao JB, Lin VC, Huang CN, Chang TY, Lan YH, Lu TL, Lee HZ, Chen LM, Ting WC, Hsieh CJ, Huang SP (2010) Individual and cumulative association of prostate cancer susceptibility variants with clinicopathologic characteristics of the disease. Clin Chim Acta 411(17–18):1232–1237

Bel Hadj Jrad B, Mahfouth W, Bouaouina N, Gabbouj S, Ahmed SB, Ltaief M et al (2006) A polymorphism in FAS gene promoter associated with increased risk of nasopharyngeal carcinoma and correlated with anti-nuclear autoantibodies induction. Cancer Lett 233:21–27

Berndt SI, Chatterjee N, Huang WY, Chanock SJ, Welch R, Crawford ED et al (2007) Variant in sex hormone-binding globulin gene and the risk of prostate cancer. Cancer Epidemiol Biomarkers Prev 16:165

Beuten J, Gelfond JA, Franke JL, Weldon KS, Crandall AC, Johnson-Pais TL, Thompson IM, Leach RJ (2009) Single and multigenic analysis of the association between variants in 12 steroid hormone metabolism genes and risk of prostate cancer. Cancer Epidemiol Biomarkers Prev 18(6):1869–1880

Bonilla C, Mason T, Long L et al (2006) E-cadherin polymorphisms and haplotypes influence risk for prostate cancer. Prostate 66:546–556

Boyer JC, Umar A, Risinger JI et al (1995) Microsatellite instability mismatch repair deficiency, and genetic defects in human cancer cell lines. Cancer Res 55:6063–6070

Burmester JK, Suarez BK, Lin JH, Jin CH, Miller RD, Zhang KQ, Salzman SA, Reding DJ, Catalona WJ (2004) Analysis of candidate genes for prostate cancer. Hum Hered 57(4):172–178

Carter BS, Beaty TH, Steinberg GD, Childs B, Walsh PC (2006) Mendelian inheritance of familial prostate cancer. Proc Natl Acad Sci USA 89:3367–3371

Chang BL, Zheng SL, Hawkins GA, Isaacs SD, Wiley KE, Turner A et al (2002) Joint effect of HSD3B1 and HSD3B2 genes is associated with hereditary and sporadic prostate cancer susceptibility. Cancer Res 62:1784

Chang BL, Spangler E, Gallagher S, Haiman CA, Henderson B, Isaacs W, Benford ML, Kidd LR, Cooney K, Strom S, Ingles SA, Stern MC, Corral R, Joshi AD, Xu J, Giri VN, Rybicki B, Neslund-Dudas C, Kibel AS, Thompson IM, Leach RJ, Ostrander EA, Stanford JL, Witte J, Casey G, Eeles R, Hsing AW, Chanock S, Hu JJ, John EM, Park J, Stefflova K, Zeigler-Johnson C, Rebbeck TR (2011) Validation of genome-wide prostate cancer associations in men of African descent. Cancer Epidemiol Biomarkers Prev 20(1):23–32

Chen Y, Wang J, Fraig MM et al (2001) Defects of DNA mismatch repair in human prostate cancer. Cancer Res 61(10):4112–4121

Chen Y, Wang J, Fraig MM et al (2003) Alterations in PMS2, MSH2 and MLH1 expression in human prostate cancer. Int J Oncol 22:1033–1043

Chen H, Hernandez W, Shriver MD, Ahaghotu CA, Kittles RA (2006) ICAM gene cluster SNPs and prostate cancer risk in African Americans. Hum Genet 120:69–76

Cheng I, Plummer SJ, Neslund-Dudas C, Klein EA, Casey G, Rybicki BA, Witte JS (2010) Prostate cancer susceptibility variants confer increased risk of disease progression. Cancer Epidemiol Biomarkers Prev 19(9):2124–2132

Chung CC, Magalhaes WC, Gonzalez-Bosquet J, Chanock SJ (2010) Genome-wide association studies in cancer – current and future directions. Carcinogenesis 31:111–120

Cooperberg MR, Lubeck DP, Mehta SS, Carroll PR (2003) Time trends in clinical risk stratification for prostate cancer: implications for outcomes (data from CaPSURE). J Urol 170:S21–S25

Cornu JN, Drouin S, Cancel-Tassin G, Bigot P, Azzouzi AR, Koutlidis N, Cormier L, Gaffory C, Rouprêt M, Sèbe P, Bitker MO, Haab F, Cussenot O (2011) Impact of genotyping on outcome of prostatic biopsies: a multicenter prospective study. Mol Med 17(5–6):473–477

Coughlin SS, Hall IJ (2002) A review of genetic polymorphisms and prostate cancer risk. Ann Epidemiol 12(3):182–196

Crawford ED (2006) Epidemiology of prostate cancer. Urology 62:3–12

Cunningham JM, Hebbring SJ, McDonnell SK, Cicek MS, Christensen GB, Wang L et al (2007) Evaluation of genetic variations in the androgen and estrogen metabolic pathways as risk factors for sporadic and familial prostate cancer. Cancer Epidemiol Biomarkers Prev 16:969

Dianat SS, Margreiter M, Eckersberger E, Finkelstein J, Kuehas F, Herwig R, Ayati M, Lepor H, Djavan B (2009) Gene polymorphisms and prostate cancer: the evidence. BJU Int 104(11):1560–1572

Douglas JA, Zuhlke KA, Beebe-Dimmer J, Levin AM, Gruber SB, Wood DP et al (2005) Identifying susceptibility genes for prostate cancer – a family-based association study of polymorphisms in CYP17, CYP19, CYP11A1, and LH-beta. Cancer Epidemiol Biomarkers Prev 14:2035

Edwards SM, Eeles RA (2004) Unravelling the genetics of prostate cancer. Am J Med Genet Part C Sem Med Genet 129C:65–73
Elledge SJ (1996) Cell cycle checkpoints: preventing an identity crisis. Science 274(5293): 1664–1672
Ferrara N, Davis-Smyth T (1997) The biology of vascular endothelial growth factor. Endocr Rev 18(1):4–25
Fitzgerald LM, Kwon EM, Koopmeiners JS, Salinas CA, Stanford JL, Ostrander EA (2009) Analysis of recently identified prostate cancer susceptibility loci in a population-based study: associations with family history and clinical features. Clin Cancer Res 15:3231–3237
Fredriksson H, Ikonen T, Autio V et al (2006) Identification of germline MLH1 alterations in familial prostate cancer. Eur J Cancer 42(16):2802–2806
Gallagher DJ, Vijai J, Cronin AM, Bhatia J, Vickers AJ, Gaudet MM, Fine S, Reuter V, Scher HI, Halldén C, Dutra-Clarke A, Klein RJ, Scardino PT, Eastham JA, Lilja H, Kirchhoff T, Offit K (2010) Susceptibility loci associated with prostate cancer progression and mortality. Clin Cancer Res 16(10):2819–2832
Ghoussaini M, Song H, Koessler T, Al Olama AA, Kote-Jarai Z, Driver KE et al (2008) Multiple loci with different cancer specificities within the 8q24 gene desert. J Natl Cancer Inst 100: 962–966
Gudmundsson J, Sulem P, Manolescu A, Amundadottir LT, Gudbjartsson D, Helgason A et al (2007) Genome-wide association study identifies a second prostate cancer susceptibility variant at 8q24. Nat Genet 39:631–637
Haiman CA, Patterson N, Freedman ML, Myers SR, Pike MC, Waliszewska A et al (2007) Multiple regions within 8q24 independently affect risk for prostate cancer. Nat Genet 39:638–644
Hamburg MA, Collins FS (2010) The path to personalized medicine. N Engl J Med 363(4):301–304, Epub 2010 Jun 15. Erratum in: N Engl J Med. 2010 Sep 9;363(11):1092
Helfand BT, Fought AJ, Loeb S, Meeks JJ, Kan D, Catalona WJ (2010) Genetic prostate cancer risk assessment: common variants in 9 genomic regions are associated with cumulative risk. J Urol 184(2):501–505
Hirata H, Hinoda Y, Tanaka Y, Okayama N, Suehiro Y, Kawamoto K, Kikuno N, Majid S, Vejdani K, Dahiya R (2007) Polymorphisms of DNA repair genes are risk factors for prostate cancer. Eur J Cancer 43(2):231–237
Hirata H, Hinoda Y, Kikuno N, Suehiro Y, Shahryari V, Ahmad AE, Tabatabai ZL, Igawa M, Dahiya R (2009) Bcl-2–938C/A polymorphism carries increased risk of biochemical recurrence after radical prostatectomy. J Urol 181(4):1907–1912
Hsing AW, Chen C, Chokkalingam AP, Gao YT, Dightman DA, Nguyen HT et al (2001) Polymorphic markers in the SRD5A2 gene and prostate cancer risk: a population-based case–control study. Cancer Epidemiol Biomarkers Prev 10:1077
Huppi K, Chandramouli GV (2004) Molecular profiling of prostate cancer. Curr Urol Rep 5:45–51
Jacobs EJ, Hsing AW, Bain EB, Stevens VL, Wang Y, Chen J, Chanock SJ, Zheng SL, Xu J, Thun MJ, Calle EE, Rodriguez C (2008) Polymorphisms in angiogenesis-related genes and prostate cancer. Cancer Epidemiol Biomarkers Prev 17(4):972–977
Jakobsson J, Palonek E, Lorentzon M, Ohlsson C, Rane A, Ekstrom L (2007) A novel polymorphism in the 17beta-hydroxysteroid dehydrogenase type 5 (aldo-keto reductase 1C3) gene is associated with lower serum testosterone levels in Caucasian men. Pharmacogenomics J 7:282
Johns LE, Houlston RS (2003) A systematic review and meta-analysis of familial prostate cancer risk. BJU Int 91:789–794
Kamoto T, Isogawa Y, Shimizu Y et al (2005) Association of a genetic polymorphism of the E-cadherin gene with prostate cancer in a Japanese population. Jpn J Clin Oncol 35:158–161
Kibel AS, Isaacs WB (2000) G(1)/S cell cycle proteins as markers of aggressive prostate carcinoma. Urology 55(3):316–322
Kibel AS, Jin CH, Klim A, Luly J, A Roehl K, Wu WS (2008) Association between polymorphisms in cell cycle genes and advanced prostate carcinoma. Prostate 68(11):1179–1186
Kirchhoff T et al (2004) BRCA mutations and risk of prostate cancer in Ashkenazi Jews. Clin Cancer Res 10:2918–2921

Kote-Jarai Z, Easton DF, Stanford JL, Ostrander EA, Schleutker J, Ingles SA et al (2008) Multiple novel prostate cancer predisposition loci confirmed by an international study: the PRACTICAL consortium. Cancer Epidemiol Biomarkers Prev 17:2052–2061

Kumazawa T, Tsuchiya N, Wang L, Sato K, Kamoto T, Ogawa O et al (2004) Microsatellite polymorphism of steroid hormone synthesis gene CYP11A1 is associated with advanced prostate cancer. Int J Cancer 110:140

Kwon EM, Salinas CA, Kolb S et al (2011) Genetic polymorphisms in inflammation pathway genes and prostate cancer risk. Cancer Epidemiol Biomarkers Prev 20:923–933

Lai HC, Sytwu HK, Sun CA, Yu MH, Yu CP, Liu HS et al (2003) Single nucleotide polymorphism at Fas promoter is associated with cervical carcinogenesis. Int J Cancer 103:221–225

Lai HC, Lin WY, Lin YW, Chang CC, Yu MH, Chen CC et al (2005) Genetic polymorphisms of FAS and FASL (CD95/CD95L) genes in cervical carcinogenesis: an analysis of haplotype and gene–gene interaction. Gynecol Oncol 99:113–118

Langeberg WJ, Kwon EM, Koopmeiners JS, Ostrander EA, Stanford JL (2010) Population-based study of the association of variants in mismatch repair genes with prostate cancer risk and outcomes. Cancer Epidemiol Biomarkers Prev 19(1):258–264

Langsenlehner T, Langsenlehner U, Renner W et al (2008) Single nucleotide polymorphisms and haplotypes in the gene for vascular endothelial growth factor and risk of prostate cancer. Eur J Cancer 44:1572–1576

Leach FS, Velasco A, Hsieh JT, Sagalowsky AI, McConnell JD (2000) The mismatch repair gene hMSH2 is mutated in the prostate cancer cell line LNCaP. J Urol 164(5):1830–1833

Lee KM, Kang D, Park SK et al (2009) Nitric oxide synthase gene polymorphisms and prostate cancer risk. Carcinogenesis 30:621–625

Li C, Wu W, Liu J, Qian L, Li A, Yang K et al (2006a) Functional polymorphisms in the promoter regions of the FAS and FAS ligand genes and risk of bladder cancer in south China: a case – control analysis. Pharmacogenet Genomics 16:245–251

Li C, Larson D, Zhang Z, Liu Z, Strom SS, Gershenwald JE et al (2006b) Polymorphisms of the FAS and FAS ligand genes associated with risk of cutaneous malignant melanoma. Pharmacogenet Genomics 16:253–263

Licastro F, Bertaccini A, Porcellini E et al (2008) Alpha 1 antichymotrypsin genotype is associated with increased risk of prostate carcinoma and PSA levels. Anticancer Res 28:395–399

Lima L, Morais A, Lobo F, Calais-da-Silva FM, Calais-da-Silva FE, Medeiros R (2008) Association between FAS polymorphism and prostate cancer development. Prostate Cancer Prostatic Dis 11(1):94–98

Lin CC, Wu HC, Tsai FJ, Chen HY, Chen WC (2003) Vascular endothelial growth factor gene-460 C/T polymorphism is a biomarker for prostate cancer. Urology 62:374–377

Lindstrom S, Zheng SL, Wiklund F, Jonsson BA, Adami HO, Balter KA et al (2006) Systematic replication study of reported genetic associations in prostate cancer: strong support for genetic variation in the androgen pathway. Prostate 66:1729

Linja MJ, Visakorpi T (2004) Alterations of androgen receptor in prostate cancer. J Steroid Biochem Mol Biol 92:255

Liu X, Cheng I, Plummer SJ, Suarez BK, Casey G, Catalona WJ, Witte JS (2011) Fine-mapping of prostate cancer aggressiveness loci on chromosome 7q22-35. Prostate 71(7):682–689

Lu Q, Dobbs LJ, Gregory CW et al (2005) Increased expression of delta-catenin/neural plakophilin-related armadillo protein is associated with the downregulation and redistribution of Ecadherin and p120ctn in human prostate cancer. Hum Pathol 36:1037–1048

Lughezzani G et al (2010) Predictive and prognostic models in radical prostatectomy candidates: a critical analysis of the literature. Eur Urol 58:687–700

Makridakis NM, Ross RK, Pike MC, Crocitto LE, Kolonel LN, Pearce CL et al (1999) Association of mis-sense substitution in SRD5A2 gene with prostate cancer in African-American and Hispanic men in Los Angeles, USA. Lancet 354:975

Margiotti K, Kim E, Pearce CL, Spera E, Novell G, Reichardt JK (2002) Association of the G289S single nucleotide polymorphism in the HSD17B3 gene with prostate cancer in Italian men. Prostate 53:65

Medeiros RM, Morais A, Vasconcelos A et al (2002) Outcome in prostate cancer: association with endothelial nitric oxide synthase Glu-Asp298 polymorphism at exon 7. Clin Cancer Res 8:3433–3437

Modugno F, Weissfeld JL, Trump DL, Zmuda JM, Shea P, Cauley JA, Ferrell RE (2001) Allelic variants of aromatase and the androgen and estrogen receptors: toward a multigenic model of prostate cancer risk. Clin Cancer Res 7(10):3092–3096

Mononen N, Schleutker J (2009) Polymorphisms in genes involved in androgen pathways as risk factors for prostate cancer. J Urol 181(4):1541–1549

Mononen N, Seppälä EH, Duggal P, Autio V, Ikonen T, Ellonen P, Saharinen J, Saarela J, Vihinen M, Tammela TL, Kallioniemi O, Bailey-Wilson JE, Schleutker J (2006) Profiling genetic variation along the androgen biosynthesis and metabolism pathways implicates several single nucleotide polymorphisms and their combinations as prostate cancer risk factors. Cancer Res 66(2):743–747

Murphy G, Gavrilovic J (1999) Proteolysis and cell migration: creating a path? Curr Opin Cell Biol 11:614–621

Nam RK, Zhang WW, Trachtenberg J, Seth A, Klotz LH, Stanimirovic A, Punnen S, Venkateswaran V, Toi A, Loblaw DA, Sugar L, Siminovitch KA, Narod SA (2009) Utility of incorporating genetic variants for the early detection of prostate cancer. Clin Cancer Res 15(5):1787–1793

Nicholson B, Theodorescu D (2004) Angiogenesis and prostate cancer growth. J Cell Biochem 91:125–150

Norris AM, Woodruff RD, D'Agostino RB Jr, Clodfelter JE, Scarpinato KD (2007) Elevated levels of the mismatch repair protein Pms2 are associated with prostate cancer. Prostate 67(2): 214–225

Ntais C, Polycarpou A, Ioannidis JP (2003) SRD5A2 gene polymorphisms and the risk of prostate cancer: a meta-analysis. Cancer Epidemiol Biomarkers Prev 12:618

Paris PL, Kupelian PA, Hall JM, Williams TL, Levin H, Klein EA et al (1999) Association between a CYP3A4 genetic variant and clinical presentation in African-American prostate cancer patients. Cancer Epidemiol Biomarkers Prev 8:901

Parkin DM, Whelan SL, Ferlay J, Teppo L, Thomas DB (eds) (2003) Cancer incidence in five continents. IARC Scientific Publications No. 155. Lyons: International Agency for Research on Cancer

Qiu LX, Li RT, Zhang JB et al (2009) The E-cadherin (CDH1) −160C/A polymorphism and prostate cancer risk: a metaanalysis. Eur J Hum Genet 17:244–249

Ravindranath N, Wion D, Brachet P, Djakiew D (2001) Epidermal growth factor modulates the expression of vascular endothelial growth factor in the human prostate. J Androl 22:432–443

Rebbeck TR, Jaffe JM, Walker AH, Wein AJ, Malkowicz SB (1998) Modification of clinical presentation of prostate tumors by a novel genetic variant in CYP3A4. J Natl Cancer Inst 90:1225

Ries LA et al (2000) The annual report to the nation on the status of cancer, 1973–1997, with a special section on colorectal cancer. Cancer 88:2398–2424

Ritchey JD, Huang WY, Chokkalingam AP, Gao YT, Deng J, Levine P, Stanczyk FZ, Hsing AW (2005) Genetic variants of DNA repair genes and prostate cancer: a population-based study. Cancer Epidemiol Biomarkers Prev 14(7):1703–1709

Salinas CA, Koopmeiners JS, Kwon EM, FitzGerald L, Lin DW, Ostrander EA, Feng Z, Stanford JL (2009) Clinical utility of five genetic variants for predicting prostate cancer risk and mortality. Prostate 69(4):363–372

Schaid DJ (2004) The complex genetic epidemiology of prostate cancer. Hum Mol Genet 13(Spec No. 1):R103–R121

Sfanos KS, De Marzo AM (2012) Prostate cancer and inflammation: the evidence. Histopathology 60(1):199–215, Review

Sfar S, Saad H, Mosbah F, Chouchane L (2009) Combined effects of the angiogenic genes polymorphisms on prostate cancer susceptibility and aggressiveness. Mol Biol Rep 36(1): 37–45

Shariat SF, Karakiewicz PI, Roehrborn CG, Kattan MW (2008) An updated catalog of prostate cancer predictive tools. Cancer 113:3075–3099

Siegel R, Naishadham D, Jemal A (2012) Cancer statistics. CA Cancer J Clin 62(1):10–29

Stetler-Stevenson WG, Aznavoorian S, Liotta LA (1993) Tumor cell interactions with the extracellular matrix during invasion and metastasis. Annu Rev Cell Biol 9:541–573

Stolze IP, Tian YM, Appelhoff RJ et al (2004) Genetic analysis of the role of the asparaginyl hydroxylase factor inhibiting hypoxia-inducible factor (HIF) in regulating HIF transcriptional target genes. J Biol Chem 279:42719–42725

Stone A, Ratnasinghe LD, Emerson GL, Modali R, Lehman T, Runnells G et al (2005) CYP3A43 Pro(340)Ala polymorphism and prostate cancer risk in African Americans and Caucasians. Cancer Epidemiol Biomarkers Prev 14:1257

Sun T, Miao X, Zhang X, Tan W, Xiong P, Lin D (2004) Polymorphisms of death pathway genes FAS and FASL in esophageal squamouscell carcinoma. J Natl Cancer Inst 96:1030–1036

Suzuki K, Matsui H, Nakazato H, Koike H, Okugi H, Hasumi M, Ohtake N, Nakata S, Takei T, Hatori M, Ito K, Yamanaka H (2003) Association of the genetic polymorphism in cytochrome P450 (CYP) 1A1 with risk of familial prostate cancer in a Japanese population: a case–control study. Cancer Lett 195(2):177–183

Teixeira AL, Ribeiro R, Cardoso D et al (2008) Genetic polymorphism in EGF is associated with prostate cancer aggressiveness and progression-free interval in androgen blockade-treated patients. Clin Cancer Res 14:3367–3371

Thompson IM, Goodman PJ, Tangen CM, Lucia MS, Miller GJ, Ford LG, Lieber MM, Cespedes RD, Atkins JN, Lippman SM, Carlin SM, Ryan A, Szczepanek CM, Crowley JJ, Coltman CA Jr (2003) The influence of finasteride on the development of prostate cancer. N Engl J Med 349:215–224

Tomlins SA, Mehra R, Rhodes DR, Cao X, Wang L, Dhanasekaran SM, Kalyana-Sundaram S, Wei JT, Rubin MA, Pienta KJ, Shah RB, Chinnaiyan AM (2007) Integrative molecular concept modeling of prostate cancer progression. Nat Genet 39(1):41–51

Ueda M, Terai Y, Kanda K, Kanemura M, Takehara M, Yamaguchi H et al (2006) Fas gene promoter −670 polymorphism in gynecological cancer. Int J Gynecol Cancer 16(Suppl 1): 179–182

Verhage BA, van Houwelingen K, Ruijter TE, Kiemeney LA, Schalken JA (2002) Single-nucleotide polymorphism in the E-cadherin gene promoter modifies the risk of prostate cancer. Int J Cancer 100:683–685

Wang LE, Cheng L, Spitz MR, Wei Q (2003) Fas A670G polymorphism, apoptotic capacity in lymphocyte cultures, and risk of lung cancer. Lung Cancer 42:1–8

Wray NR, Goddard ME, Visscher PM (2008) Prediction of individual genetic risk of complex disease. Curr Opin Genet Dev 18:257–263

Yeager M, Xiao N, Hayes RB, Bouffard P, Desany B, Burdett L et al (2008) Comprehensive resequence analysis of a 136 kb region of human chromosome 8q24 associated with prostate and colon cancers. Hum Genet 124:161–170

Yeager M, Chatterjee N, Ciampa J, Jacobs KB, Gonzalez-Bosquet J, Hayes RB et al (2009) Identification of a new prostate cancer susceptibility locus on chromosome 8q24. Nat Genet 41:1055–1057

Yeh CC, Lee C, Dahiya R (2001) DNA mismatch repair enzyme activity and gene expression in prostate cancer. Biochem Biophys Res Commun 285:409–413

Yoon SO, Park SJ, Yun CH, Chung AS (2003) Roles of matrix metalloproteinases in tumor metastasis and angiogenesis. J Biochem Mol Biol 36:128–137

Zabaleta J, Lin H-Y, Sierra RA et al (2008) Interactions of cytokine gene polymorphisms in prostate cancer risk. Carcinogenesis 29:573–578

Zabaleta J, Su LJ, Lin H-Y et al (2009) Cytokine genetic polymorphisms and prostate cancer aggressiveness. Carcinogenesis 30:1358–1362

Zeegers MP, Kiemeney LA, Nieder AM, Ostrer H (2004) How strong is the association between CAG and GGN repeat length polymorphisms in the androgen receptor gene and prostate cancer risk? Cancer Epidemiol Biomarkers Prev 13:1765

Zhang X, Miao X, Sun T, Tan W, Qu S, Xiong P et al (2005) Functional polymorphisms in cell death pathway genes FAS and FASL contribute to risk of lung cancer. J Med Genet 42:479–484

Zhang Z, Wang LE, Sturgis EM, El-Naggar AK, Hong WK, Amos CI et al (2006) Polymorphisms of FAS and FAS ligand genes involved in the death pathway and risk and progression of squamous cell carcinoma of the head and neck. Clin Cancer Res 12:5596–5602

Zheng SL, Sun J, Cheng Y, Li G, Hsu FC, Zhu Y et al (2007) Association between two unlinked loci at 8q24 and prostate cancer risk among European Americans. J Natl Cancer Inst 99: 1525–1533

Chapter 10
Expression Signature

Stefania Staibano and Angela Celetti

Abstract The prostate gland can be the site of multiple neoplastic transformation events, many of which give rise only to latent prostate cancer that does not progress to clinically detectable disease.

While evidence of major subtypes of prostate cancer is lacking at the histopathological level, recent genomic analyses have provided increasing evidence for molecularly defined subtypes (Tomlins et al., Neoplasia 10(2):177–188, 2008; Palanisamy et al., Nat Med 16(7):793–798, 2010; Taylor et al., Cancer Cell 18(1):11–22, 2010) but expression profiling analyses of tumor specimens have not strictly defined molecular signatures associated with distinct subtypes that specifically correlate with disease outcome (Singh et al., J Androl 23(5):652–660, 2002a; Singh et al., Cancer Cell 1: 203–209, 2002b; Lapointe et al., Proc Natl Acad Sci USA 101(3):811–886, 2004; Tomlins et al., Nat Genet 39(1):41–51, 2007a; Tomlins et al., Nature 448(7153), 595–599, 2007b). However, oncogenomic pathway analyses that integrate analyses of gene expression, copy number alterations, and exon resequencing may provide a unified approach for distinguishing prostate cancer subtypes and stratifying patient outcome (Taylor et al., Cancer Cell 18(1):11–22, 2010).

Integrating "omics" analyses with epigenetics will probably allow the identification of true different subtypes of prostate cancers characterized by divergent biological behavior and/or response to therapy.

S. Staibano (✉)
Department of Advanced Biomedical Sciences, Pathology Section, Faculty of Medicine and Surgery, University of Naples "Federico II", via S. Pansini, n.5, Naples, Italy
e-mail: staibano@unina.it

A. Celetti
Department of Cellular and Molecular Biology and Pathology, CNR, Institute of Experimental Endocrinology and Oncology 'G. Salvatore' EOS-CNR, Naples, Italy
e-mail: celetti@unina.it

S. Staibano (ed.), *Prostate Cancer: Shifting from Morphology to Biology*,
DOI 10.1007/978-94-007-7149-9_10,

This chapter aims to summarize the most exciting data emerging from recent genetic and translational studies on prostate cancer, potentially shedding new light on surprising aspects concerning its biology and extremely promising for the generation of more effective and safe new molecular therapies.

10.1 Advances in Prostate Cancer Genomics

Stefania Staibano (✉)

It is hard to summarize the spectacular advances made in cancer genomics in the last few years.

The emergence of Next Generation DNA and exome sequencing of malignant tumors is revealing thousands of mutations in every tumor type, many of which seem unique to each prostate cancer patient. This confirms that the word "cancer" is a figurative umbrella covering incredible spectra of diseases. This biological complexity justify the extraordinary hurdle to translate the results of basic research into real benefit for the single cancer patient (Barbieri et al. 2012).

A group of genes strongly correlates with prostate tumor differentiation stage, according to the Gleason score (Singh et al. 2002a). The gene expression data generated by DNA micro-arrays profiles predict with accuracy the patient evolution after prostatectomy. These data support the notion that the PCa clinical behavior is related to specific differences in gene expression profile that are detectable at the time of diagnosis.

This looks particularly promising also for the identification of new targets for therapy.

As an example, it has been found that the transmembrane serine protease called hepsin is specifically over-expressed in non-metastatic carcinoma cells, and PCa cell lines overexpressing hepsin show a dramatic reduction in growth and invasion, and increase of apoptosis. It has been then hypothesized that the decrease/loss of hepsin expression could be related with a poor prognosis of PCa and then hepsin could represent a potential target for prostate cancer gene therapy (Magee et al. 2001).

An integrated analysis of 218 primary and metastatic prostate cancers, 12 cell lines and xenografts, performed by assessment of DNA copy number, mRNA expression, and focused exon resequencing identified as expected, changes in the PI3K and androgen receptor (AR) pathways in nearly all metastatic samples and in a number of primary cancer tissue (Taylor et al. 2010).

Unexpectly, the nuclear receptor coactivator NCOA2 gene on the 8q13 was found mutated and acting as an oncogene in 11 % of primary tumors. NCOA2 and other regulators of nuclear receptor function such as NCOR2, are involved in AR pathway molecular signaling. This finding is of relevance, because it extends the potential importance of AR pathway perturbation even to disease initiation, while AR gene amplification or mutation is generally restricted to metastatic, castration-resistant disease (Tomlins et al. 2007a, b).

Several other emerging candidate cancer genes are *SPTA1* and *ADAM18*. *ADAM18* encodes a disintegrin and metalloprotease domain family member involved in sperm function. ADAM proteins exert key cell–cell and cell–matrix interactions.

In addition, *HSPA2*, *HSPA5* and *HSP90AB1*, heat shock genes encoding Hsp70 and Hsp90 isoforms, which form a chaperone complex, and the potassium channel genes *KCNQ3* and *KCNT1, with putative negative* tumor cell growth regulating activity, have been found to harbor point-mutation in a percentage of prostate cancer. Their functional significance, however, is still to be determined (Barbieri et al. 2012).

Anyhow, it has emerged that overall somatic point mutations and protein-altering point mutations are uncommon in prostate cancer if compared with other malignant tumor types, such as glioblastoma, lung cancer and melanoma (Barbieri et al. 2012; Taylor et al. 2010; Kumar et al. 2011; Gimba and Barcinski 2003; Greenman et al. 2007; Pleasance et al. 2010a, b).

In addition, no single gene emerged as commonly mutated. TP53 and PTEN, which act as prostate cancer tumor suppressors (Dong 2006; Pourmand et al. 2007), showed preferentially copy-number loss rather than point mutation.

The genomic and clinical outcome data from one of this study population are made available as a public resource, with the aim that it may contribute to define clusters of low- and high-risk disease beyond Gleason score of tumors (Taylor et al. 2010).

Novel adaptive clinical trial designs, linking *oncogenomic* (genomic and proteomic) alterations to treatment response and survival, are needed to translate molecular advances into clinical practice.

Nowaday, they have already changed our understanding of prostate cancer, with a progressive shift to a omics-based disease stratification approach and to molecularly guided therapeutic intervention modalities.

Definition of genetic and translational context will provide the data sets required to derive new classification schemes and the generation of a "biological road map" of prostate cancer, favoring the formulation of treatments tailored on patient specific tumor biology (Johnston and Lawler 2012). The end-point of this process will be the transition from the poorly understood, clinically heterogeneous prostate cancer superfamily to a collection of homogeneous molecular subtypes with the development of biomarkers able to distinguishing aggressive from indolent disease (Barbieri et al. 2012).

This approach holds promise as a way to maximize the benefit of targeted treatments while minimizing unnecessary side effects, with a predictable positive implications also for health economics (Johnston and Lawler 2012).

10.2 Interplay Between Genetic and Epigenetic Events in Prostate Cancer

Angela Celetti (✉)

The interplay between genetic and epigenetic events has a causative role in the development and progression of prostate cancer. In fact, loss or gains of several chromosomes have been reported at chromosomes 8p and 8q, loss at 5q, 6q,10q, 13q, 16q, 18 and gains at 1q, 3q, 7 and Xq12, as indicated in Fig. 10.1 (Ribeiro et al. 2006; Sun et al. 2007). Which genes might be affected by these genetic events on each chromosome is still object of investigation. For example gains and amplification at chromosome 8q may lead to overexpression of myc with increase in the proliferation of the epithelial prostate neoplastic cells. On the other hand loss of genes at 8p may determine the loss of the NKX3A gene whose activity consists in the regulation of prostate epithelial development (Fig. 10.1).

In the most aggressive histotypes, the loss of function of PTEN, RB1 and TP53 tumor suppressors, by allelic loss or mutation, has been found in advanced stage of the disease. Alterations of autocrine and paracrine growth factor signaling pathways are also very common, even if RAS mutations have been rarely reported, so far.

In more than half of the prostate cancers, chromosomal rearrangements involving oncogenic transcription factors of the ETS family have been reported (Kumar-Sinha et al. 2008; Tomlins et al 2005).

The major translocation reported involves chromosome 21 and creates a fusion gene, in which the androgen-responsive TMPRSS2 promoter induces the expression of the ERG transcription factor (Tomlins et al. 2005). Two different mechanisms, an internal deletion within the chromosome or a chromosomal rearrangement, in which a fragment of the chromosome 21, separating the two genes, is translocated elsewhere, could be envisaged at the basis of the translocation.

One of the genes involved in the chromosomal translocation, TMPRSS2 (androgen-regulated trans-membrane protease, serine 2), encodes for a serine

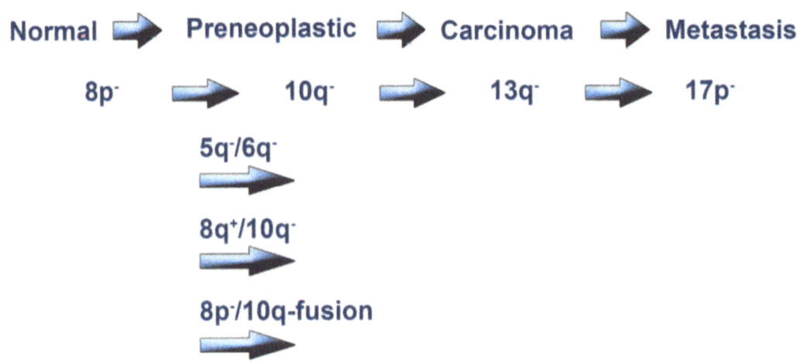

Fig. 10.1 Models of prostate cancer progression

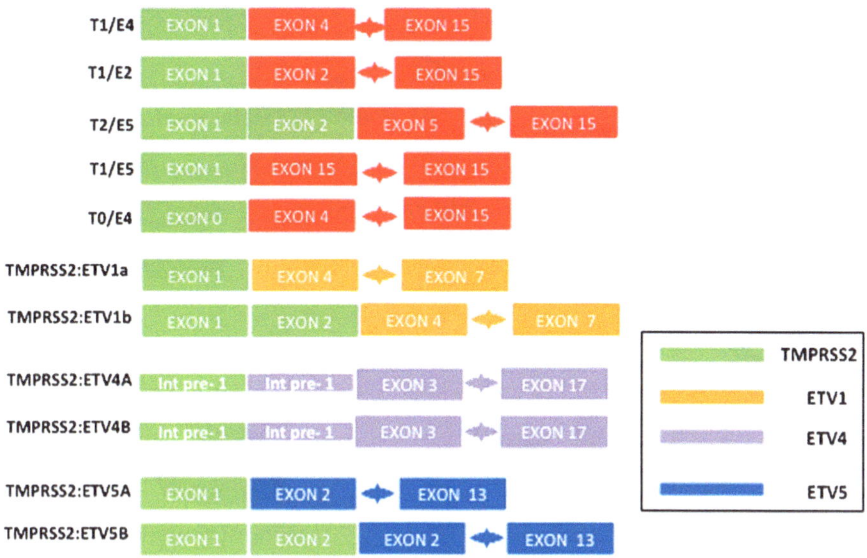

Fig. 10.2 Isoforms described to date of the TMRSS2-ERG fusion genes that give the idea of great instability of this rearrangement

protease secreted by prostate epithelial cells in an androgen-dependent manner (Afar et al. 2001), the other, ERG or ETV1, identified as member of the ETS family of oncogenes (Tomlins et al. 2005), has been previously classified as the most commonly overexpressed proto-oncogene in prostate cancer (72 % of all prostate cancer) (Petrovics et al. 2005). Both intra-chromosomal and inter-chromosomal genetic rearrangements create a fusion transcript involving ETS family members, whose activity is regulated by post-translational modifications.

The TMPRSS2 and ERG genes are roughly 3 megabases (Mb) distant on chromosome 21. In more than half of samples, fusion is the result of the deletion of the intervening DNA sequence, but fusion may also occur by a translocation (Yoshimoto et al. 2006; Tu et al. 2007). The exact points of DNA rupture, and the exons conserved in the fusion product, vary between patients and more than 20 TMPRSS2:ERG variants have been reported so far (Tomlins et al. 2005, 2006; Clark et al. 2007; Liu et al. 2007). Then, a nomenclature lists the variant transcripts, depending on which exons of the genes are involved (Clark et al. 2007). The most frequent variants result from the recombination between either exon 1 or exon 2 of TMPRSSR2 and exon 4 of ERG genes. Rarely, exons 2–5 have been reported. The fusion transcript including exon 1 of TMPRSS2 and exon 4 of ERG is one of the most described and identified as the T1/E4 following the above mentioned nomenclature (Clark et al. 2007) with a rate of up to 86 % among the reported fusions (Wang et al. 2006) (Fig. 10.2).

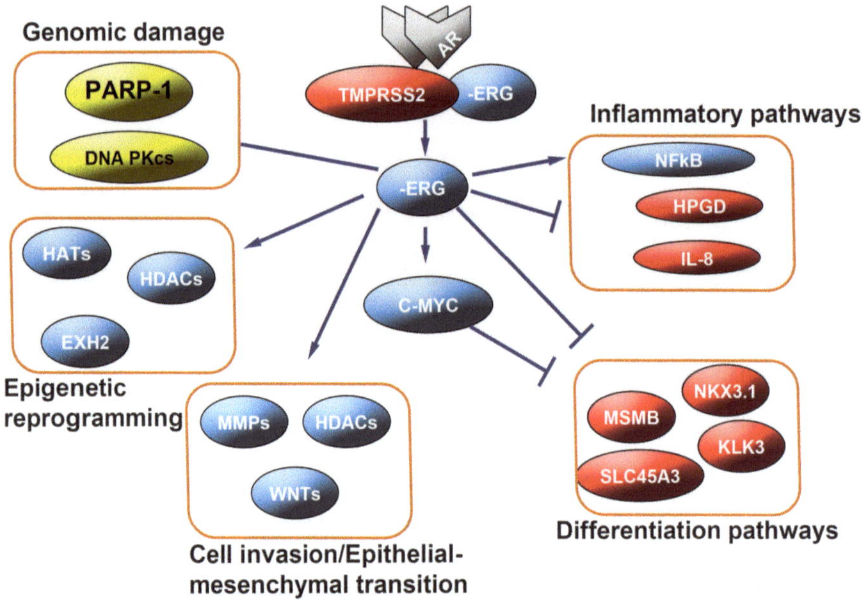

Fig. 10.3 ERG regulated prostate cancer pathways

The list of genes and the variants, involved in fusion transcripts, is continuously enlarging. In fact, new members of the ETS gene family (ETV4 and ETV5) have been reported in a few cases of prostate cancer (Tomlins et al. 2006; Helgeson et al. 2008). On the 5′ side of the translocation, new partners have also been described. A chimeric product derived from a variant isoform of TMPRSS2, which mapped 4 kb upstream of the more common start site has been reported (Lapointe et al. 2007). About 50 fusion partners for ETV1, comprising SLC5A3, HERV-K22q11.23, C15orf21, and HNRPA2B1 have been involved (Tomlins et al. 2007a, b; Helgeson et al. 2008). SLC5A3 recombine to ETV5, as well as to ETV1, but not to ERG (Tomlins et al. 2007a, b; Helgeson et al. 2008). Two additional fusion partners of ETV4 kallikrein 2 (KLK2) and calcium-activated nucleotidase 1 (CANT1) have been also reported (Hermans et al. 2008) (Fig. 10.3).

Overall, members of the ETS family are overexpressed in most prostate cancers and alternative mechanisms to gene fusions can be also envisaged. In fact, overexpression of ERG, in absence of fusion has been reported as well, but the underlying genetic mechanisms was not determined (Petrovics et al. 2005; Cai et al. 2007). Interestingly, androgen-dependent cases have been reported where the expression of androgen receptor and PSA levels are associated to the presence of TMPRSS2:ERG fusion transcript and to overexpression of the ERG gene. However, some androgen-independent cancers were found to harbour the TMPRSS2:ERG fusion transcript, in absence of the androgen receptor. Nevertheless, these tumors might have been dependent from androgens at the beginning of the transformation process. FlI-1 and ETV4 have been found overexpressed in androgen-independent advanced prostate tumors.

Among the ETV1 fusion partners originally reported (Tomlins et al. 2007a, b) three of them, TMPRSS2, SLC5A3, and HERV-K22q11.23, appear to be androgen-responsive and two, C15orf21 and HNRPA2B1, drive the constitutive overexpression of ETV1 in the absence of androgen stimulation. In the next future, the interplay between clinical studies and the molecular biology understanding of the tumor should help to distinguish the course of the disease, in cases of cancer with different fusion proteins, and should help to correlate the response to androgen ablation treatments.

Fusion oncogenes of this type may explain how androgens come to drive cell proliferation in prostate cancers, instead of promoting cell differentiation, favouring cell survival and maintaining regulating secretory function as in the normal prostate gland.

Nevertheless, several parallel pathways of genetic alterations may exist in prostate cancer and key genetic changes may determine the aggressiveness of the single tumor. Even if it is true that prostate cancers develop through several steps, a better understanding of the sequential genetic events could help us to perform an early diagnosis and to select a personalized therapy.

10.2.1 Characterisation of TMPRSS2erg Protein

The TMPRSS2-ERG gene fusion generates a chimeric transcript that combine the prostate-specific promoter of the TMPRSS2 gene to the ERG oncogene open reading frame (ORF). Thus, the protein sequences have been predicted from the sequence of the fusion ORFs. Among the various fusion transcripts identified from the cDNA sequence, some are predicted to generate premature stop codons and to encode for a truncated protein, not functional. In some other cases, non-aminoacid sequence derived from TMPTSS2 is integrated in the hybrid ORF and therefore a fusion protein is not created (Clark et al. 2007).

10.2.2 Prevalence of Fusion Product Among Unselected Prostate Cancer Cases

The presence of a gene fusion product can be determined with different methods, like RT-PCR, that detect the level of RNA expression, like FISH, which measure the inappropriate juxtaposition of non-adjacent sequences or the breakage of a single gene and fusion to different chromosome sites, or like the array technology that reveal the imbalance expression of individual exons. The assay used, the volume of cancer, the number of foci analysed and the number of chimeric variants studied in the screening panel may affect the rate and quality of the fusions reported so far. Moreover, a single cancer may have distinct foci that harbour different rearrangements involving separate genes, or no rearrangement at all. These data suggest that most of prostate cancers (more than 70 %) carry a fusion product (Hermans et al. 2006; Perner et al. 2006; Soller et al. 2006; Rajput et al. 2007; Tu et al. 2007; Nam et al. 2007). Since the number of variant species is continuously enlarging and the detection methods become always more sensitive, the proportion of prostate cancer samples containing more than one variant is predicted to increase progressively. Moreover, the heterogeneity of TMPRSS2:ERG gene fusion may account for the distinct foci of cancer that occur within a multifocal prostate cancer, which might represent different malignant clones and could, then, limit and delay the transfer to clinical use of the fusion products as putative biomarkers. Even if a complete characterization of the fusion products identified so far is still missing, an aggressive clinical behaviour has been reported together with the presence of blue-tinged mucin, a cribriform growth pattern, macronucleoli, intraductal tumor spread, and signet-ring cell features. Nevertheless, Gleason grade or stage, or PSA levels has not been associated with a particular type of fusion gene, yet (Perner et al. 2006; Wang et al. 2006; Lapointe et al. 2007; Rajput et al. 2007; Tu et al. 2007).

10.2.3 Clinical Significance of TMPRSS2:Erg Gene Fusion

Histologic grade (measured by the Gleason scoring system), tumor stage and PSA level at diagnosis are considered reliable prognostic factors for men with localised prostate cancer, so far. Men with tumors of higher grade (Gleason 8–10), stage (T3–T4), or PSA level (420 ng/ml) experience relatively high rates of progression to metastasis, when compared with men with tumors of lower grade, local stage, or low PSA level. Novel biomarkers are urgently needed in order to help to select specific treatments for individuals.

In conclusion, the original discovery by Tomlins et al. in 2005 of a frequent genetic event in prostate cancers has highlighted the role of chromosomal rearrangements in the aetiologies of common solid tumors. The importance of this genetic fusion have been confirmed and the classes of fusion genes, that are now considered among the most frequent recurrent rearrangements in cancer, have been enlarging. The consequence of the various chimeric transcripts is the overexpression of a member of the ETS family of oncogenes that tend to lose the androgen dependence in advanced disease after an initial phase of androgen control, lost later in advanced disease. The activation of this pathway may be causative to prostate carcinogenesis, but the clinical implication of the various fusion products is still under characterization. All the efforts are, in fact, now focalized to classify patients with different risk, identify a screening test and finally target the ETS family oncogene to open the way to novel molecular therapies.

References

Afar DE, Vivanco I, Hubert RS, Kuo J, Chen E, Saffran DC, Raitano AB, Jakobovits A (2001) Catalytic cleavage of the androgen-regulated TMPRSS2 protease results in its secretion by prostate and prostate cancer epithelia. Cancer Res 61:1686–1692

Barbieri CE, Baca SC, Lawrence MS, Demichelis F, Blattner M, Theurillat JP, White TA, Stojanov P, Van Allen E, Stransky N, Nickerson E, Chae SS, Boysen G, Auclair D, Onofrio RC, Park K, Kitabayashi N, MacDonald TY, Sheikh K, Vuong T, Guiducci C, Cibulskis K, Sivachenko A, Carter SL, Saksena G, Voet D, Hussain WM, Ramos AH, Winckler W, Redman MC, Ardlie K, Tewari AK, Mosquera JM, Rupp N, Wild PJ, Moch H, Morrissey C, Nelson PS, Kantoff PW, Gabriel SB, Golub TR, Meyerson M, Lander ES, Getz G, Rubin MA, Garraway LA (2012) Exome sequencing identifies recurrent SPOP, FOXA1 and MED12 mutations in prostate cancer. Nat Genet 44(6):685–689

Cai C, Hsieh CL, Omwancha J, Zheng Z, Chen SY, Baert JL, Shemshedini L (2007) ETV1 is a novel androgen receptor-regulated gene that mediates prostate cancer cell invasion. Mol Endocrinol 21:1835–1846

Clark J, Merson S, Jhavar S, Flohr P, Edwards S, Foster CS, Eeles R, Martin FL, Phillips DH, Crundwell M, Christmas T, Thompson A, Fisher C, Kovacs G, Cooper C (2007) Diversity of TMPRSS2 – ERG fusion transcripts in the human prostate. Oncogene 26:2667–2673

Dong JT (2006) Prevalent mutations in prostate cancer. J Cell Biochem 7(3):433–447. Review

Gimba ER, Barcinski MA (2003) Molecular aspects of prostate cancer: implications for future directions. Int Braz J Urol 29(5):401–410; discussion 411

Greenman C, Stephens P, Smith R, Dalgliesh GL, Hunter C, Bignell G, Davies H, Teague J, Butler A, Stevens C, Edkins S, O'Meara S, Vastrik I, Schmidt EE, Avis T, Barthorpe S, Bhamra G, Buck G, Choudhury B, Clements J, Cole J, Dicks E, Forbes S, Gray K, Halliday K, Harrison R, Hills K, Hinton J, Jenkinson A, Jones D, Menzies A, Mironenko T, Perry J, Raine K, Richardson D, Shepherd R, Small A, Tofts C, Varian J, Webb T, West S, Widaa S, Yates A, Cahill DP, Louis DN, Goldstraw P, Nicholson AG, Brasseur F, Looijenga L, Weber BL, Chiew YE, DeFazio A, Greaves MF, Green AR, Campbell P, Birney E, Easton DF, Chenevix-Trench G, Tan MH, Khoo SK, Teh BT, Yuen ST, Leung SY, Wooster R, Futreal PA, Stratton MR (2007) Patterns of somatic mutation in human cancer genomes. Nature 446(7132):153–158

Helgeson BE, Tomlins SA, Shah N, Laxman B, Cao Q, Prensner JR, Cao X, Singla N, Montie JE, Varambally S, Mehra R, Chinnaiyan AM (2008) Characterization of TMPRSS2:ETV5 and SLC45A3:ETV5 gene fusions in prostate cancer. Cancer Res 68:73–80

Hermans KG, van Marion R, van Dekken H, Jenster G, van Weerden WM, Trapman J (2006) TMPRSS2ERG fusion by translocation or interstitial deletion is highly relevant in androgen dependent prostate cancer, but is bypassed in late-stage androgen receptor-negative prostate cancer. Cancer Res 66:10658–10663

Hermans KG, Bressers AA, van der Korput HA, Dits NF, Jenster G, Trapman J (2008) Two unique novel prostate-specific and androgen-regulated fusion partners of ETV4 in prostate cancer. Cancer Res 68:3094–3098

Johnston PG, Lawler M (2012) Expert opinion: future frontiers and challenges in cancer medicine. Oncologist 17(5):e3–e5

Kumar A et al (2011) Exome sequencing identifies a spectrum of mutation frequencies in advanced and lethal prostate cancers. Proc Natl Acad Sci USA 108:17087–17092

Kumar-Sinha C, Tomlins SA, Chinnaiyan AM (2008) Recurrent gene fusions in prostate cancer. Nat Rev Cancer 8(7):497–511

Lapointe J et al (2004) Gene expression profiling identifies clinically relevant subtypes of prostate cancer. Proc Natl Acad Sci USA 101(3):811–816

Lapointe J, Li C, Giacomini CP, Salari K, Huang S, Wang P et al (2007) Genomic profiling reveals alternative genetic pathways of prostate tumorigenesis. Cancer Res 67:8504–8510

Liu W, Ewing CM, Chang BL, Li T, Sun J, Turner AR, Dimitrov L, Zhu Y, Sun J, Kim JW, Zheng SL, Isaacs WB, Xu J (2007) Multiple genomic alterations on 21q22 predict various TMPRSS2/ERG fusion transcripts in human prostate cancers. Genes Chromosomes Cancer 46(11):972–980

Magee JA, Araki T, Patil S, Shrig T, True L, Humphrey PA et al (2001) Expression profile reveals hepsin overexpression in prostate cancer. Cancer Res 61:5692–5696

Nam RK, Sugar L, Yang W, Srivastava S, Klotz LH, Yang LY et al (2007) Expression of the TMPRSS2:ERGfusion gene predicts cancer recurrence after surgery for localised prostate cancer. Br J Cancer 97:1690–1695

Palanisamy N et al (2010) Rearrangements of the RAF kinase pathway in prostate cancer, gastric cancer and melanoma. Nat Med 16(7):793–798

Perner S, Demichelis F, Beroukhim R, Schmidt FH, Mosquera JM, Setlur S, Tchinda J, Tomlins SA, Hofer MD, Pienta KG, Kuefer R, Vessella R, Sun XW, Meyerson M, Lee C, Sellers WR, Chinnaiyan AM, Rubin MA (2006) TMPRSS2:ERG fusion-associated deletions provide insight into the heterogeneity of prostate cancer. Cancer Res 66:8337–8341

Petrovics G, Liu A, Shaheduzzaman S, Furasato B, Sun C, Chen Y, Nau M, Ravindranath L, Chen Y, Dobi A, Srikantan V, Sesterhenn IA, McLeod DG, Vahey M, Moul JW, Srivastava S (2005) Frequent overexpression of ETS-related gene-1 (ERG1) in prostate cancer transcriptome. Oncogene 24:3847–3852

Pleasance ED, Cheetham RK, Stephens PJ, McBride DJ, Humphray SJ, Greenman CD, Varela I, Lin ML, Ordóñez GR, Bignell GR, Ye K, Alipaz J, Bauer MJ, Beare D, Butler A, Carter RJ, Chen L, Cox AJ, Edkins S, Kokko-Gonzales PI, Gormley NA, Grocock RJ, Haudenschild CD, Hims MM, James T, Jia M, Kingsbury Z, Leroy C, Marshall J, Menzies A, Mudie LJ, Ning Z, Royce T, Schulz-Trieglaff OB, Spiridou A, Stebbings LA, Szajkowski L, Teague J, Williamson D, Chin L, Ross MT, Campbell PJ, Bentley DR, Futreal PA, Stratton MR (2010a) A comprehensive catalogue of somatic mutations from a human cancer genome. Nature 463(7278):191–196

Pleasance ED, Stephens PJ, O'Meara S, McBride DJ, Meynert A, Jones D, Lin ML, Beare D, Lau KW, Greenman C, Varela I, Nik-Zainal S, Davies HR, Ordoñez GR, Mudie LJ, Latimer C, Edkins S, Stebbings L, Chen L, Jia M, Leroy C, Marshall J, Menzies A, Butler A, Teague JW, Mangion J, Sun YA, McLaughlin SF, Peckham HE, Tsung EF, Costa GL, Lee CC, Minna JD, Gazdar A, Birney E, Rhodes MD, McKernan KJ, Stratton MR, Futreal PA, Campbell PJ (2010b) A small-cell lung cancer genome with complex signatures of tobacco exposure. Nature 463(7278):184–190

Pourmand G, Ziaee AA, Abedi AR, Mehrsai A, Alavi HA, Ahmadi A, Saadati HR (2007) Role of PTEN gene in progression of prostate cancer. Urol J 4(2):95–100

Rajput AB, Miller MA, De Luca A, Boyd N, Leung S, Hurtado-Coll A, Fazli L, Jones EC, Palmer JB, Gleave ME, Cox ME, Huntsman DG (2007) Frequency of the TMPRSS2:ERG gene fusion is increased in moderate to poorly differentiated prostate cancers. J Clin Pathol 60:1238–1243

Ribeiro FR, Henrique R, Hektoen M, Berg M, Jeronimo C, Teixeira MR et al (2006) Comparison of chromosomal and array-based comparative genomic hybridization for the detection of genomic imbalances in primary prostate carcinomas. Mol Cancer 5:33

Singh D, Febbo PG, Ross K, Jackson DG, Manola J, Ladd C et al (2002a) Gene expression correlates of clinical prostate behavior. Cancer Cell 1:203–209

Singh J et al (2002b) Prostate epithelial expression of a novel androgen target gene. J Androl 23(5):652–660

Soller MJ, Isaksson M, Elfving P, Soller W, Lundgren R, Panagopoulos I (2006) Confirmation of the high frequency of the TMPRSS2/ERG fusion gene in prostate cancer. Genes Chromosomes Cancer 45:717–719

Sun J, Liu W, Adams TS, Sun J, Li X, Turner AR et al (2007) DNA copy number alterations in prostate cancers: a combined analysis of published CGH studies. Prostate 67:692–700

Taylor BS, Schultz N, Hieronymus H, Gopalan A, Xiao Y, Carver BS, Arora VK, Kaushik P, Cerami E, Reva B, Antipin Y, Mitsiades N, Landers T, Dolgalev I, Major JE, Wilson M, Socci

ND, Lash AE, Heguy A, Eastham JA, Scher HI, Reuter VE, Scardino PT, Sander C, Sawyers CL, Gerald WL (2010) Integrative genomic profiling of human prostate cancer. Cancer Cell 18(1):11–22

Tomlins SA, Rhodes DR, Perner S, Dhanasekaran SM, Mehra R, Sun XW, Varambally S, Cao X, Tchinda J, Kuefer R, Lee C, Montie JE, Shah RB, Pienta KJ, Rubin MA, Chinnaiyan AM (2005) Recurrent fusion of TMPRSS2 and ETS transcription factor genes in prostate cancer. Science 310:644–648

Tomlins SA, Mehra R, Rhodes DR, Smith LR, Roulston D, Helgeson BE, Cao X, Wei JT, Rubin MA, Shah RB, Chinnaiyan AM (2006) TMPRSS2:ETV4 gene fusions define a third molecular subtype of prostate cancer. Cancer Res 66:3396–3400

Tomlins SA, Laxman B, Dhanasekaran SM, Helgeson BE, Cao X, Morris DS, Menon A, Jing X, Cao Q, Han B, Yu J, Wang L, Montie JE, Rubin MA, Pienta KJ, Roulston D, Shah RB, Varambally S, Mehra R, Chinnaiyan AM (2007a) Distinct classes of chromosomal rearrangements create oncogenic ETS gene fusions in prostate cancer. Nature 448(7153): 595–599

Tomlins SA et al (2007b) Integrative molecular concept modeling of prostate cancer progression. Nat Genet 39(1):41–51

Tomlins SA et al (2008) Role of the TMPRSS2-ERG gene fusion in prostate cancer. Neoplasia 10(2):177–188

Tu JJ, Rohan S, Kao J, Kitabayashi N, Mathew S, Chen YT (2007) Gene fusions between TMPRSS2 and ETS family genes in prostate cancer: frequency and transcript variant analysis by RT–PCR and FISH on paraffin-embedded tissues. Mod Pathol 20:921–928

Wang J, Cai Y, Ren C, Ittmann M (2006) Expression of variant TMPRSS2/ ERG fusion messenger RNAs is associated with aggressive prostate cancer. Cancer Res 66:8347–8351

Yoshimoto M, Joshua AM, Chilton-Macneill S, Bayani J, Selvarajah S, Evans AJ, Zielenska M, Squire JA (2006) Three-color FISH analysis of TMPRSS2/ERG fusions in prostate cancer indicates that genomic microdeletion of chromosome 21 is associated with rearrangement. Neoplasia 8:465–469

Chapter 11
Mapping Prostate Cancer Aggressiveness Loci

Maria Siano, Silvia Varricchio, and Gennaro Ilardi

Abstract Non-metastatic primary prostate cancers frequently contain multiple independent histologic foci of cancer, and appear as truly multifocal tumors, since these different foci are often genetically distinct (Aihara et al., Urology 43:60–66, 1994; Bostwick et al., Cancer 83:1995–2002,1998; Macintosh et al., Cancer Res 58:23–28, 1998; Mehra et al., Cancer Res 67:7991–7995, 2007; Clark et al., Oncogene 27:1993–2003, 2008). By converse, multiple metastases in the same patient are clonally related, indicating that advanced prostate cancer is monoclonal both at molecular and cytogenetic level (Mehra et al., Cancer Res 68:3584–3590, 2008; Liu et al. 2009).

Genomics will hopefully allow the sub-typing of prostate cancer for diagnostic purposes, overcoming the limits of morphology to prognostically evaluate this tumor (Taylor et al., Cancer Cell 18(1):11–22, 2010). The major part of studies searching for genetic variants correlated with prostate cancer, have yielded preferentially results indicating that several genetic loci impact early stages of prostate cancer development. Few data exist, to date, on the existence of loci unequivocally correlated with prostate cancer progression (Liu et al., Front Endocrinol (Lausanne) 3:72, 2012).

Recently, however, integrative genomic analysis techniques identified copy number variation as a biomarker predictive of prostate cancer outcome (Ding et al., Nature 470(7333):269–273, 2011), and comparative oncogenomics have derived a four-gene signature and an additional pathway-representative fourteen-gene panel that resulted better prognostic biomarkers with respect to PSA (Ding et al., Nature 470(7333):269–273, 2011). Moreover, the use of a five-SNP panel was found useful to predict the aggressive behavior of prostate cancer (Lin et al., Cancer Epidemiol

M. Siano (✉) • S. Varricchio • G. Ilardi
Department of Advanced Biomedical Sciences, Pathology Section, Faculty of Medicine and Surgery, University of Naples "Federico II", via S. Pansini, n.5, Naples, Italy
e-mail: dysian@tin.it; silvia.varricchio@gmail.com; gennaro.ilardi@gmail.com

S. Staibano (ed.), *Prostate Cancer: Shifting from Morphology to Biology*, DOI 10.1007/978-94-007-7149-9_11, © Springer Science+Business Media Dordrecht 2013

Biomarkers Prev 20:954–961, 2011). Thus, the advancement of genetic techniques has opened up the promising scenario of a next coming mapping of prostate cancer aggressiveness loci.

According to the most recent scientific data, prostate cancer can be regarded as a collection of cancer subtypes, each characterized by a different set of molecular and/or genetic alterations (Barbieri et al. 2012). Numerous genomic rearrangements are frequent in tumor cells, and are frequently responsible for a growth advantage. In addition, inherited genetic variants both in the form of SNPs and CNVs can predispose to disease progression, potentially favouring specific somatic events. Prostate cancer has resulted attributable to hereditary factors more frequently that most of the other solid tumors, so the germline risk factors may predispose to prostate cancer development or even to its aggressive form (Lichtenstein et al. 2000).

Specifically, genetic defects was estimated as 42 %, and a number of them likely may have a role in modulating prostate cancer progression.

Single nucleotide polymorphisms (SNPs) represent the most common source of genetic variation among humans. Several variants are associated with disease progression and adverse outcome (Cheng et al. 2010; Holt et al. 2008, 2010; Huang et al. 2010; Nguyen et al. 2010; Wright et al. 2010; Gallagher et al. 2010).

Lin. et al., in 2011, reported for the first time the independent association with PCa-specific mortality for five SNPs involving ARVCF, LEPR, CRY1, RNASEL and IL4 genes (Lin et al. 2011).

A series of previous studies, both genome-wide association studies (GWAS) and family linkage analyses, reported on multiple independent (SNPs) as PCa risk markers. In 2006, Amundadottir et al. (2006) and Freedman et al. (2006) detected PCa risk SNPs in three regions of 8q24 using linkage analysis followed by fine-mapping in an Icelandic family (Amundadottir et al. 2006) and using an admixture scan approach in West African ancestry men. Multiple other loci have subsequently been identified (Eeles et al. 2008, 2009; Thomas et al. 2008) and replicated (Zheng et al. 2008). On one hand, the data clearly reflect the strong genetic component involved in PCa incidence. On the other hand, the modest reported effects of the risk SNPs diminish their suitability in disease detection applications. In addition, it has been extremely challenging to demonstrate a functional role for these risk SNPs, which are most often outside gene coding areas.

Most of the studies on the correlation between inherited genetic variants and prostate cancer biology, have indicated then that several genetic loci impact early stages of prostate cancer development, rather than its metastasizing phase. In some instances, a clusterization of several genetic loci has been shown in prostate cancer patients, but its biological effects are still unclarified.

However, in the last few years, several reports have indicated that this eveniences could be more frequent than it has been previously thought.

8q24 is the most frequently involved chromosomal region in prostate tumors and the rs1447295 is its most reported SNP. The genes closest to 8q24 are FAM84B, a breast cancer membrane-associated protein, and the oncogene c-MYC. In PCa c-MYC is over-expressed while its reduction inhibits tumor growth; it has been hypothesized that the risk variant in 8q24 modifies c-MYC regulation, predisposing to genomic instability associated with aggressive tumors, hormone independence and poor prognosis (Sturge et al. 2011).

Nine out twenty-six prostate cancer susceptibility loci have been recently associated with prostate cancer progression after accounting for known predictors of prostate cancer outcomes like PSA, Gleason score, stage, and primary treatment (Sturge et al. 2011).

Five loci (ITGA6, NUDT10/NUDT11, KIAA1211, SLCC22A3, HNF1B/TCF) were found as independent factors predictive for progression. The strongest association was observed at the ITGA6 locus, that encodes for alpha 6 integrin, involved in important features of tumor invasion, cell adhesion, migration, and signaling (Xu et al. 2008). The predictive relevance of this locus has been recently outlined in a follow-up study, and in a further study, outlining its association with a poor prognosis of prostate cancer (Eeles et al. 2009). This association may relate to pleiotropy in the biological effects of alpha 6 integrin. Alteration in integrin function may affect, as a result of genetic variations, cell adhesion and migration activity (Eeles et al. 2009). Prostate tumors persistently express alpha 6 integrin, which in preclinical studies has been linked to increased tumor cell invasion, migration, and metastasis, supporting the role of ITGA6 locus in prostate cancer progression (Abhijit et al. 2011).

The SNP rs10486567 on chromosome 7 occurs in the second intron of the JAZF zinc finger 1 gene (JAZF1) that encodes for a transcriptional repressor of NR2C2 (highly expressed in prostate tissue, interacts with androgen receptor to repress its target gene expression). In a study from Gallagher et al. (2010), this SNP resulted associated with biochemical recurrence and castration-resistant metastases.

On chromosome 10 two SNPs (rs10993994 and rs7920517) have been identified in the region containing the microsemino-protein beta gene (MSMB) that encodes PSP94, a member of the immunoglobulin binding factor family synthesized by epithelial cells in prostate and secreted in seminal plasma. Loss of PSP94 is associated with recurrence after prostatectomy, while, in androgen-insensitive PCa, MSMB is silenced by EZH2. Leading to the conclusion that variants in MSMB gene may represent a genetic risk marker for PCa.

A SNP (rs2735839) is located between the KLK2 and KLK3genes, that have been reported to influence the PCa risk. SNPs in the promoter region of KLK3 have been associated with PSA concentrations, and, in some cases, with PCa risk and stage of disease (Kader et al. 2009).

rs4962416 falls in the fifth intron of the CTBP2 gene, that encodes for a transcriptional co-repressor activated under metabolic stress, highly expressed in prostate tissue. Its expression is associated with decreased PTEN expression and posphatidylinositol3-kinase pathway activation.

rs9623117 and rs7291619 are on 22q13 and are in linkage disequilibrium. They occurs within the TNRC6B gene that encode a protein that mediate miRNA-guided mRNA cleavage. Its expression is suppressed in hormone-refractory metastatic PCa compared to prostate carcinoma and genetic variation may alter the levels of mRNA species under its control and, therefore, contribute to carcinogenesis(for an extensive discussion on these topics see Chap. 9).

The second most common source of variation among humans regards copy number variants (CNVs), (Feuk et al. 2006; Freeman et al. 2006) defined as copy number changes – gains or losses – of stretches of DNA between a few hundred bases to several Mb wide. Similar to SNPs, CNVs are seen commonly in the genome of healthy individuals (Redon et al. 2006; McCarroll et al. 2008) and confer susceptibility to diseases such as Alzheimer's disease, Parkinson's disease, mental retardation, autism, bipolar disorder and schizophrenia (Stankiewicz and Lupski 2010; Zhang et al. 2009).

The occurrence rates of SNPs and CNVs are different, where CNVs have higher rates of occurrences, suggesting that these two types of polymorphisms potentially carry different information (Korbel et al. 2008).

Among recurrent non-synonymous mutations, the most common in prostate cancer involves SPOP, a gene encoding for the substrate-recognition component of a Cullin3-based E3-ubiquitin ligase (Nagai et al. 1997; Hernandez-Munoz et al. 2005). Mutations in SPOP have been reported originally in three recent sequencing studies (Kan et al. 2010; Berger et al. 2011; Zhuang et al. 2009). With an incidence ranging from 6 to 13 % of human prostate cancers.

Structural analysis suggests that these mutations will inactivate SPOP function by disrupting SPOP–substrate interaction, and this in prostate cell lines resulted in increased invasion and altered gene expression; evidence of this expression signature was identified also in primary tumor displaying characteristic somatic copy number aberrations. This could be the starting point for the identification of a new distinct molecular class of PCa.

Overall, the chance to incorporate genomic analysis into prostate cancer screening could lead in the next future to the formulation of risk-stratified population screening, including a polygenic risk profile. Basing on the emerging data, it may allow us to more efficiently subtyping prostate cancer by a prognostic point-of-view with respect to the current age-stratified screening (Chowdhury et al. 2013).

References

Abhijit MG, Vincent C, Njar O (2011) Prostate cancer: current and emerging therapies. In: Ozdemir O (ed) Current cancer treatment: novel beyond conventional approaches. InTech. Rijeka, Croatia. ISBN: 978-953-307-397-2. doi: 10.5772/23893

Aihara M, Wheeler TM, Ohori M, Scardino PT (1994) Heterogeneity of prostate cancer in radical prostatectomy specimens. Urology 43:60–66; discussion 6–7

Amundadottir LT, Sulem P, Gudmundsson J et al (2006) A common variant associated with prostate cancer in European and African populations. Nat Genet 38:652–658

Barbieri CE, Baca SC, Lawrence MS, Demichelis F, Blattner M, Theurillat JP, White TA, Stojanov P, Van Allen E, Stransky N, Nickerson E, Chae SS, Boysen G, Auclair D, Onofrio RC, Park K, Kitabayashi N, MacDonald TY, Sheikh K, Vuong T, Guiducci C, Cibulskis K, Sivachenko A, Carter SL, Saksena G, Voet D, Hussain WM, Ramos AH, Winckler W, Redman MC,

Ardlie K, Tewari AK, Mosquera JM, Rupp N, Wild PJ, Moch H, Morrissey C, Nelson PS, Kantoff PW, Gabriel SB, Golub TR, Meyerson M, Lander ES, Getz G, Rubin MA, Garraway LA (2012) Exome sequencing identifies recurrent SPOP, FOXA1 and MED12 mutations in prostate cancer. Nat Genet 44(6):685–689

Berger MF, Lawrence MS, Demichelis F et al (2011) The genomic complexity of primary human prostate cancer. Nature 470:214–220

Bostwick DG, Shan A, Qian J, Darson M, Maihle NJ, Jenkins RB et al (1998) Independent origin of multiple foci of prostatic intraepithelial neoplasia: comparison with matched foci of prostate carcinoma. Cancer 83:1995–2002

Cheng I, Plummer SJ, Neslund-Dudas C, Klein EA, Casey G, Rybicki BA, Witte JS (2010) Prostate cancer susceptibility variants confer increased risk of disease progression. Cancer Epidemiol Biomarkers Prev 19(9):2124–2132

Chowdhury S, Dent T, Pashayan N, Hall A, Lyratzopoulos G, Hallowell N, Hall P, Pharoah P, Burton H (2013) Incorporating genomics into breast and prostate cancer screening: assessing the implications. Genet Med 15(6):423–432, Epub 2013 Feb 14

Clark J, Attard G, Jhavar S, Flohr P, Reid A, De-Bono J et al (2008) Complex patterns of ETS gene alteration arise during cancer development in the human prostate. Oncogene 27:1993–2003

Ding Z, Wu CJ, Chu GC, Xiao Y, Ho D, Zhang J, Perry SR, Labrot ES, Wu X, Lis R, Hoshida Y, Hiller D, Hu B, Jiang S, Zheng H, Stegh AH, Scott KL, Signoretti S, Bardeesy N, Wang YA, Hill DE, Golub TR, Stampfer MJ, Wong WH, Loda M, Mucci L, Chin L, DePinho RA (2011) SMAD4-dependent barrier constrains prostate cancer growth and metastatic progression. Nature 470(7333):269–273

Eeles RA, Kote-Jarai Z, Giles GG et al (2008) Multiple newly identified loci associated with prostate cancer susceptibility. Nat Genet 40:316–321

Eeles RA, Kote-Jarai Z, Al Olama AA et al (2009) Identification of seven new prostate cancer susceptibility loci through a genomewide association study. Nat Genet 41:1116–1121

Feuk L, Marshall CR, Wintle RF, Scherer SW (2006) Structural variants: changing the landscape of chromosomes and design of disease studies. Hum Mol Genet 15(Spec. no. 1):R57–R66

Freedman ML, Haiman CA, Patterson N et al (2006) Admixture mapping identifies 8q24 as a prostate cancer risk locus in African-American men. Proc Natl Acad Sci USA 103: 14068–14073

Freeman JL, Perry GH, Feuk L et al (2006) Copy number variation: new insights in genome diversity. Genome Res 16:949–961

Gallagher DJ, Vijai J, Cronin AM, Bhatia J, Vickers AJ, Gaudet MM, Fine S, Reuter V, Scher HI, Halldén C, Dutra-Clarke A, Klein RJ, Scardino PT, Eastham JA, Lilja H, Kirchhoff T, Offit K (2010) Susceptibility loci associated with prostate cancer progression and mortality. Clin Cancer Res 16(10):2819–2832

Hernandez-Munoz I, Lund AH, van der Stoop P et al (2005) Stable X chromosome inactivation involves the PRC1 Polycomb complex and requires histone MACROH2A1 and the CULLIN3SPOP ubiquitin E3 ligase. Proc Natl Acad Sci USA 102:7635–7640

Holt SK, Karyadi DM, Kwon EM, Stanford JL, Nelson PS, Ostrander EA (2008) Association of megalin genetic polymorphisms with prostate cancer risk and prognosis. Clin Cancer Res 14:3823–3831

Holt SK, Kwon EM, Koopmeiners JS et al (2010) Vitamin D pathway gene variants and prostate cancer prognosis. Prostate 70:1448–1460

Huang SP, Ting WC, Chen LM, Huang LC, Liu CC, Chen CW, Hsieh CJ, Yang WH, Chang TY, Lee HZ, Bao BY (2010) Association analysis of Wnt pathway genes on prostate-specific antigen recurrence after radical prostatectomy. Ann Surg Oncol 17(1):312–322

Kader AK, Sun J, Isaacs SD et al (2009) Individual and cumulative effect of prostate cancer risk-associated variants on clinicopathologic variables in 5,895 prostate cancer patients. Prostate 69:1195–1205

Kan Z, Jaiswal BS, Stinson J et al (2010) Diverse somatic mutation patterns and pathway alterations in human cancers. Nature 466:869–873

Korbel JO, Kim PM, Chen X et al (2008) The current excitement about copy-number variation: how it relates to gene duplications and protein families. Curr Opin Struct Biol 18:366–374

Lichtenstein P, Holm NV, Verkasalo PK et al (2000) Environmental and heritable factors in the causation of cancer – analyses of cohorts of twins from Sweden, Denmark, and Finland. N Engl J Med 343:78–85

Lin DW, Fitzgerald LM, Fu R et al (2011) Genetic variants in the LEPR CRY1, RNASEL, IL4, and ARVCF genes are prognostic markers of prostate cancer-specific mortality. Cancer Epidemiol Biomarkers Prev 20:954–961

Liu Y, Hegde P, Zhang F, Hampton G, Jia S (2012) Prostate cancer – a biomarker perspective. Front Endocrinol (Lausanne) 3:72

Liu W, Laitinen S, Khan S, Vihinen M, Kowalski J, Yu G et al (2009) Copy number analysis indicates monoclonal origin of lethal metastatic prostate cancer. Nat Med 15:559–565

Macintosh CA, Stower M, Reid N, Maitland NJ (1998) Precise microdissection of human prostate cancers reveals genotypic heterogeneity. Cancer Res 58:23–28

McCarroll SA, Kuruvilla FG, Korn JM et al (2008) Integrated detection and population-genetic analysis of SNPs and copy number variation. Nat Genet 40:1166–1174

Mehra R, Han B, Tomlins SA, Wang L, Menon A, Wasco MJ et al (2007) Heterogeneity of TMPRSS2 gene rearrangements in multifocal prostate adenocarcinoma: molecular evidence for an independent group of diseases. Cancer Res 67:7991–7995

Mehra R, Tomlins SA, Yu J, Cao X, Wang L, Menon A et al (2008) Characterization of TMPRSS2-ETS gene aberrations in androgen-independent metastatic prostate cancer. Cancer Res 68:3584–3590

Nagai Y, Kojima T, Muro Y et al (1997) Identification of a novel nuclear speckle-type protein, SPOP. FEBS Lett 418:23–26

Nguyen PL, Ma J, Chavarro JE et al (2010) Fatty acid synthase polymorphisms, tumor expression, body mass index, prostate cancer risk, and survival. J Clin Oncol 28:3958–3964

Redon R, Ishikawa S, Fitch KR et al (2006) Global variation in copy number in the human genome. Nature 444:444–454

Stankiewicz P, Lupski JR (2010) Structural variation in the human genome and its role in disease. Annu Rev Med 61:437–455

Sturge J, Caley MP, Waxman J (2011) Bone metastasis in prostate cancer: emerging therapeutic strategies. Nat Rev Clin Oncol 8(6):357–368

Taylor BS, Schultz N, Hieronymus H, Gopalan A, Xiao Y, Carver BS, Arora VK, Kaushik P, Cerami E, Reva B, Antipin Y, Mitsiades N, Landers T, Dolgalev I, Major JE, Wilson M, Socci ND, Lash AE, Heguy A, Eastham JA, Scher HI, Reuter VE, Scardino PT, Sander C, Sawyers CL, Gerald WL (2010) Integrative genomic profiling of human prostate cancer. Cancer Cell 18(1):11–22

Thomas G, Jacobs KB, Yeager M et al (2008) Multiple loci identified in a genome-wide association study of prostate cancer. Nat Genet 40:310–315

Wright JL, Kwon EM, Lin DW et al (2010) CYP17 polymorphisms and prostate cancer outcomes. Prostate 70:1094–1101

Xu J, Isaacs SD et al (2008) Association of prostate cancer risk variants with clinicopathologic characteristics of the disease. Clin Cancer Res 14:5819–5824

Zhang F, Gu W, Hurles ME, Lupski JR (2009) Copy number variation in human health, disease, and evolution. Annu Rev Genomics Hum Genet 10:451–481

Zheng SL, Sun J, Wiklund F et al (2008) Cumulative association of five genetic variants with prostate cancer. N Engl J Med 358:910–919

Zhuang M, Calabrese MF, Liu J et al (2009) Structures of SPOP–substrate complexes: insights into molecular architectures of BTB-Cul3 ubiquitin ligases. Mol Cell 36:39–50

Chapter 12
Epigenetic Mechanisms: Histone Acetylation, DNA Methylation, miRNA, Chromatin Modifiers

Angela Celetti

Abstract In prostate cancer DNA methylation have been recorded as one of the main epigenetic event. More than 50 genes, in fact, have been found hypermethylated in prostate cancer (Li et al., Biochim Biophys Acta 1704:87–102, 2004; Nelson et al., Front Biosci 12:4254–4266, 2007). However, only a small number of genes, e.g. RASSF1A, RARB2, APC, and GSTP1 (Florl et al., Br J Cancer 91:985–994, 2004) or GSTP1, APC and MDR1 (Enokida et al., Clin Cancer Res 11:6582–6588, 2005) can help to discriminate between benign and cancerous changes in the prostate, by hypermethylation assay. The assays for genes hypermethylated in a fraction of the cases may help to distinguish different subgroups of prostate cancer, even if the low sensitivity of the assay is mostly dependent on the amount of DNA obtained by the primary sample (biopsy, serum, urine). Then, the reliability of this tests, unfortunately, is still low (Li et al., Biochim Biophys Acta 1704:87–102, 2004).

Hypermethylation at some genes may be dynamic during tumor progression, as in the case of the ESR2 gene encoding the estrogen receptor (Zhu et al. 2004). Otherwise, methylation at some genes is varying, affecting the methylation-specific PCR assays. Hypermethylation of genes like GSTP1, detected in high percentage of prostate cancers, occur at an early stage of cancer development and is detected in the best-established precursor lesion of prostate cancer, high-grade prostatic intraepithelial neoplasia (HG-PIN). The specificity of genes methylation in carcinoma is confirmed by the weak methylation of the same genes in morphologically benign adjacent areas. A "field effect" which may promote the onset of carcinomas has also been reported (Florl et al. 2004; Aitchinson et al. 2007).

A. Celetti (✉)
Institute of Experimental Endocrinology and Oncology 'G. Salvatore' IEOS, CNR, c/o Department of Molecular Medicine and Medical Biotechnology, University of Naples, "Federico II", Naples, Italy
e-mail: celetti@unina.it

S. Staibano (ed.), *Prostate Cancer: Shifting from Morphology to Biology*,
DOI 10.1007/978-94-007-7149-9_12, © Springer Science+Business Media Dordrecht 2013

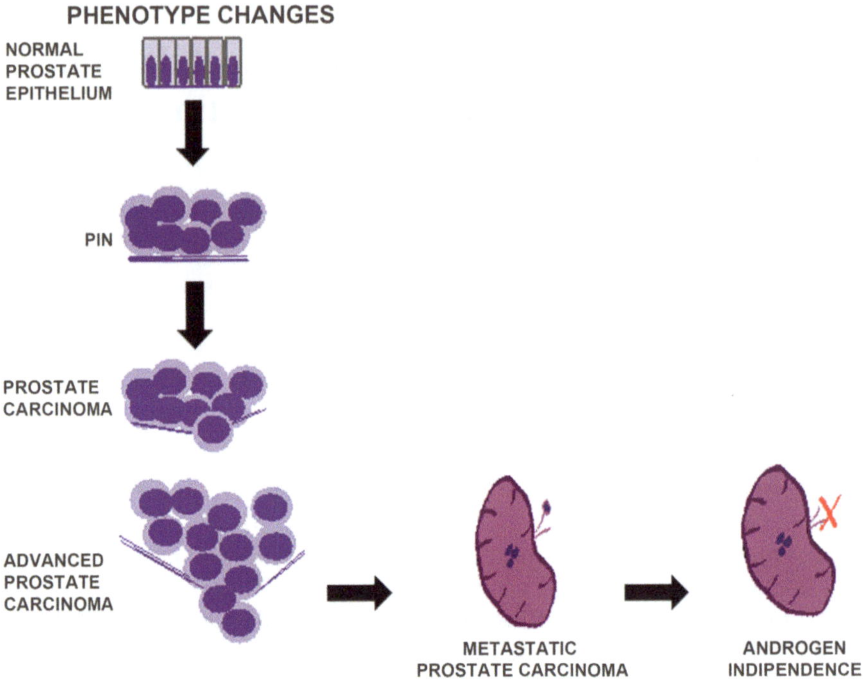

Fig. 12.1 Epigenetic contribution to prostate initiation, progression and metastasis

Interestingly, methylation of DLC1 gene is more frequent in samples from older men and is associated with decreased expression (Guan et al. 2006; Hornstein et al. 2008). Genes with this expression pattern may contribute to the first step of prostate transformation. Understanding the mean of coordinate hypermethylation of a group of genes at the onset of prostate cancer will reveal important insights into the etiology of the disease. A model that mimic the hypermethylation pattern of a defined group of genes in human prostate cancers is still in need of characterization. However, cadmium, a known prostate carcinogen, has been reported to induce hypermethylation of several genes (Benbrahim-Tallaa et al. 2007) and also viruses have been implicated in hypermethylation of few genes in prostate cancer (Liu et al. 2005; Morey et al. 2008).

Thus, early phase of prostate cancer development are associated with hypermethylation of specific genes, whereas global DNA hypomethylation has been linked to cancer progression. Indeed, metastatic cases are characterized by pronounced decreases in methylcytosine content and in particular hypomethylation of LINE-1 retrotransposons, usually methylated in normal cells (Florl et al. 2004). Hypomethylation is mainly recorded in metastatic cases suggesting that the mechanisms maintaining methylation at retroelements and other normally methylated sequences are affected during prostate cancer progression (Fig. 12.1).

12.1 Polycombs and DNA Methylation in Prostate Cancer

EZH2 high levels were detected in metastatic prostate cancers suggesting a link between polycombs and prostate cancer (Varambally et al. 2002; Hoffman et al. 2007; Bracken et al. 2003; Saramaki et al. 2006; Berezovska et al. 2006). Loss of negative regulators of EZH2, like RB1, microRNAs miR26 (Sander et al. 2008) and miR101 (Varambally et al. 2002), or gene amplification in metastatic cases (Saramaki et al. 2006; Berezovska et al. 2006) may account for the EZH2 increase.

The polycomb group (PcG) proteins have originally been discovered in Drosophila melanogaster as repressors of homeotic genes. Polycomb proteins are organized in two distinct multiprotein complexes, PRC1 and PRC2 (Figure 5). PRC2 is thought to initiate silencing. The complex include EED, SUZ12, EZH1 and the histone methyltransferase (HMT) EZH2, which can trimethylate histone H3 lysine 27 (H3K27) and in some cases add two methyl groups to H3K9. Methylation of H3K27 allows recruitment of the PRC1 complex which is implicated in stable maintenance of gene repression. PRC1 components are members of the polycomb and polyhomeotic family (CBX/HPC and EDR/HPH), RING1A, RING1B, YY1 and BMI-1, which ubiquitinates histone H2A at K119 (Fig. 12.2).

A cell type-specific composition of PcG complexes (Gunster et al. 2001), with related differences in specificity, is suggested to explain differences in the expression of the various components between tissues and to identify specific cell types and developmental stages.

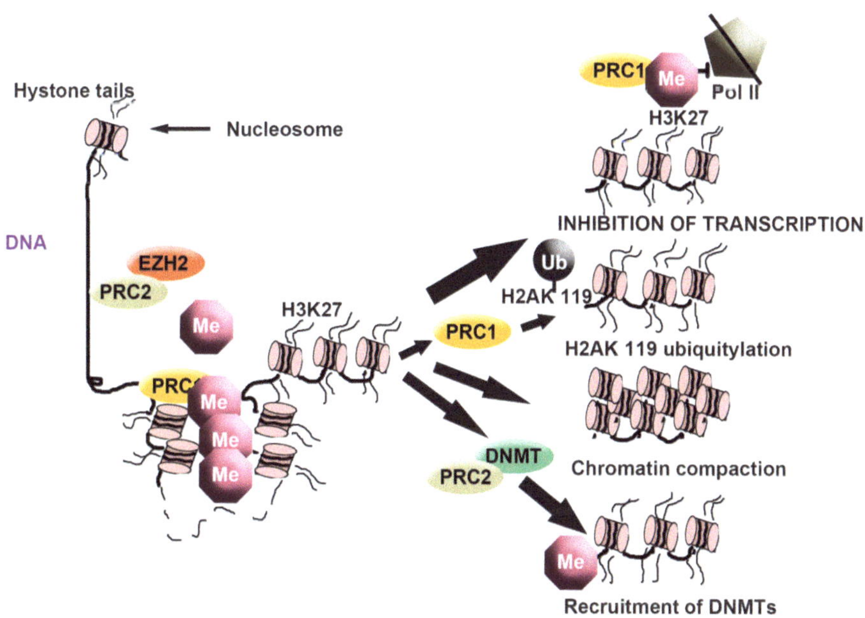

Fig. 12.2 Epigenetic gene silencing by Polycomb protein complexes

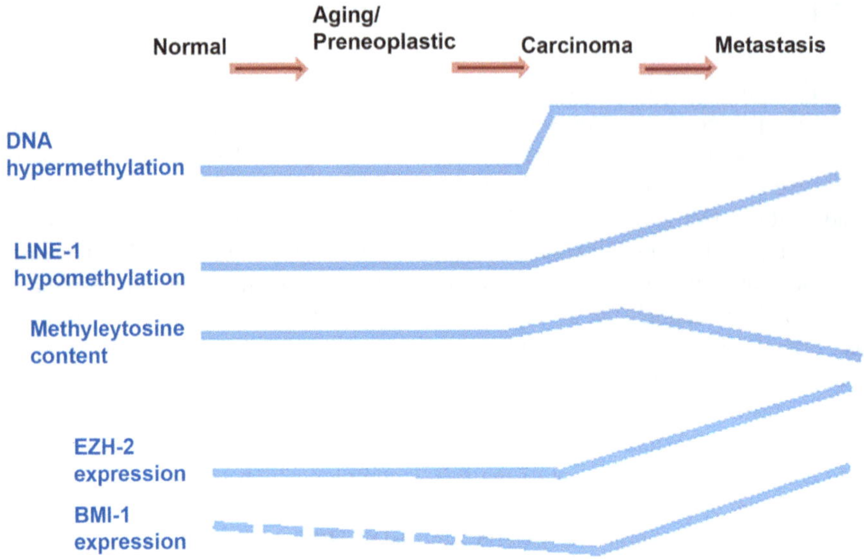

Fig. 12.3 Relationship between epigenetic changes and prostate cancer progression

For example, BMI is a marker of proliferation (Berezovska et al. 2006; van Leenders et al. 2007) and in prostate cancer is associated to higher stage of prostate cancer (Hoffman et al. 2007). Furthermore, PcG proteins like EZH2 and BMI-1 are considered crucial for the maintenance of adult multipotent stem cells (Valk-Lingbeek et al. 2004) (Fig. 12.3).

In prostate cancer, increased expression of EZH2, but not of other epigenetic regulators like BMI-1 and DNMTs, has been reported to correlate with the number of hypermethylated genes (Hoffman et al. 2007). Notably, not all hypermethylated genes are PcG targets nor do all PcG targets become hypermethylated. DNA hypermethylation in prostate cancer seems not particularly dependent from Polycomb genes. Overall, over-expression of EZH2 and modifications of polycomb complexes are associated with prostate cancer progression.

12.2 Epigenetics of the Androgen Response in Prostate Cancer

Growth of prostate cancers is dependent on androgens that signal through the androgen receptor. Fusion oncogenes, that contain androgen-responsive promoters, may be responsible of this dependency, in addition to the well known effect of androgens on function and survival of prostate epithelial cells. Therefore, anti-androgenic treatment represent a milestone of prostate cancer therapy. Unfortunately, cancers may become resistant to this therapy by several mechanisms (Feldman and Feldman 2001). Nevertheless, other oncogenic signals may account for some metastatic, hormone-refractory cancers that do not contain androgen-responsive fusion genes. Still, AR promoter hypermethylation (Jarrard et al. 1998) may respond of some advanced stage prostate cancers. Lower androgen availability may be imputated to AR overexpression for gene amplification, to AR point mutations that affect ligand binding specificity, or to increased activity of signaling pathways that allow ligand-independent AR activity. More than 100 co-regulators have been described so far for the AR (Heemers and Tindall 2007) and they, through interaction with the transcriptional activation domains, may mediate, allow or inhibit activation of transcription by AR without directly binding to DNA. In addition, specific transcription factors, like FOXA1/HNF3, HOXB13, and GATA-3, bind to DNA and influence AR activity. Finally, b-Catenin, as coregulator, may facilitate AR transport to the nucleus, whereas ARNIP, may target the receptor for degradation.

Well-known AR coregulators include chromatin remodeling complex components, such as BRG1 and BAF57, protein acetylases, such as the steroid hormone coactivators SRC1, SRC 2 and SRC 3, p300, CBP and PCAF, deacetylases, including SIRT1 and several class I and II HDACs, protein methylases, such as PRMT1, PRMT5, and G9a, and demethylases, such as LSD1 and Jumonji-domain proteins (Heemers and Tindall 2007). Of note, some of these enzymes, especially acetylases and deacetylases, may modify the AR itself in addition to histones.

Recently, histone demethylases have been recognized as AR coregulators and the mechanisms of action is reported in the Fig. 12.4.

Importantly, increased expression of LSD1 (Metzger et al. 2005) and JARID1B (Xiang et al. 2007) has been associated to prostate cancer advanced stages and could account for resistance to anti-androgenic therapy (Fig. 12.4).

In general, androgen signaling is necessary at least in the first step of the neoplastic process to induce prostate cancer proliferation and could contribute to select for further modifications in AR coregulators. Moreover, additional effects, beside the androgen signaling, could account for further advantages to the growing tumors. In the prostate epithelial cell, few genes are dependent on androgens through the AR, therefore we can believe that modifications in some AR coregulators that modify chromatin in prostate cancer cells might drive the proliferative process.

12.3 Future Developments and Present Conclusions

The epigenetic changes that occur in prostate cancer are still unknown. In the next future, genome-wide analysis will contribute to a better understanding of altered methylation in prostate cancer. An important question is to understand if

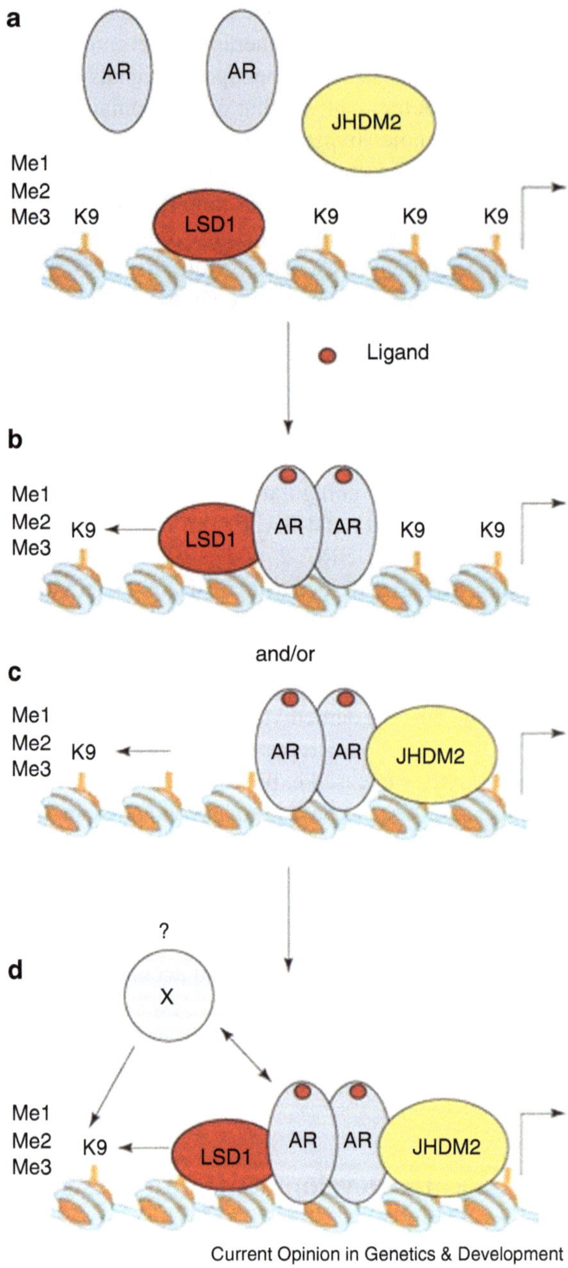

Fig. 12.4 Demethylases regulate the transcriptional activity of the androgen receptor. LSD1 specifically associates with chromatin on the promoter regions of androgen receptor (*AR*) target genes in either the absence or the presence of ligand (**a**). Chromatin association is independent of the presence of the androgen receptor. Once the ligand-activated androgen receptor translocates to the nucleus and binds to the ARE, LSD1 and AR form a transcriptionally active multi-protein complex that demethylates H3K9me1 and H3K9me2 but fail to demethylate H3K9me3 (**b**). In addition, the ligand-activated AR recruits a second demethylase, JHJDM2, which in concert with LSD1 regulates demethylation of H3K9me2, but never H3K9me1 or H3K9me3 (**c**). Given that ligand-dependent activation of AR target genes is associated with demethylation of H3K9me3, there must be additional demethylases that remove H3K9 trimethyl marks (**d**)

methylation changes in individual cancers are related or not to some particular group of chromosomal or genetic defects in prostate cancer. In addition, large-scale analyses of chromatin structure will contribute to understand which aberrant androgen action and modifications in coregulators could contribute to chromatin changes at individual genes and genome-wide.

Changes in microRNAs can be considered other epigenetic events that could contribute to prostate cancer even if they are subject to epigenetic regulation themselves. We already reported that the polycomb protein EZH2 is negatively regulated by miR-26a (Sander et al. 2008) and by miR101, which is often lost in prostate cancer (Varambally et al. 2008). Moreover, Polycomb components have been reported to bind to miR gene clusters (Guil and Esteller 2009; Marson et al. 2008). Interestingly, deregulated expression of microRNAs in prostate cancers (Ozen et al. 2008; Porkka et al. 2007; Ambs et al. 2008) which may affect important cancer-related genes like RAS, BCL-2, CCND1, WNT3A, E2F1 and CDKN1A have been reported. Moreover, differential miR expression patterns could help to discern between benign tissues and carcinoma, androgen-dependent and androgen-refractory tumors (Porkka et al. 2007) organ-confined tumors and extraprostatic disease (Ambs et al. 2008). Thus, a more accurate analysis of miR expressions in prostate tissues and identification of their targets may allow the identifications of molecular markers in order to improve personalized therapy decisions. In particular, crucial microRNAs could represent targets for small molecule therapies and re-expression of miRNA could result in growth arrest and apoptosis (Bonci et al. 2008).

Interestingly, changes in polycomb complexes, may also account for the establishment of a more embryonal-like cancer stem cell (Rajasekhar and Begemann 2007), that represent a novel issue in the prostate field research. The stem cells of normal prostate tissue are thought to give rise to essentially three cell types, basal, luminal (secretory) and neuroendocrine cells. By contrast, prostatic cancer stem cells can give rise to a limited number of aberrant phenotypes, such as luminal secretory cells, characterized by PSA expression and lack of GSTP1, a variable proportion of neuroendocrine cells, and under some conditions cells mimicking osteoblasts, which may uncover the ability of the prostate cancer cells to metastasize to bones (Koeneman et al. 1999) (Fig. 12.5).

CD133 antigen positivity identifies prostate cancer stem cells that also express the androgen receptor (Vander Griend et al. 2008). In this way we could hypothesize that androgens could support cancer stem cells also on the basis of androgen-regulated fusion oncogenes, while basal cells, which lack the AR, do not contribute to prostate cancers. The epigenetic changes in prostate cancer, i.e. consistent hypermethylation of a specific set of genes and aberrant polycomb activity, could also contribute to limit the aberrant differentiation potential of the cancer stem cells, in concert to chromosomal alterations.

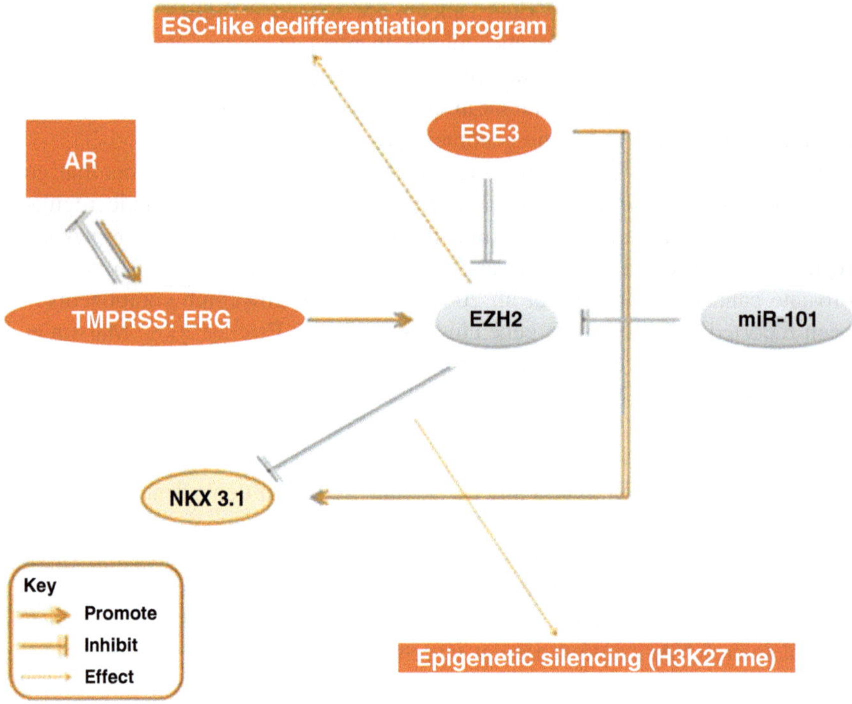

Fig. 12.5 Schematic overview of the cross-talk between ETS-polycomb group proteins and ETS-miRNA in prostate cancer. *ESC* embryonic stem cell, *AR* androgen receptor

Overall, prostate cancer development is a multifactorial and multisequential process, driven by genetic changes with the contribution of abnormal epigenetic regulation. In the next future diagnostic and therapeutic procedures that would take into considerations the genetic and the epigenetic alterations may allow an optimized treatment of each affected patient.

References

Aitchison A, Warren A, Neal D, Rabbitts P (2007) RASSF1A promoter methylation is frequently detected in both pre-malignant and non-malignant microdissected prostatic epithelial tissues. Prostate 67:638–644

Ambs S, Prueitt RL, Yi M, Hudson RS, Howe TM, Petrocca F et al (2008) Genomic profiling of microRNA and messenger RNA reveals deregulated microRNA expression in prostate cancer. Cancer Res 68:6162–6170

Benbrahim-Tallaa L, Waterland RA, Dill AL, Webber MM, Waalkes MP (2007) Tumor suppressor gene inactivation during cadmium-induced malignant transformation of human prostate cells correlates with overexpression of de novo DNA methyltransferase. Environ Health Perspect 115:1454–1459

Berezovska OP, Glinskii AB, Yang Z, Li XM, Hoffman RM, Glinsky GV (2006) Essential role for activation of the polycomb group (PcG) protein chromatin silencing pathway in metastatic prostate cancer. Cell Cycle 5:1886–1901

Bonci D, Coppola V, Musumeci M, Addario A, Giuffrida R, Memeo L et al (2008) The miR-15a-miR-16-1 cluster controls prostate cancer by targeting multiple oncogenic activities. Nat Med 14:1271–1277

Bracken AP, Pasini D, Capra M, Prosperini E, Colli E, Helin K (2003) EZH2 is down- stream of the pRB-E2F pathway, essential for proliferation and amplified in cancer. EMBO J 22:5323–5335

Enokida H, Shiina H, Urakami S, Igawa M, Ogishima T, Li LC et al (2005) Multigene methylation analysis for detection and staging of prostate cancer. Clin Cancer Res 11:6582–6588

Feldman BJ, Feldman D (2001) The development of androgen-independent prostate cancer. Nat Rev Cancer 1:34–45

Florl AR, Steinhoff C, Muller M, Seifert HH, Hader C, Engers R et al (2004) Coordinate hypermethylation at specific genes in prostate carcinoma precedes LINE-1 hypomethylation. Br J Cancer 91:985–994

Guan M, Zhou X, Soulitzis N, Spandidos DA, Popescu NC (2006) Aberrant methylation and deacetylation of deleted in liver cancer-1 gene in prostate cancer: potential clinical applications. Clin Cancer Res 12:1412–1419

Guil S, Esteller M (2009) DNA methylomes, histone codes and miRNAs: tying it all together. Int J Biochem Cell Biol 41:87–95

Gunster MJ, Raaphorst FM, Hamer KM, den Blaauwen JL, Fieret E, Meijer CJ et al (2001) Differential expression of human Polycomb group proteins in various tissues and cell types. J Cell Biochem Suppl 36:129–143

Heemers HV, Tindall DJ (2007) Androgen receptor (AR) coregulators: a diversity of functions converging on and regulating the AR transcriptional complex. Endocr Rev 28:778–808

Hoffmann MJ, Engers R, Florl AR, Otte AP, Muller M, Schulz WA (2007) Expression changes in EZH2, but not in BMI-1, SIRT1, DNMT1 or DNMT3B are associated with DNA methylation changes in prostate cancer. Cancer Biol Ther 6:1403–1412

Hornstein M, Hoffmann MJ, Alexa A, Yamanaka M, Muller M, Jung V et al (2008) Protein phosphatase and TRAIL receptor genes as new candidate tumor genes on chromosome 8p in prostate cancer. Cancer Genomics Proteomics 5:123–136

Jarrard DF, Kinoshita H, Shi Y, Sandefur C, Hoff D, Meisner LF et al (1998) Methylation of the androgen receptor promoter CpG island is associated with loss of androgen receptor expression in prostate cancer cells. Cancer Res 58:5310–5314

Koeneman KS, Yeung F, Chung LW (1999) Osteomimetic properties of prostate cancer cells: a hypothesis supporting the predilection of prostate cancer metastasis and growth in the bone environment. Prostate 39:246–261

Li LC, Okino ST, Dahiya R (2004) DNA methylation in prostate cancer. Biochim Biophys Acta 1704:87–102

Liu L, Zhang J, Bates S, Li JJ, Peehl DM, Rhim JS et al (2005) A methylation profile of *in vitro* immortalized human cell lines. Int J Oncol 26:275–285

Marson A, Levine SS, Cole MF, Frampton GM, Brambrink T, Johnstone S et al (2008) Connecting microRNA genes to the core transcriptional regulatory circuitry of embryonic stem cells. Cell 134:521–533

Metzger E, Wissmann M, Yin N, Muller JM, Schneider R, Peters AH et al (2005) LSD1 demethylates repressive histone marks to promote androgen-receptor- dependent transcription. Nature 437:436–439

Morey K Sr, Smiraglia DJ, James SR, Moser MT, Foster BA, Karpf AR (2008) Stage-specific alterations of DNA methyltransferase expression, DNA hyper- methylation, and DNA hypomethylation during prostate cancer progression in the transgenic adenocarcinoma of mouse prostate model. Mol Cancer Res 6:1365–1374

Nelson WG, Yegnasubramanian S, Agoston AT, Bastian PJ, Lee BH, Nakayama M et al (2007) Abnormal DNA methylation, epigenetics, and prostate cancer. Front Biosci 12:4254–4266

Ozen M, Creighton CJ, Ozdemir M, Ittmann M (2008) Widespread deregulation of microRNA expression in human prostate cancer. Oncogene 27:1788–1793

Porkka KP, Pfeiffer MJ, Waltering KK, Vessella RL, Tammela TL, Visakorpi T (2007) MicroRNA expression profiling in prostate cancer. Cancer Res 67:6130–6135

Rajasekhar VK, Begemann M (2007) Concise review: roles of polycomb group proteins in development and disease: a stem cell perspective. Stem Cells 25:2498–2510

Sander S, Bullinger L, Klapproth K, Fiedler K, Kestler HA, Barth TF et al (2008) MYC stimulates EZH2 expression by repression of its negative regulator miR-26a. Blood 112:4202–4212

Saramaki OR, Tammela TL, Martikainen PM, Vessella RL, Visakorpi T (2006) The gene for polycomb group protein enhancer of zeste homolog 2 (EZH2) is amplified in late-stage prostate cancer. Genes Chromosomes Cancer 45:639–645

Valk-Lingbeek ME, Bruggeman SW, van Lohuizen M (2004) Stem cells and cancer; the polycomb connection. Cell 118:409–418

van Leenders GJ, Dukers D, Hessels D, van den Kieboom SW, Hulsbergen CA, Witjes JA et al (2007) Polycomb-group oncogenes EZH2, BMI1, and RING1 are over- expressed in prostate cancer with adverse pathologic and clinical features. Eur Urol 52:455–463

Vander Griend DJ, Karthaus WL, Dalrymple S, Meeker A, DeMarzo AM, Isaacs JT (2008) The role of CD133 in normal human prostate stem cells and malignant cancer-initiating cells. Cancer Res 68:9703–9711

Varambally S, Dhanasekaran SM, Zhou M, Barrette TR, Kumar-Sinha C, Sanda MG et al (2002) The polycomb group protein EZH2 is involved in progression of prostate cancer. Nature 419:624–629

Varambally S, Cao Q, Mani RS, Shankar S, Wang X, Ateeq B et al (2008) Genomic loss of microRNA-101 leads to overexpression of histone methyltransferase EZH2 in cancer. Science 322(5908):1695–1699

Xiang Y, Zhu Z, Han G, Ye X, Xu B, Peng Z et al (2007) JARID1B is a histone H3 lysine 4 demethylase up-regulated in prostate cancer. Proc Natl Acad Sci USA 104:19226–19231

Zhu X, Leav I, Leung YK, Wu M, Liu Q, Gao Y et al (2004) Dynamic regulation of estrogen receptor-beta expression by DNA methylation during prostate cancer development and metastasis. Am J Pathol 164:2003–2012

Part IV
A Modern Approach to Therapy of Prostate Cancer: Targeting the Deadly Subset

Chapter 13
Molecular Markers for Patient Selection and Stratification: Personalized Prognostic Predictive Models

Stefania Staibano

Abstract The emerging data from US statistics on prostate cancer screening (Carlsson et al., J Clin Oncol 30(21):2581–2584, 2012; Brawley, Ann Intern Med 157(2):135–136, 2012) and the early results of the 11-year follow-up European Randomized Study of Screening for Prostate Cancer (ERSPC) involving eight countries (Belgium, Finland, France, Italy, Netherlands, Spain, Sweden and Switzerland), have evidenced that the main downside to large-scale PSA screening is over-diagnosis of biologically "indolent" cancers. They constitute about 30 % of detected cancers, and cause the facing of patients with the side effects of unnecessary treatment. Currently, the only way for men suffering for these "biologically insignificant" prostate cancers to delay unnecessary therapies is to offer them an Active Surveillance programme based upon regular check-up schedules.

It seems evident that there is an urgent need to find new reliable molecular markers able to predict the biological aggressiveness of each case of prostate cancer, in order to (Hamburg and Collins, N Engl J Med 363(11):1092, 2010) warrant the successful establishment and performance of personalized medicine. This, in turn, necessitate of substantial investments in infrastructure and standards, which may hasten a thorough understanding of the genctic basis of this disease.

Currently, clinical decision for prostate cancer mostly depends on the initial diagnosis on tumor biopsy. This approach has some weak points in terms of representativity of the mutational background of the entire tumor bulk, due to the frequent intratumor heterogeneity. Quite obviously, the primary tumor likely often lacks several genomic alterations that will characterize the metastatic deposits.

S. Staibano (✉)
Department of Advanced Biomedical Sciences, Pathology Section, Faculty of Medicine and Surgery, University of Naples "Federico II", via S. Pansini, n.5, 80131 Naples, Italy
e-mail: staibano@unina.it

S. Staibano (ed.), *Prostate Cancer: Shifting from Morphology to Biology*,
DOI 10.1007/978-94-007-7149-9_13, © Springer Science+Business Media Dordrecht 2013

For this reason, new surrogate biomarkers, such as circulating cancer cells (CCC), and microRNA, could represent biological samples which may add important "prognostic" information to histological diagnosis and PSA.

Genetic and epigenetic profiling and next-generation sequencing (NGS) of matched primary localized prostate cancers and mCRPC, are improving our knowledge of the multistage and highly heterogeneous nature of prostate cancer.

Besides AR splicing variants in CRPC, several new putative NGS-driven prostate cancer biomarker are being discovered, and sophisticated computational algorithms and systems biology approach are currently being explored to aid in the identification of the driving oncogenic events that change a subset of primary prostate cancers in the lethal bone-metastasizing CRPC. Where biomarkers will guide the clinical decision

This chapter aims to provide an up-date on existing and potential biomarkers predictive for CRPC, with the hope that they may soon aid in patients selection and prostate cancer treatment.

Screening procedures for prostate cancer detection with prostate specific antigen (PSA), have led to a profound stage migration, with the majority of prostate tumors detected today being of low grade, limited volume and clinically localized. To date, new reliable biochemical criteria for selecting men harbouring tumors most likely to progress following treatments and thus should receive the most aggressive treatment are lacking, this raising the possibility of overtreatment of patients with otherwise non-aggressive prostate cancer (Aggarwal and Ryan 2011; Merino et al. 2011).

Clinical and pathological features such as Gleason grade and tumor stage, are currently used to predict which prostate tumors have poor prognosis and the highest potential for recurrence and/or death (Merino et al. 2011).

To date, online nomograms predicting the outcomes of prostate cancer therapy following surgery and radiation therapy are available. The two most widely used are the Kattan nomograms (Kattan et al. 1997, 1998, 2000, 2001) and the Partin tables, (Partin et al. 1993, 2001) incorporating istopathological features, PSA levels, age, and treatment type to predict biochemical and overall outcome following therapy.

With such prognostic tools, most men diagnosed with this disease undergo radical prostatectomy or radiation treatment even when these treatment are unnecessary, suffering from side effects that have a relevant fall-out on quality of life.

In the last few years, the evolution of imaging in prostate cancer care has registered a rapid evolution, and new-generation FDG-PET imaging shows an increasing rate of correspondence with Gleason grade, clinical stage, and serum PSA level (Jadvar 2009).

Immuno-PET imaging for antibody drug conjugates offers exciting potential diagnostic applications (Nakajima et al. 2011). Similarly to FDG-PET for that concerning glucose methabolism, other metabolomics, as citrate, polyamines, lactate, choline, and creatine, may be detected by HHR-MAS spectroscopy of biopsy tissues and this looks very promising as reliable non-invasive prostate cancer detection based on cancer cells metabolic changes (Spratlin et al. 2009; Roberts et al. 2011), in addition to sarcosine and alanine (Tessem et al. 2008; Sreekumar et al. 2009).

The recent advancement in omics-based technology has provided unprecedented opportunities to look to the intricate signaling networks crosstalking in prostate cancer.

Moreover, the increasing availability of micro-dissected tumor tissue has sensibly clarified the different but complementary role of epithelial prostate cancer cells and tumor stromal cells in determining cancer aggressiveness and resistance to therapy (Liu et al. 2012).

Novel biomarkers that might have prognostic significance have been then identified and in the major part of cases are still under evaluation.

This is the case for the tight junctional proteins Claudins, involved either in paracellular transport and signal transduction (Szász et al. 2010). Claudins control also a broader range of processes, like cell growth, and promote carcinogenesis and cancer progression in several solid tumors. In prostate cancer they seem to have prognostic value. Szasz et al. showed in fact that different claudin expression levels may variably influence clinical behaviour of the tumor. Claudin-1 was found almost absent in patients with clinically advanced or metastatic disease, that in turn showed high levels of claudin-4. A further study on 141 prostatic adenocarcinomas, reported a variable immunohistochemical expression of claudins, with claudin-3 and -4 overexpressed in metastatic disease (Szász et al. 2010).

Another study showed instead the expression of claudin-1 in benign prostatic hyperplasia and prostatic intraepithelial neoplasia, and has been proposed as a novel marker for benign prostatic lesions. Further studies on more representative series of patients are needed before translating these findings in clinical practice (Fu et al. 2012).

The possible role of several cytokines specifically found elevated both in local and metastatic prostate cancer samples, like interleukin-4 (IL-4), IL-6 and IL-10, transforming growth factor beta (TGFβ) serum levels, and TGFβ1 have been found elevated (Schroten et al. 2012) has been discussed elsewhere in this book.

Particular interest has been generated by the reports signaling that the activation of Telomerase enzyme looks as a critical step in cell immortality and aggressiveness of prostate cancer cells. As it is well known, Telomerase is a ribonucleoprotein comprised of an integral RNA template (hTR) and a reverse transcriptase protein component (hTERT). The progressive shortening of telomeres is considered a major cause of cellular senescence. PCa show a relevant increased telomerase activity from the stage of in situ neoplasia (Gimba and Barcinski 2003).

This indicates that malignant prostate cells become able of bypassing senescence, reactivate telomerase. This leads to stabilization of telomere ends and continued cellular proliferation.

For this reason, the finding of high telomerase expression (about tenfold with respect to normal epithelial prostate cells) in prostate biopsies is considered strongly suggestive for malignant conversion, in addition to PSA levels, and is detected in 90 % of prostate carcinomas (Nanni et al. 2002; Akalin et al. 2001).

By converse, the absence of telomerase expression may be considered as a strong indicator of benign lesions (Akalin et al. 2001).

Most of the proposed new markers for prostate cancer patient selection and stratification are constituted by secreted proteins, detectable at tumor tissue level and/or in the bloodstream of patients.

However, the National Cancer Institute defines a biomarker as "a biological molecule found in blood, other body fluids, or tissues that is a sign of a normal or abnormal process, or of a condition or disease." Based on these postulates, then, also cells, molecules involved in epigenetic regulation, DNA or RNA, may be considered as candidate biomarkers, which can influence therapy selection and/or may be determinant for the generation of new drugs (Liu et al. 2012).

In line with this idea, methylated genes eventually present in tumor biopsy samples and urine samples from prostate cancer patients. including GSTP1, DNMT3A2, and EZH2, reflecting alteration in epigenetic regulation of cancer cells, have been proposed as attractive biomarkers for prostate cancer (Liu et al. 2012; Lofton-Day et al. 2008; Cairns et al. 2001; Rosenbaum et al. 2005). Assays for these are commercially available, although they awaits definitive approval by regulatory agencies (Fernandez et al. 2011).

As well, attempts to stratify prostate cancers on the basis of RNA signature and histology have been recently made, but they yielded disappointing results (Hegde et al. 2007).

However, Tomlins et al. (2011) demonstrated that urine PCA3, a prostate cancer-specific non-coding mRNA, when co-found with urine TMPRSS2:ERG fusion transcript, enhances the predictive value of serum PSA for prostate cancer risk and clinically relevant cancer on biopsy. As it has been reported elsewhere in this book, non-coding microRNAs (miRNAs) are single-stranded, highly conserved short RNA molecules, that repress gene expression in a sequence- dependent manner (Tomlins et al. 2011).

Deregulated miRNA expression correlates in several cases with clinically aggressive or metastatic cancers (Tavazoie et al. 2008; Bryant et al. 2012). Changes in circulating, and are also detected in exosomes, membrane vesicles secreted by normal and neoplastic cells and detectable in almost all biological physiological and malignant fluids (Valadi et al. 2007; Mitchell et al. 2008).

miRNA levels have been demonstrated to associate with prostate cancer. Br. J. Cancer 106, 768–774., and circulating miR-141 has been found to identify prostate cancer patients from healthy men (Mitchell et al. 2008).

miR-21 predicted the resistance to docetaxel-based chemotherapy in patients with mCRPC (Zhang et al. 2011) and increased levels of exosomes–associated miRNA were found in blood of patients with late-stage prostate cancer (Duijvesz et al. 2011).

Currently, microRNA gene signatures are being evaluated in clinical trials, and the research on these molecules is actively on, also because they resist to formalin-fixation and paraffin-embedding of archival prostate tissue, appearing then as an ideal candidate as a cancer biomarker to be used in a diagnostic daily context.

Lastly, also circulating cancer cells (CCCs) Alix-Panabières (2012) have been proposed as a new biomarker for prostate cancer. Evidence exists that when the

CTC number is solid cancers is reduced to fewer than five cells per 7.5 ml of blood, survival outcomes often improve, and this finding may influence treatment decisions.

The quantification and typization of CCCs in the peripheral blood and disseminated cancer cells (DCCs) in bone marrow may then provide important prognostic information and might help to monitor response to therapy of patients with metastatic castration-resistant prostate cancer (Delacruz 2012).

The challenges of interpreting such a number of highly complex data sets has lead several research groups to generate mathematical models of the individual natural history of prostate cancer, to elaborate with a more objective detail the individual natural history of cancer and retrospectively assess the effects of treatment (Hanin 2013).

That a stratified screening for prostate cancer incorporating genomic may be weighted by scientific, ethical, and logistical implications is a matter of fact. For instance, the use of genetic data may lead to discrimination against high-risk individuals by insurers, or constitute a hard ethical burden for the tutors in the case of genetic testing of minors (Chowdhury et al. 2013).

For these considerations, it will be imperative to unequivocally recognize the pros and cons of such a delicate personalized screening, before we choose it instead of existing screening methods.

Even considering these problems, it seems evident that sending off a tumor sample for a broad screening of genetic aberrations, instead of just a single test, may increase the chance of finding the right therapy for the right patient, likely expanding his survival. The search for new prognostic and/or diagnostic biomarkers is currently in an extraordinary phase of intense investigation, fueled by the increasing knowledge of the molecular processes that underlie prostate cancer progression.

References

Aggarwal R, Ryan CJ (2011) Castration-resistant prostate cancer: targeted therapies and individualized treatment. Oncologist 16(3):264–275

Akalin A, Elmore LW, Forsythe HL, Amaker BA, McCollum ED, Nelson PS et al (2001) A novel mechanism for Chaperone-mediated telomerase regulation during prostate cancer progression. Cancer Res 61:4791–4796

Alix-Panabières C (2012) EPISPOT assay: detection of viable DTCs/CTCs in solid tumor patients. Recent Results Cancer Res 195:69–76

Brawley OW (2012) Prostate cancer screening: what we know, don't know, and believe. Ann Intern Med 157(2):135–136

Bryant RJ, Pawlowski T, Catto JW, Marsden G, Vessella RL, Rhees B, Kuslich C, Visakorpi T, Hamdy FC (2012) Changes in circulating microRNA levels associated with prostate cancer. Br J Cancer 106(4):768–774

Cairns P et al (2001) Molecular detection of prostate cancer in urine by GSTP1 hypermethylation. Clin Cancer Res 7:2727–2730

Carlsson S, Vickers AJ, Roobol M, Eastham J, Scardino P, Lilja H, Hugosson J (2012) Prostate cancer screening: facts, statistics, and interpretation in response to the US Preventive Services Task Force review. J Clin Oncol 30(21):2581–2584

Chowdhury S, Dent T, Pashayan N, Hall A, Lyratzopoulos G, Hallowell N, Hall P, Pharoah P, Burton H (2013) Incorporating genomics into breast and prostate cancer screening: assessing the implications. Genet Med 15(6):423–32, Epub 2013 Feb 14

Delacruz A (2012) Using circulating tumor cells as a prognostic indicator in metastatic castration-resistant prostate cancer. Clin J Oncol Nurs 16(2):E44–E47

Duijvesz D, Luider T, Bangma CH, Jenster G (2011) Exosomes as biomarker treasure chests for prostate cancer. Eur Urol 59(5):823–831

Fernandez AF et al (2011) A DNA methylation fingerprint of 1628 human samples. Genome Res 22:407–419

Fu W, Madan E, Yee M, Zhang H (2012) Progress of molecular targeted therapies for prostate cancers. Biochim Biophys Acta 1825(2):140–152

Gimba ER, Barcinski MA (2003) Molecular aspects of prostate cancer: implications for future directions. Int Braz J Urol 29(5):401–410, discussion 411

Hamburg MA, Collins FS (2010) The path to personalized medicine. N Engl J Med 363(4):301–304, Epub 2010 Jun 15. Erratum in: N Engl J Med. 2010 Sep 9;363(11):1092

Hanin L (2013) Seeing the invisible: how mathematical models uncover tumor dormancy, reconstruct the natural history of cancer, and assess the effects of treatment. Adv Exp Med Biol 734:261–282

Hegde PS, Rusnak D, Bertiaux M, Alligood K, Strum J, Gagnon R, Gilmer TM (2007) Delineation of molecular mechanisms of sensitivity to lapatinib in breast cancer cell lines using global gene expression profiles. Mol Cancer Ther 6(5):1629–1640

Jadvar H (2009) FDG PET, in prostate cancer. PET Clin 4(2):155–161

Kattan MW, Stapleton AM, Wheeler TM et al (1997) Evaluation of a nomogram used to predict the pathologic stage of clinically localized prostate carcinoma. Cancer 79:528–537

Kattan MW, Eastham JA, Stapleton AMF et al (1998) A preoperative nomogram for disease recurrence following radical prostatectomy for prostate cancer. J Natl Cancer Inst 90:766–771

Kattan MW, Zelefsky MJ, Kupelian PA et al (2000) Pretreatment nomogram for predicting the outcome of three-dimensional conformal radiotherapy in prostate cancer. J Clin Oncol 18:3352–3359

Kattan MW, Potters L, Blasko JC et al (2001) Pretreatment nomogram for predicting freedom from recurrence after permanent prostate brachytherapy in prostate cancer. Urology 58:393–399

Liu Y, Hegde P, Zhang F, Hampton G, Jia S (2012) Prostate cancer – a biomarker perspective. Front Endocrinol (Lausanne) 3:72

Lofton-Day C et al (2008) DNA methylation biomarkers for blood-based colorectal cancer screening. Clin Chem 54:414–423

Merino M, Pinto A, González R, Espinosa E (2011) Antiangiogenic agents and endothelin antagonists in advanced castration resistant prostate cancer. Eur J Cancer 47(12):1846–1851

Mitchell PS, Parkin RK, Kroh EM, Fritz BR, Wyman SK, Pogosova-Agadjanyan EL, Peterson A, Noteboom J, O'Briant KC, Allen A, Lin DW, Urban N, Drescher CW, Knudsen BS, Stirewalt DL, Gentleman R, Vessella RL, Nelson PS, Martin DB, Tewari M (2008) Circulating microRNAs as stable blood-based markers for cancer detection. Proc Natl Acad Sci U S A 105(30):10513–10518

Nakajima T, Mitsunaga M, Bander NH, Heston WD, Choyke PL, Kobayashi H (2011) Targeted, activatable, in vivo fluorescence imaging of prostate-specific membrane antigen (PSMA) positive tumors using the quenched humanized J591 antibody-indocyanine green (ICG) conjugate. Bioconjug Chem 22(8):1700–1705

Nanni S, Narduci M, Pietra LD, Moretti F, Grasselli A, De Carli O et al (2002) Siganling through estrogen receptors modulates telomerase activity in human prostate cancer. J Clin Invest 11:219–227

Partin AW, Yoo J, Carter HB et al (1993) The use of prostate specific antigen, clinical stage and Gleason score to predict pathological stage in men with localized prostate cancer. J Urol 150:110–114

Partin AW, Mangold LA, Lamm DM et al (2001) Contemporary update of prostate cancer staging nomograms (Partin tables) for the new millennium. Urology 58:843–848

Roberts MJ, Schirra HJ, Lavin MF, Gardiner RA (2011) Metabolomics: a novel approach to early and noninvasive prostate cancer detection. Korean J Urol 52(2):79–89

Rosenbaum E et al (2005) Promoter hypermethylation as an independent prognostic factor for relapse in patients with prostate cancer following radical prostatectomy. Clin Cancer Res 11:8321–8325

Schroten C, Dits NF, Steyerberg EW, Kranse R, van Leenders AG, Bangma CH, Kraaij R (2012) The additional value of TGFβ1 and IL-7 to predict the course of prostate cancer progression. Cancer Immunol Immunother 61(6):905–910

Spratlin JL, Serkova NJ, Eckhardt SG (2009) Clinical applications of metabolomics in oncology: a review. Clin Cancer Res 15(2):431–440

Sreekumar A, Poisson LM, Rajendiran TM, Khan AP, Cao Q, Yu J, Laxman B, Mehra R, Lonigro RJ, Li Y, Nyati MK, Ahsan A, Kalyana-Sundaram S, Han B, Cao X, Byun J, Omenn GS, Ghosh D, Pennathur S, Alexander DC, Berger A, Shuster JR, Wei JT, Varambally S, Beecher C, Chinnaiyan AM (2009) Metabolomic profiles delineate potential role for sarcosine in prostate cancer progression. Nature 457(7231):910–914

Szász AM, Nyirády P, Majoros A, Szendrõi A, Szûcs M, Székely E, Tõkés AM, Romics I, Kulka J (2010) Beta-catenin expression and claudin expression pattern as prognostic factors of prostatic cancer progression. BJU Int 105(5):716–722

Tavazoie SF, Alarcon C, Oskars-son T, Padua D, Wang Q, Bos PD, Gerald WL, Mas-sague J (2008) Endogenous human microRNAs that suppress breast cancer metastasis. Nature 451:147–152

Tessem MB, Swanson MG, Keshari KR, Albers MJ, Joun D, Tabatabai ZL, Simko JP, Shinohara K, Nelson SJ, Vigneron DB, Gribbestad IS, Kurhanewicz J (2008) Evaluation of lactate and alanine as metabolic biomarkers of prostate cancer using 1H HR-MAS spectroscopy of biopsy tissues. Magn Reson Med 60(3):510–516

Tomlins SA, Aubin SM, Siddiqui J, Lonigro RJ, Sefton-Miller L, Miick S, Williamsen S, Hodge P, Meinke J, Blase A, Penabella Y, Day JR, Varambally R, Han B, Wood D, Wang L, Sanda MG, Rubin MA, Rhodes DR, Hollenbeck B, Sakamoto K, Silberstein JL, Fradet Y, Amberson JB, Meyers S, Palanisamy N, Rittenhouse H, Wei JT, Groskopf J, Chinnaiyan AM (2011) Urine TMPRSS2: ERG fusion transcript stratifies prostate cancer risk in men with elevated serum PSA. Sci Transl Med 3(94):94ra72

Valadi H, Ekström K, Bossios A, Sjöstrand M, Lee JJ, Lötvall JO (2007) Exosome-mediated transfer of mRNAs and microRNAs is a novel mechanism of genetic exchange between cells. Nat Cell Biol 9(6):654–659

Zhang HL, Yang LF, Zhu Y, Yao XD, Zhang SL, Dai B, Zhu YP, Shen YJ, Shi GH, Ye DW (2011) Serum miRNA-21: elevated levels in patients with metastatic hormone-refractory prostate cancer and potential predictive factor for the efficacy of docetaxel-based chemotherapy. Prostate 71(3):326–331

Chapter 14
Targeting Tumor Angiogenesis

Stefania Staibano and Paolo Antonio Ascierto

Abstract Four decades after the seminal work of Judah Folkman, in 1971, cancer therapies based on the suppression of neo-angiogenesis (Folkman, N Engl J Med 285:1182–1186, 1971) are becoming a reality (Verheul et al., Clin Cancer Res 14(11):3589–3597, 2008).

The shift toward the up-regulation of pro-angiogenic factors secretion from both tumor and stroma, results from the interplay between endothelial cell activation, proliferation, extracellular matrix degradation, migration, canalization. It leads to the generation of a chaotic vascular vessels network in prostate cancer tissue (Ahmed and Bicknell, Method Mol Biol 467:3–24, 2009), which can be detected also by modern imaging techniques based on magnetic resonance, ultrasound, and nuclear imaging through targeting of key angiogenic factors (Russo et al., BJU Int 110(11 Pt C):E794–E808, 2012).

This hopefully will lead to further improvements in prostate cancer diagnosis and staging. Preclinical evidence indicates that angiogenesis inhibitors can improve the efficacy of conventional cytotoxic agents mainly by normalizing tumor blood flow, thus improving drug delivery. Although significant biological activity of most vascular growth factors-interfering agents is demonstrated in preclinical models, single-agent activity is almost universally poor (Aragon-Ching et al., J Oncol 2010:361836, 2010). Due to the redundancy within the signalling pathways that promote angiogenesis, combining anti-angiogenic agents with different

S. Staibano (✉)
Department of Advanced Biomedical Sciences, Pathology Section, Faculty of Medicine and Surgery, University of Naples "Federico II", via S. Pansini, n.5, Naples, Italy
e-mail: staibano@unina.it

P.A. Ascierto
Unit of Melanoma, Cancer Immunotherapy and Innovative Therapy, Department of Melanoma, National Institute of Tumors Fondazione "G. Pascale", Via Mariano Semmola 1, 80131 Naples, Italy
e-mail: paolo.ascierto@gmail.com

S. Staibano (ed.), *Prostate Cancer: Shifting from Morphology to Biology*,
DOI 10.1007/978-94-007-7149-9_14,

mechanisms of action seems likely to significatively potentiate their therapeutic efficacy (Corcoran and Gleave 2012; Ellis and Hicklin, Nat Rev Cancer 8: 579–591, 2008; Verheul et al., Cancer Chemother Pharmacol 60:29–39, 2007).

14.1 Prostate Tumor Microenvironment, Hypoxia and Tumor Neoangiogenesis

Stefania Staibano (✉)

Prostate cancer-associated angiogenesis is a well recognised process (Russo et al. 2012).

Microvessel density (MVD) usually is higher in primary tumors of patients with metastatic disease compared with localised prostate cancers (Weidner et al. 1993; Fregene et al. 1993; Strohmeyer et al. 2000; Gravdal et al. 2009). As well, higher MVD correlates with advanced pathological stage (Lee et al. 2004), increased PSA levels (Lee et al. 2001a, b), higher tumor grade (Park et al. 2007), increased metastatic potential (Aragon-Ching and Dahut 2009; Park et al. 2007), and decreased survival of patients (Park et al. 2007; Lee et al. 2001a, b).

Moreover, tumor blood vessels show multiple structural and functional abnormalities (Russo et al. 2012), increased tortuosity, blind ends and high cellular proliferation rate, leading to dysfunctional and heterogeneous tumor tissue microcirculation, with frequent avascular tumor areas, hypoxia, acidosis, and glucose deprivation (Shannon et al. 2003; Teicher et al. 1990; Vaupel et al. 1989; Airley et al. 2000; Brown 1999; Folkman 1971).

This results in a net efflux of fluid into the interstitial space, devoted of functional lymphatics, so that it distends the extracellular matrix and increases the interstitial pressure (An et al. 1998).

All these features are associated with metastatic risk (Siim et al. 1996; Wang et al. 1992; Peters et al. 2001).

VEGF is a 46-kDa dimeric protein also known as a vascular permeability factor (VPF), and represents the most potent growth factor acting in stimulating cell proliferation, angiogenesis and lymphangiogenesis (Ferrer et al. 1998; Shweiki et al. 1995). VEGF, in prostate cancer progression, is regulated by hypoxia (Shweiki et al. 1992, 1995; Minchenko et al. 1994; Walker et al. 1994), cytokines, and androgens; moreover, several oncogenes, as Ras-, Raf-, and Src, the inactivation of tumor-suppressor genes as p53 and von Hippel – Lindau (Ravi et al. 2000) concur to its modulation.

VEGF immunohistochemical expression is highest in metastatic prostatic cancer tissue, but it does not predict prostate cancer progression (Botelho et al. 2010), nor correlates with VEGF expression and clinical and pathological features of tumors (Shariat et al. 2004).

VEGF acts in a paracrine manner by binding its receptors (VEGF-R1 and VEGF-R2) expressed on the surface of endothelial cells. VEGF binding activates the receptor's tyrosine kinase activity, and via the stimulation of several molecular pathways, as ERK and Akt, leads to vasodilatation, increased vascular permeability, cell proliferation, degradation and invasion of the underlying stroma (Aragon-Ching and Dahut 2009).

However, these vascular growth factors have multiple functions. As an example, prostate cancer cells overexpress also the VEGF receptor (VEGFR), so VEGF and VEGFR reciprocally act in an autocrine manner promoting, besides neo-angiogenesis, prostate cancer cell proliferation and survival (Jackson et al. 2002). In addition, it inhibits tumor cell apoptosis by inducing the expression of the anti-apoptotic protein Bcl-2 (Pidgeon et al. 2001).

Addition of VEGF inhibitors to antiandrogen therapy results in increased oxygen delivery to hypoxic tumors areas and thus further potentiates radiation therapy (Zhu and Kyprianou 2008).

Prostate tumor cells respond to hypoxia with the over-transcription of the hypoxia-inducible factor-1a (HIF-1) (Rang et al. 1999), which in turn overstimulates VEGF production and leads to neo-angiogenesis (Cvetkovic et al. 2001). VEGF production and signaling is partly dependent on mTOR induced expression of HIF-1a (Treins et al. 2002).

HIF-1a is hydroxylated at the proline residue and degraded by interaction with the von Hipple-Lindau protein complex and proteosome machinery (Semenza 2003; Forsythe et al. 1996) in normoxic conditions. In prostate cells, androgens can activate HIF-1 through an autocrine loop, and HIF-1 interacts with AR on PSA gene promoter, thereby activating its expression (Zhong et al. 2008).

Under hypoxic conditions, as in advanced prostate cancer, HIF-1a protein is stabilized and translocated into the nucleus for specific gene expression regulation including VEGF, and regulates intracellular pH, metabolism, cell invasion and autophagy, preventing death of aggressive cancer cells (Pouyssegur et al. 2006).

HIF-1a is then a preferential target for the development of anticancer drugs (Pili and Donehower 2003).

Histone deacetylase (HDAC) inhibitors have shown an anti-angiogenic activity mediated in part by HIF-1a down-regulation in both tumor and endothelial cells, with the consequent down-regulation of VEGF and other HIF-1a regulated angiogenesis-related genes (Qian et al. 2004). Class II HDAC are important modifiers of HIF-1a. Recently, it has been reported that the HDAC inhibitor LBH589 reduced tumor growth and angiogenesis in a preclinical prostate cancer model (Qian et al. 2006a, b).

Prostate cancer cells overexpress also TGF β, which promotes either extracellular matrix production and angiogenesis (Russell et al. 1998), favouring also osteoblastic bone metastases in experimental systems. The increase of TGF β RI is associated with high-grade and higher clinical stage of prostate cancer.

TGFβ RI expression correlates with tumor vascularity, tumor grade, and metastasis (Wikström et al. 2001). On the opposite side, TGF β RIII expression is decreased or lost in most human prostate cancers, where it correlates with advanced tumor stage and high risk of PSA recurrence.

The occurrence of intraepithelial prostate cancer correlates instead with the loss of TGF β RII responsiveness in stromal fibroblasts. Thus, partially blocking TGFβ through angiogenesis inhibitors, e.g. angiostatin and endostatin could potentially reverse the continuous stimulation of tumor angiogenesis (Deryugina and Quigley 2006).

Prostate cancer has the ability to produce MMPs, TGFβ; and cyclooxygenase 2 (COX-2).

Several endogenous inhibitors of angiogenesis have also been described in prostate cancer, namely angiostatin, endostatin, PSA, TSP1, interleukin 8, and interferons.

Overall, the microenvironment of prostate cancer is a critical determinant in cancer genesis (Chung et al. 2005).

14.2 Targeting the Angiogenic Pathways in Castration-resistant Prostate Cancer

Paolo Antonio Ascierto (✉)

Monoclonal antibody Bevacizumab is a recombinant humanized IgG1 monoclonal antibody with high affinity and specificity for all VEGF-A isoforms. Upon binding to soluble VEGF-A, bevacizumab limits ligand binding to EC receptors VEGFR-1 and VEGFR-2, thus blocking pro-angiogenic intracellular signals transduction.

In a phase II study, Reese et al. evaluated bevacizumab at 10 mg/kg every 14 days for 6 cycles in 15 chemotherapy-naïve patients with CRPC. No objective responses were showed.

But there was a PSA decline (less than 50 %) in 27 % of patients. Antibodies to VEGF slow tumor proliferation in prostate cancer xenograft models, especially when combined with chemotherapy (Gross et al. 2009; Antonarakis and Armstrong 2011).

Despite strong preclinical rationale, a phase III randomized study in men with chemotherapy-untreated CRPC (CALGB 90401) failed to show a survival advantage with the anti-VEGF antibody bevacizumab when combined with docetaxel compared with docetaxel used alone (22.6 vs 21.5 months), although significant improvements were seen with respect to PSA responses (70 vs 58 %) and radiographic responses (53 vs 42 %), as well as progression-free survival (9.9 vs 7.5 months) (George et al. 2011).

However, these results do not indicate that antiangiogenic therapies may never have a role in the treatment of CRPC, as much of this failure may be explained by an imbalance of treatment-related toxicities (cardiovascular events, neutropenic complications) in this older population with multiple co-morbidities. To this end, it was reported that the presence and number of co-morbidities (for example, cardiovascular disease, hypertension, diabetes, renal disease, liver disease) among patients in the CALGB 90401 trial significantly correlated with survival, and that there was an increase in the average number of co-morbidities in the docetaxel-bevacizumab arm (Wu et al. 2005).

Future development of this and other antiangiogenic agents may rely on combinations with other classes of angiogenesis inhibitors or other chemotherapeutic drugs whose toxicities do not overlap, and will require careful patient selection for those men most likely to benefit and not be harmed by this class of agents.

Combinations of bevacizumab with other agents were also evaluated. The phase II trial CALGB 90006 enrolled 79 patients with metastatic CRPC patients who were treated with docetaxel (70 mg/mq every 21 days), bevacizumab (15 mg/kg every 21 days) and estramustine (280 mg on days 1–5 of the 21 day cycle).

Promising results were showed with a median PFS of 8 months and a median OS of 24 months, with a PSA decline (higher than 50 %) in 75 % of patients (epub.theoncologist.com) (Sturge et al. 2011).

In a phase III trial with metastatic CRPC, 1,050 patients were randomized to receive docetaxel (75 mg/mq every 21 days), prednisone (5 mg twice daily) and either bevacizumab (15 mg/kg every 21 days) or placebo.

This study showed, despite an improvement in the secondary endpoints of progression-free survival (PFS), measurable disease response and post-therapy PSA decline, but the combination with bevacizumab was not statistically significant for OS (22.6 vs 21.5 months). Furthermore, there was higher toxicity in the experimental arm (epub.theoncologist.com).

Other combinations of bevacizumab with other drugs (cytotoxin agents and immunotherapy) didn't show results and further studies are needed. Sunitinib is an oral TKI targeted to all three VEGFR isoforms as well as PDGFβ and KIT, currently approved for renal cell carcinoma (Powles et al. 2011) and gastrointestinal stromal tumor. Several phase II studies of sunitinib were conducted in patients with metastatic CRPC, both in chemotherapy-naïve patients and in post-docetaxel setting (epub.theoncologist.com).

Also a multicenter, randomized, double-blind phase III trial comparing sunitinib plus prednisone versus prednisone alone (NCT00676650) in patients with post docetaxel progressive metastatic CRPC was conducted. But This trial has been recently interrupted prematurely since the combination of sunitinib with prednisone didn't improve OS when compared to prednisone alone.

Aflibercept is a recombinant fusion protein of the extracellular domain of human VEGF-R1 and VEGF-R2 and the Fc portion of human IgG. It acts as a 'VEGF trap' or decoy receptor, binding free ligand and preventing it from interacting with and activating membrane-bound receptor. As expected, it potently binds all naturally occurring ligands of VEGF-R1 and VEGF-R2, including VEGF-A, VEGF-B, and placental growth factor, and so may be anticipated to possess greater anti-angiogenic activity than bevacizumab. In phase I trials, the combination of aflibercept and docetaxel was shown to be safe and well tolerated, and the combination is now under evaluation in a large phase III trial (VENICE study) in patients with mCRPC.51 This study has completed accrual of approximately 1,200 patients, and is expected to report in mid-2012, with overall survival as the primary endpoint (Corcoran and Gleave 2012).

Other immunomodulatory-antiangiogenic agents like Thalidomide have been studied. The best data on thalidomide were in combination with other cytotoxic agents. Weekly docetaxel (30 mg/mq weekly for 3 out every 4 weeks) with or without thalidomide (200 mg/day) has been evaluated in chemotherapy-naïve metastatic CRPC. The combination arm was favored in terms of PSA decline (53 versus 37 % experiencing >50 % decrease in PSA) and PFS (5.9 versus 3.7 months). The most frequent adverse events were fatigue, peripheral neuropathy and constipation. Furthermore the combination arm may increase thromboembolic events and requires prophylactic anticoagulant therapy. Lenalidomide is an analog of thalidomide that has been evaluated in CRPC patients showing lower toxicity than thalidomide and a better antiangiogenic effect (Merino et al. 2011).

A phase I–II trial evaluating efficacy and tolerability of lenalidomide has been conducted and it compared lenalidomide 5 versus 25 mg/day, administered during 6 months, or until progression, in 60 patients, without hormonal therapy, after PSA relapse. Main toxicity was neutropenia, thrombotic events, asthenia and rash, with more grades 3–4 events in the 25 mg dose arm. Despite higher toxicity, PSA decline curve was favourable to patients receiving the 25 mg/day dose. The first results of a phase II trial combining bevacizumab, lenalidomide, docetaxel and prednisone in CRPC patients were presented at ASCO meeting 2011.

Among 24 patients who had completed four or more cycles, 22 patients had a >50 % PSA decline, and 20 patients had >75 % PSA decline, 14 patients, with measurable disease, showed 2 RC, 9 PR and 3 SD (overall response rate of 78.6 %). Therefore this combination seems to be associated with a high response rate with manageable toxicity. A phase III trial comparing different doses of lenalidomide combined with docetaxel-prednisone versus placebo is, however, currently underway (Merino et al. 2011).

Although one might conclude from these studies that antiangiogenic therapies are ineffective in mCRPC, we believe these negative data highlight an important biologic principle in prostate cancer angiogenesis that should inform the design of future trials (Antonarakis and Armstrong 2011). Specifically, the bone marrow microenvironment contains multiple proangiogenic factors in addition to VEGF including PDGF, basic fibroblast growth factor (bFGF), interleukin 8, and other soluble cytokines. This multiplicity of angiogenic pathways creates "redundancy" and the potential for "tumor escape" from antiangiogenic therapies and suggests that blocking multiple pathways simultaneously, rather than VEGF alone, may be necessary to effectively block angiogenesis in mCRPC. In support of this, our experience with clinical trials suggests that blocking PDGF and VEGF simultaneously (with sunitinib) is more potent in eliciting PSA responses in patients with mCRPC than blocking either VEGF alone (with bevacizumab) or PDGF alone (with imatinib) (Chi et al. 2005). Reflecting these data, studies are currently

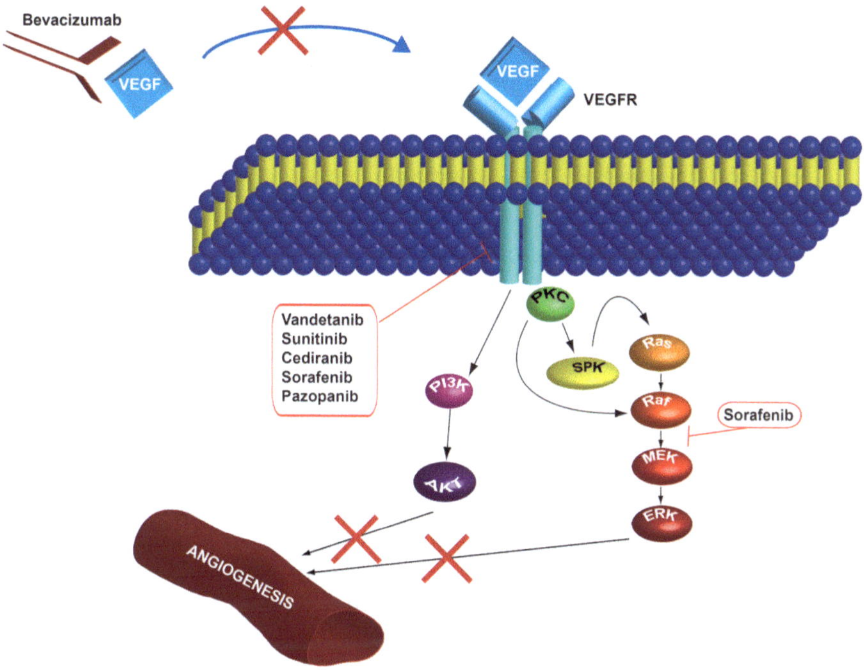

Fig. 14.1 Therapies targeting angiogenic pathway. Vascular endothelial growth factor (VEGF) and its receptors (VEGFR-1, VEGFR-2 and VEGFR-3) provide for new vessel formation and their maintenance. VEGF expression is markedly higher in prostate cancer specimens compared to non-neoplastic prostatic tissue controls and plasma VEGF levels are significantly higher with metastatic versus localized disease. Bevacizumab is a recombinant humanized IgG1 monoclonal antibody with high affinity and specificity for all isoforms of VEGF-A. It binds to soluble VEGF-A limiting ligand binding to EC receptors VEGFR-1 and VEGFR-2 and blocking the transduction of proangiogenic intracellular signals. Sorafenib, sunitinib and Cediranib are multitargeted receptor tyrosine kinase inhibitors (TKI), that exerts their antiangiogenic effect targeting, respectively, RAF kinase, VEGFR-2 and platelet-derived growth factor receptor (PDGFR-β), the three VEGFR isoforms, PDGFβ and KIT, and VEGFR 1 and 2

underway using tyrosine kinase inhibitors that target multiple angiogenic pathways (e.g., TKI258, which potently blocks VEGF, PDGF, and bFGF) (ClinicalTrials.gov identifier: NCT00831792), or alternatively, combine agents that block angiogenesis through different mechanisms (e.g., combining bevacizumab plus lenalidomide). In addition, in a recent phase I/II study combining sunitinib and docetaxel for the treatment of mCRPC in the frontline setting, patients demonstrated reductions in both PSA levels and tumor burden that were more substantial than a historical cohort of patients receiving docetaxel alone (Fig. 14.1) (Sowery et al. 2008).

The observation that both bevacizumab and sunitinib have shown prolongation of progression-free survival without differences in overall survival also raises the possibility that sustained suppression of angiogenesis is required to affect overall survival (Antonarakis and Armstrong 2011; Dayyani et al. 2011). Enhanced tumor growth following cessation of antiangiogenic therapy has been described, a "rebound" phenomenon that could influence overall survival (Chi et al. 2009). To address these limitations, it may be necessary to continue antiangiogenic therapy beyond standard definitions of disease progression to observe a beneficial impact on overall survival.

There are many question to be answered to optimize antiangiogenic therapy for advanced prostate cancers.

- The role of several angiogenic regulator factors is still poor understood. As an example, we currently know that the prostate-specific membrane antigen (PSMA) expression in tumor-associated neovasculature is necessary for angiogenesis and endothelial cell invasion, but we are unaware of its real role in angiogenesis (Gordon et al. 2008).
- As well, VEGF activation is probably mediated by other still unknown transcription factors such as the Activator protein 1 (AP-1) transcription factor complex (Shih and Claffey 1998).
- Further studies will also address the predictive role of expression of HIF-1 alpha, VEGF, and other angiogenic growth factors in patients treated with radiotherapy alone. These patients, in fact, lack the beneficial effect on tumor vascularization exerted by a neoadjuvant androgen deprivation. Therefore, the angiogenic markers may be even more important in this subgroup of patients (Vergis et al. 2008).

Anyhow, it is a matter of fact that almost all the key regulators of angiogenesis are upregulated in prostate cancer, particularly in the castration-resistant setting, and this undoubtedly has a great relevance for the gain of prostate cancer aggressiveness. This strongly stimulates the search for new reliable marks for effectively targeting prostate cancer angiogenesis.

References

Ahmed Z, Bicknell R (2009) Angiogenic signalling pathways. Methods Mol Biol 467:3–24

Airley RE, Monaghan JE, Stratford IJ (2000) Hypoxia and disease: opportunities for novel diagnostic and therapeutic prodrug strategies. Pharm J 264:666–673

An WG, Kanekal M, Simon MC, Maltepe E, Blagosklonny MV, Neckers LM (1998) Stabilization of wild-type p53 by hypoxiainducible factor-1a. Nature 392:405–408

Antonarakis ES, Armstrong AJ (2011) Emerging therapeutic approaches in the management of metastatic castration-resistant prostate cancer. Prostate Cancer Prostatic Dis 14(3):206–218

Aragon-Ching JB, Dahut WL (2009) VEGF inhibitors and prostate cancer therapy. Curr Mol Pharmacol 2:161–168

Aragon-Ching JB, Madan RA, Dahut WL (2010) Angiogenesis inhibition in prostate cancer: current uses and future promises. J Oncol 2010:361836

Botelho F, Pina F, Lunet N (2010) VEGF and prostate cancer: a systematic review. Eur J Cancer Prev 19:385–392

Brown JM (1999) The hypoxic cell: a target for selective cancer therapy – eighteenth Bruce F. Cain Memorial Award Lecture. Cancer Res 59:5863–5870

Chi KN, Eisenhauer E, Fazli L et al (2005) A phase I pharmacokinetic and pharmacodynamic study of OGX-011, a 2′-methoxyethyl antisense oligonucleotide to clusterin, in patients with localized prostate cancer. J Natl Cancer Inst 97:1287–1296
Chi KN, Hotte SJ, Yu EY et al (2009) Mature results of a randomized phase II study of OGX-011 in combination with docetaxel/prednisone versus docetaxel/prednisone in patients with metastatic castration-resistant prostate cancer [abstract 5012]. J Clin Oncol 27(15 suppl):238s
Chung LW, Baseman A, Assikis V, Zhau HE (2005) Molecular insights into prostate cancer progression: the missing link of tumor microenvironment. J Urol 1(73):10–20
Corcoran NM, Gleave ME (2012) Targeted therapy in prostate cancer. Histopathology 60(1): 216–231
Cvetkovic D, Movsas B, Dicker AP, Hanlon AL, Greenberg RE, Chapman JD, Hanks GE, Tricoli JV (2001) Increased hypoxia correlates with increased expression of the angiogenesis marker vascular endothelial growth factor in human prostate cancer. Urology 57(4):821–825
Dayyani F, Gallick GE, Logothetis CJ, Corn PG (2011) Novel therapies for metastatic castrate-resistant prostate cancer. J Natl Cancer Inst 103(22):1665–1675
Deryugina EI, Quigley JP (2006) Matrix metalloproteinases and tumor metastasis. Cancer Metastasis Rev 25:9–34
Ellis LM, Hicklin DJ (2008) VEGF-targeted therapy: mechanisms of antitumour activity. Nat Rev Cancer 8:579–591
Ferrer FA, Miller LJ, Andrawis RI et al (1998) Angiogenesis and prostate cancer: *in vivo* and *in vitro* expression of angiogenesis factors by prostate cancer cells. Urology 51:161–167
Folkman J (1971) Tumour angiogenesis: therapeutic implications. N Engl J Med 285:1182–1186
Forsythe JA, Jiang BH, Iyer NV et al (1996) Activation of vascular endothelial growth factor gene transcription by hypoxia-inducible factor 1. Mol Cell Biol 16:4604–4613
Fregene TA, Khanuja PS, Noto AC et al (1993) Tumor-associated angiogenesis in prostate cancer. Anticancer Res 13:2377–2381
George DJ, Armstrong AJ, Creel P et al (2011) A phase II study of RAD001 in men with hormone-refractory metastatic prostate cancer (HRPC) [abstract 181]. Presented at the 2008 American Society of Clinical Oncology Genitourinary Cancers symposium, San Francisco, 14–16 Feb 2008. Available at: http://www.asco.org/ASCOv2/Meetings/Abstracts. Accessed 8 Feb 2011
Gordon IO, Tretiakova MS, Noffsinger AE, Hart J, Reuter VE, Al-Ahmadie HA (2008) Prostate-specific membrane antigen expression in regeneration and repair. Mod Pathol 12:1421–1427
Gravdal K, Halvorsen OJ, Haukaas SA, Akslen LA (2009) Proliferation of immature tumor vessels is a novel marker of clinical progression in prostate cancer. Cancer Res 69:4708–4715
Gross ME, Soscia J, Sakowsky S et al (2009) Phase I trial of RAD001, bevacizumab, and docetaxel for castration-resistant prostate cancer [abstract 5154]. J Clin Oncol 27(15 suppl):272s
Jackson MW, Roberts JS, Heckford SE et al (2002) A potential autocrine role for vascular endothelial growth factor in prostate cancer. Cancer Res 62:854–859
Lee LF, Guan J, Qiu Y, Kung HJ (2001a) Neuropeptide-induced androgen independence in prostate cancer cells: roles of nonreceptor tyrosine kinases Etk/Bmx, Src, and focal adhesion kinase. Mol Cell Biol 21:8385–8397
Lee SE, Chung WJ, Kwak HB et al (2001b) Tumor necrosis factor-alpha supports the survival of osteoclasts through the activation of Akt and ERK. J Biol Chem 276:49343–49349
Lee LF, Louie MC, Desai SJ et al (2004) Interleukin-8 confers androgenin dependent growth and migration of LNCaP: differential effects of tyrosine kinases Src and FAK. Oncogene 23: 2197–2205
Merino M, Pinto A, González R, Espinosa E (2011) Antiangiogenic agents and endothelin antagonists in advanced castration resistant prostate cancer. Eur J Cancer 47(12):1846–1851
Minchenko A, Bauer T, Salceda S, Caro J (1994) Hypoxic stimulation of vascular endothelial growth factor expression in vitro and in vivo. Lab Invest 71:374–379
Park SI, Shah AN, Zhang J, Gallick GE (2007) Regulation of angiogenesis and vascular permeability by Src family kinases: opportunities for therapeutic treatment of solid tumors. Expert Opin Ther Targets 11:1207–1217

Peters KB, Wang H, Brown JM, Iliakis G (2001) Inhibition of DNA replication by Tirapazamine. Cancer Res 61:5425–5431

Pidgeon GP, Barr MP, Harmey JH, Foley DA, Bouchier-Hayes DJ (2001) Vascular endothelial growth factor (VEGF) up regulates Bcl-2 and inhibits apoptosis in human and murine mammary adenocarcinoma cells. Br J Cancer 85:273–278

Pili R, Donehower RC (2003) Is HIF-1a a valid therapeutic target? J Natl Cancer Inst 95:498–499

Pouyssegur J, Dayan F, Mazure NM (2006) Hypoxia signalling in cancer and approaches to enforce tumour regression. Nature 441:437–443

Powles T, Chowdhury S, Jones R et al (2011) Sunitinib and other targeted therapies for renal cell carcinoma. Br J Cancer 104:741–745

Qian DZ, Wang X, Kachhap SK et al (2004) The histone deacetylase inhibitor NVP-LAQ824 inhibits angiogenesis and has a greater antitumor effect in combination with the vascular endothelial growth factor receptor tyrosine kinase inhibitor PTK787/ZK222584. Cancer Res 64:6626–6634

Qian DZ, Kachhap SK, Collis SJ et al (2006a) Class II histone deacetylases are associated with VHL-independent regulation of hypoxia-inducible factor 1a. Cancer Res 66:8814–8821

Qian DZ, Kato Y, Shabbeer S et al (2006b) Targeting tumor angiogenesis with histone deacetylase inhibitors: the hydroxamic acid derivative LBH589. Clin Cancer Res 12:634–642

Rang HP, Dall MM, Ritter JM (1999) Pharmacology, 4th edn. Churchill Livingstone, Edinburgh, pp 670–677

Ravi R, Mookerjee B, Bhujwalla ZM et al (2000) Regulation of tumor angiogenesis by p53-induced degradation of hypoxia inducible factor 1alpha. Genes Dev 14:34–44

Russell J, Bennett S, Stricker P (1998) Growth factor involvement in progression of prostate cancer. Clin Chem 44:705–723

Russo G, Mischi M, Scheepens W, De la Rosette JJ, Wijkstra H (2012) Angiogenesis in prostate cancer: onset, progression and imaging. BJU Int 110(11 Pt C):E794–E808

Semenza GL (2003) Targeting HIF-1for cancer therapy. Nat Rev Cancer 3:721–732

Shannon AM, Bouchier-Hayes DJ, Condron CM, Toomey D (2003) Tumour hypoxia, chemotherapeutic resistance and hypoxia-related therapies. Cancer Treat Rev 29(4):297–307, Review

Shariat SF, Anwuri VA, Lamb DJ, Shah NV, Wheeler TM, Slawin KM (2004) Association of preoperative plasma levels of vascular endothelial growth factor and soluble vascular cell adhesion molecule – 1 with lymph node status and biochemical progression after radical prostatectomy. J Clin Oncol 22:1655–1663

Shih SC, Claffey KP (1998) Hypoxia-mediated regulation of gene expression in mammalian cells. Int J Exp Pathol 79:347–357

Shweiki D, Itin A, Soffer D, Keshet E (1992) Vascular endothelial growth factor induced by hypoxia may mediate hypoxia initiated angiogenesis. Nature 359:843–845

Shweiki D, Neeman M, Itin A, Keshet E (1995) Induction of vascular endothelial growth factor expression by hypoxia and by glucose deficiency in multicell spheroids: implications for tumour angiogenesis. Proc Natl Acad Sci USA 92:768–772

Siim BG, van Zijl PL, Brown JM (1996) Tirapazamine-induced DNA damage measured using the comet assay correlates with cytotoxicity towards hypoxic tumour cells in vitro. Br J Cancer 73:952–960

Sowery RD, Hadaschik BA, So AI et al (2008) Clusterin knockdown using the antisense oligonucleotide OGX-011 re-sensitizes docetaxel-refractory prostate cancer PC-3 cells to chemotherapy. BJU Int 102:389–397

Strohmeyer D, Rossing C, Strauss F, Bauerfeind A, Kaufmann O, Loening S (2000) Tumor angiogenesis is associated with progression after radical prostatectomy in pT2/pT3 prostate cancer. Prostate 42:26–33

Sturge J, Caley MP, Waxman J (2011) Bone metastasis in prostate cancer: emerging therapeutic strategies. Nat Rev Clin Oncol 8(6):357–368

Teicher BA, Holden SA, Al-Achi A, Herman TS (1990) Classification of antineoplastic treatments by their differential toxicity toward putative oxygenated and hypoxic tumour subpopulation in vivo in the FSaIIC murine fibrosarcoma. Cancer Res 50:3339–3344

Treins C, Giorgetti-Peraldi S, Murdaca J, Semenza GL, Van Obberghen E (2002) Insulin stimulates hypoxia inducible factor 1 through a phosphatidylinositol 3-kinase/target of rapamycin-dependent signaling pathway. J Biol Chem 277:27975–27981
Vaupel P, Kallinowski F, Okunieff P (1989) Blood flow, oxygen and nutrient supply, and metabolic microenvironment of human tumours: a review. Cancer Res 49:6449–6465
Vergis R, Corbishley CM, Norman AR et al (2008) Intrinsic markers of tumour hypoxia and angiogenesis in localised prostate cancer and outcome of radical treatment: a retrospective analysis of two randomised radiotherapy trials and one surgical cohort study. Lancet Oncol 9:342–351
Verheul HM, Qian DZ, Carducci MA, Pili R (2007) Sequence-dependent antitumor effects of differentiation agents in combination with cell cycle-dependent cytotoxic drugs. Cancer Chemother Pharmacol 60:29–39
Verheul HM, Salumbides B, Van Erp K, Hammers H, Qian DZ, Sanni T, Atadja P, Pili R (2008) Combination strategy targeting the hypoxia inducible factor-1 alpha with mammalian target of rapamycin and histone deacetylase inhibitors. Clin Cancer Res 14(11):3589–3597
Walker LJ, Craig RB, Harris AL, Hickson ID (1994) A role for the human DNA-repair enzyme HAP1 in cellular-protection against DNA-damaging agents and hypoxic stress. Nucleic Acids Res 22:4884–4889
Wang J, Biedermann KA, Brown JM (1992) Repair of DNA and chromosome breaks in cells exposed to SR 4233 under hypoxia or to ionizing radiation. Cancer Res 52:4473–4477
Weidner N, Carroll R, Flax J, Blumenfeld W, Folkman J (1993) Tumor angiogenesis correlates with metastasis in invasive prostate carcinoma. Am J Pathol 143:401–409
Wikström P, Damber J, Bergh A (2001) Role of transforming growth factor-beta1 in prostate cancer. Microsc Res Tech 52:411–419
Wu L, Birle DC, Tannock IF (2005) Effects of the mammalian target of rapamycin inhibitor CCI-779 used alone or with chemotherapy on human prostate cancer cells and xenografts. Cancer Res 65:2825–2831
Zhong D, Liu X, Khuri FR, Sun SY, Vertino PM, Zhou W (2008) LKB1 is necessary for Akt-mediated phosphorylation of proapoptotic proteins. Cancer Res 68(18):7270–7277
Zhu ML, Kyprianou N (2008) Androgen receptor and growth factor signaling cross-talk in prostate cancer cells. Endocr Relat Cancer 15:841–849

Chapter 15
Efficacy of Signal Transduction Inhibition in Advanced Prostate Cancer

Stefania Staibano

Abstract The overall survival of patients with metastatic Castration-resistant Prostate Cancer (CRPC) is discouraging low (Attar et al., Clin Cancer Res 15:3251–3255, 2009).

CRPC exhibit tremendous heterogeneity and complexity, reflecting the dysregulation of multiple patterns, mutations, and pathways, combined in a different manner in each patient. Of course, the impact of this heterogeneity on outcome and response to therapy is tremendous. It is therefore an urgent need to identify the multiple cellular pathways cooperatively promoting progression of the single cases of CRPC for successfully therapeutically target them. Several molecular pathways have been implicated in prostate cancer progression from localized androgen-sensitive disease to lethal CRPC.

In this article, we will review some of the recent findings on signal transduction studies performed to identify novel targets and alternative chances of therapeutic intervention for advanced prostate cancer.

A wide range of prostate cancer models are currently available, including a new orthotopic prostate model involving the bioluminescent cell line PC3M, which allows the visualization of injected prostate cancer cells in vivo, in nude mice, by means of a bioluminescent spectrum imaging system (Kumari 2012).

However, despite these resources and the extensive research, the molecular mechanism underlying the progression of CRPC is still poorly understood.

S. Staibano (✉)
Department of Advanced Biomedical Sciences, Pathology Section, Faculty of Medicine and Surgery, University of Naples "Federico II", via S. Pansini, n.5, 80131 Naples, Italy
e-mail: staibano@unina.it

S. Staibano (ed.), *Prostate Cancer: Shifting from Morphology to Biology*,
DOI 10.1007/978-94-007-7149-9_15,

An increasing number of molecules belonging to the classical and "alternative" pathways have been proposed for new target therapies or are currently matter for active research. It may be then useful to take a wide shot over the current research on this topic (Sarwar and Persson 2011).

Growth and differentiation of normal prostate epithelial cells are largely driven by androgens, acting through the androgen receptor (AR), encoded by the AR gene, located on chromosome Xq11-q13. In its inactive form, the AR protein resides in the cytoplasm of epithelial prostate cells, bound to chaperone molecules, such as the heat-shock proteins (HSP) (Edwards and Bartlett 2005). Upon androgens binding, conformational changes occur that allow to the androgen-AR complex to dissociate from the HSP-complex, form a homodimer, and rapidly reach the nucleus, where it binds to specific nuclear DNA-sequences (androgen-responsive elements) to stimulate transcription of androgen-regulated genes. This transcription is then modulated by the action of several co-activator and co-repressor factors (Edwards and Bartlett 2005). The growth and differentiation of benign prostate epithelial cells are also indirectly dependent on the stromal compartment. The binding of androgen to AR to stromal cells induces, in fact, the release of soluble peptides (andromedins) which in turn goes across the basement membrane and reaches the prostatic epithelial secretory compartment (the "paracrine" pathway) (Cunha 2008).

In androgen-dependent prostate cancer, AR promotes cell proliferation through the regulation of G1/S transition, only in the presence of androgens (Comstock and Knudsen 2007). Malignant prostate cancer cells can grow instead in androgen-depleted conditions, irrespective of the availability of paracrine stromal growth factors, passing to an "autocrine" pathway of growth factor synthesis and secretion (Gao et al. 2001; Vander Griend et al. 2010).

In other terms, the testosterone-AR pathway is bypassed, and prostate cancer cells find alternative ways to continue the AR-mediated functions.

After an initial response to ADT, then, prostate cancer cell growth becomes independent of the plasma testosterone level, and the onset of CRPC is followed by treatment failure on chemotherapy-based therapy and further clinical progression of disease in the vast majority of cases. In androgen-independent prostate cancer cells, AR up-regulates M-phase cell cycle genes, as the UBE2C gene, which inactivates the M-phase checkpoint (Wang et al. 2009). UBE2C activation is driven also by the concurrent histone H3K4 methylation and FoxA1 transcription factor binding (Wang et al. 2009).

Thus, AR continue to drive prostate cancer growth and progression even in androgen-independent CRPC (Scher and Sawyers 2005; Attard et al. 2008; Ang et al. 2009). Common changes in AR signaling during progression to CRPC comprise AR overexpression despite low circulating androgens (Chen et al. 2004; Haag et al. 2005; Linja et al. 2001; Holzbeierlein et al. 2004) coupled with high Intratumoral levels of androgens due the ability of tumor cells to independently synthesize androgens *de novo* (Locke et al. 2008; Montgomery et al. 2008).

In addition, AR-overexpressing CRPC cells may become hypersensitive to reduced levels of androgens (Waltering et al. 2009; Kawata et al. 2010) and/or mutations in the AR (Taplin et al. 2003) splice variants, or changes in coregulatory proteins may occur (Sun et al. 2010; Urbanucci et al. 2008). A plethora of molecular pathways cooperate to the development of prostate cancer metastasizing ability.

Epidermal growth factor (EGF), interleukin-6 (IL-6), the neuropeptide including bombesin and gastrin releasing peptides can activate AR in the absence of androgen in prostate cancer cell lines (Malinowska et al. 2009; Kung and Evans 2009; Grossmann et al. 2001). Accumulating evidence suggests that deregulation of multiple AR-independent pathways, such as the phosphatidylinositol 3-kinase (PI3-K)-Akt/mTOR pathway, may contribute to the progression to CRPC (Li et al. 2005; Catz and Johnson 2003; Nelson et al. 2007).

15.1 PI3K/Akt/Pathway

Upregulation of the phosphoinositide-3-kinase (PI3K)–Akt–mammalian target of rapamycin (mTOR) pathway has been detected in various tumors, including prostate cancer (Morgan et al. 2009) PI3K is activated by several extracellular receptors, including EGF receptor and insulin-like growth factor-1 receptor (IGF-1R), in addition to intracellular oncogenes such as RAS (LoPiccolo et al. 2008). In turn, activated PI3K induces Akt to phosphorylate and activate mTOR, which promotes cell division. PI3K activation is regulated by tumor suppressor phosphatase and tensin homolog (PTEN), and loss of PTEN function has been detected in prostate cancer (Cairns et al. 1997; McMenamin et al. 1999). Preclinical studies suggest that loss of PTEN function and/or activation of the PI3K–Akt–mTOR pathway can result in androgen-independent prostate cancer growth (Shen and Abate-Shen 2007; Jiao et al. 2007). Furthermore, deletion of PTEN has been associated with earlier disease progression in patients with prostate cancer (Schmitz et al. 2007; Yoshimoto et al. 2007) and greater AR expression and cancer-associated mortality in patients with CRPC (Sircar et al. 2009; Antonarakis and Armstrong 2011).

Several mTOR inhibitors have been developed. mTOR inhibitors show modest single-agent activity in advanced CRPC; however, their use combined with docetaxel was shown to be able to reverse or delay chemotherapy resistance in prostate cancer cell lines, particularly in patients who have activation of the Akt pathway as a result of PTEN mutation/loss or other genetic alteration) (Antonarakis and Armstrong 2011).

Limitations of the use of single-agent mTOR inhibitors include feedback up-regulation of upstream survival signals (such as PI3K) and the lack of induction of apoptosis or prolonged cytostasis due to the activation of other oncogenic pathways.

Several mTOR inhibitors have then entered human clinical testing in combination with other agents. Among these, everolimus (RAD001) was shown to inhibit the growth of prostate cancer cells in mouse bone, and this effect was potentiated by combination treatment with docetaxel and zoledronic acid. In a phase I dose-escalation trial on everolimus plus docetaxel chemotherapy-naïve patients with metastatic CRPC and a positive fluorodeoxyglucose positron emission tomography scan, no dose-limiting toxicities have been found. In a phase I trial on everolimus plus docetaxel and bevacizumab metastatic chemotherapy-naïve CRPC patients, 50 % PSA declines were seen in 10 patients over 12. In a phase II, single-arm, Simon two-stage study on 19 patients with CRPC of everolimus monotherapy, most of whom docetaxel refractory, the median TTP was 85 days and no PSA or tumor responses were recorded. In preclinical studies, temsirolimus (CCI-779) inhibited the growth of prostate cancer cell lines and xenografts, and had greater activity in combination with docetaxel. In addition, phase I studies of ridaforolimus (AP23573) on patients with advanced solid tumors, have successfully been completed. A single-arm, phase II trial of ridaforolimus monotherapy in taxane-resistant CRPC patients has completed enrollment and results are pending (ClinicalTrials.gov Identifier, NCT00110188) (Antonarakis and Armstrong 2011). Toxicities of mTOR agents include maculopapular rash, hypertriglyceridemia, hyperglycemia, allergic reactions, mucositis, pneumonitis and thrombocytopenia.

15.2 Chaperone Proteins

Chaperone (heat-shock) proteins have antiapoptotic properties and are an established target for anticancer therapy. Although heat-shock protein 90 (HSP90) was an early focus for study, no HSP90 inhibitor has so far proved to be therapeutically viable for prostate cancer, although work is still ongoing . Clusterin, an alternative chaperone protein, is a novel target. In prostate cancer cell lines, clusterin overexpression resulted in androgen-independent growth and clusterin gene silencing, induced apoptosis and significantly reduced growth. Clusterin expression is upregulated in patients with prostate cancer who have undergone androgen-deprivation therapy (ADT). Custirsen (OGX-011) is an antisense inhibitor of clusterin that suppresses clusterin expression in tumor tissue when administered to patients with localized prostate cancer. In vitro, instead, custirsen was found to resensitize docetaxel-refractory prostate cancer cell lines to docetaxel. A randomized phase II study of docetaxel plus prednisone with or without custirsen in patients with metastatic CRPC (n=82) has been completed, and showed a longer median overall survival time in the custirsen arm (24 versus 17 months; HR, 0.61; p=.06), although PSA rates and tumor response were similar 58. Based on these findings, phase III trials of OGX-011 plus docetaxel and prednisone are planned (Antonarakis and Armstrong 2011).

15.3 IGF-1R Pathway

Insulin-like growth factor receptor-1 (IGF-1R) and its ligands have antiapoptotic and transforming activities, and IGF-1R–mediated signaling can be detected during several stages of metastasis, including adhesion, migration, and invasion. In vitro models suggest that increased IGF-1R expression in prostate cancer cells can lead to androgen independence. In a recent study, using frozen tissue specimens, IGF-1R was more frequently expressed in stromal tissue surrounding malignant than in surrounding nonmalignant tissue and in high-grade tumors than in low-grade ones. Studies of IGF-1R ligands have provided further evidence of the oncogenic role of IGF signaling. In transgenic mice expressing human IGF-1 in the basal prostate epithelium, spontaneous tumorigenesis was seen. In a study of prostatic tumor tissue, IGF-1 and IGF-2 expression was higher in high-grade than in low-grade tumors. Furthermore, in a meta-analysis of clinical studies, elevated circulating IGF-1 concentrations were associated with a greater risk for prostate cancer (Antonarakis and Armstrong 2011).

Therapeutic monoclonal antibodies that bind to the extracellular domain of IGF-1R can potently inhibit the function of this receptor. In prostate cancer cell lines and in xenograft models, such antibodies have been shown to inhibit growth of both androgen-dependent and -independent tumors.

Three monoclonal antibodies against IGF-1R, cixutumumab (IMC-A12), figitumumab (CP-751,871), and AMG-479, are being assessed in CRPC patients and have demonstrated good tolerability in phase I studies. Further studies on IGF-1R antibodies are in progress (Antonarakis and Armstrong 2011). Cixutumumab specifically targets IGF-1R, inhibiting ligand binding and IGF signaling. Toxicities with this agent included fatigue, hyperglycemia, thrombocytopenia, hyperkalemia and muscle spasms. The development of figitumumab was suspended after an unexpected finding of a higher treatment-related mortality rate when this agent was added to standard chemotherapy (Antonarakis and Armstrong 2011).

15.4 MET

Aberrant activation or overexpression of MET is a common event in castration-resistant bone metastasizing prostate cancer, where it is associated with proliferation, invasion and angiogenesis. MET protein is a transmembrane receptor with a only one known ligand, the hepatocyte growth factor. Androgen suppression induces increased MET expression. Cabozantinib (XL184) is an oral potent inhibitor of MET and VEGFR. It has shown antiangiogenic, antiproliferative and anti-invasive effects in preclinical systems, and improvements in bone scans in 95 % of men with osseous metastases from CRPC.

Toxic side-effects from Cabozantinib include fatigue, diarrhea, anorexia, emesis and hypertension. Confirmatory controlled trials to assess the overall clinical benefit of this agent as well as the appropriate schedule for long-term use are needed, as well as further investigation of its intra-bone activity through novel imaging techniques (18F-positron emission tomography) or pharmacodynamic studies (Antonarakis and Armstrong 2011).

15.5 Src Pathway

Src and other members of the Src-family kinases (SFKs) are nonreceptor tyrosine kinases that transduce signals from a range of upstream proteins, including receptors for epidermal growth factor (EGF), platelet-derived growth factor (PDGF), and vascular endothelial growth factor (VEGF) (Chang et al. 2007). In addition to the established role of growth factor receptors in prostate cancer oncogenesis, preclinical studies have shown that Src and SFKs are highly active and/or overexpressed during prostate tumor growth and metastasis (Fizazi 2007). Src is also required during osteoclast functioning (Araujo and Logothetis 2009). In a recent study of tumor sampling from patients with CRPC, SFK activity was elevated in approximately 30 % of cases and patients with greater SFK activity had a significantly shorter overall survival. Dasatinib is a potent inhibitor of Src and SFKs that suppress proliferation of prostate cancer cell lines, and inhibit adhesion, migration and invasion (Antonarakis and Armstrong 2011) (Fig. 15.1).

It has shown preclinical antitumor and antimetastatic activity against prostate cancer cells and antiosteoclast activity (Nam et al. 2005; Vandyke et al. 2009). Dasatinib is an oral inhibitor of multiple oncogenic kinases including Src. In experimental models, dasatinib In addition, dasatinib reduced tumor growth and lymph node involvement in a prostate cancer mouse xenograft model (Antonarakis and Armstrong 2011).

In a phase II trial of dasatinib monotherapy on 47 patients metastatic chemotherapy-naïve CRPC, 6 % had a 50 % reduction in PSA, 12 of 23 patients with RECIST-evaluable disease had SD, and 23 of 41 patients with bone metastases at baseline had no new bone lesions at week 12 (Yu et al. 2009). In a phase I/II study of dasatinib plus docetaxel and prednisone in chemotherapy-naïve patients with CRPC, 49 % of evaluable patients had a 50 % PSA decline and 58 % had a RECIST PR. Bone scans showed size and/or number of lesions reduction in 28 % of patients and no new lesions in 69 % of patients (Araujo et al. 2009). These findings led to a randomized, placebo-controlled phase III trial of dasatinib plus docetaxel therapy. Adverse effects of dasatinib include diarrhea, nausea, fatigue and fluid retention.

Saracatinib (AZD0530) is another oral Src inhibitor in clinical development. In preclinical studies, saracatinib blocked proliferation and migration in a range of prostate cancer cell lines, including androgen-independent xenografts (Chang et al. 2008; Yang et al. 2009). Saracatinib has also shown antiosteoclast activity in vitro and in vivo (de Vries et al. 2009; Evans et al. 2011). In an initial phase II, single-arm, Simon two-stage trial of saracatinib monotherapy in patients with advanced CRPC, 5 of the 28 patients evaluated had a slight decline in PSA, though no patient achieved a 30 % decline and the median progression-free survival interval was 8 weeks.

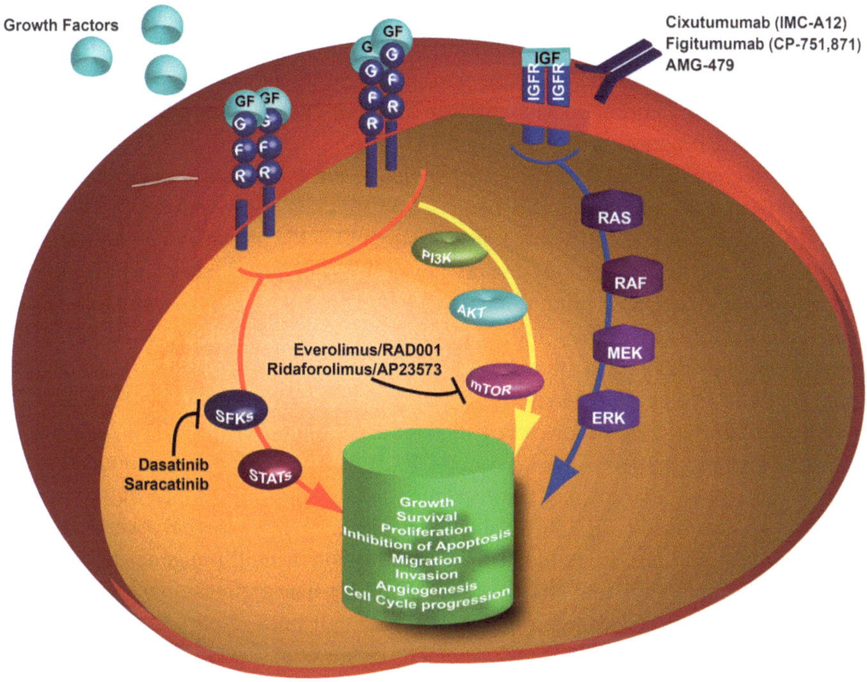

Fig. 15.1 Novel therapies designed to target non-AR mediated pathways. In addition to AR-mediated pathways, evidence suggests that several alternative signaling pathways may also be involved in prostate cancer disease progression. In this picture we report three pathways: Src-family kinases (SFKs) (*left side*), phosphoinositide-3-kinase (PI3K)–Akt–mammalian target of rapamycin (mTOR) (*middle*) and IGF-1R (*right side*). The Src-family kinases (SFKs) are nonreceptor tyrosine kinases that transduce signals from receptors for several growth factors and are highly active and/or overexpressed during prostate tumor growth and metastasis. This pathway is blocked by dasatinib and saracatinib (AZD0530). Upregulation of PI3K-Akt-mTOR pathway has been detected in various tumors. PI3K is activated by several extracellular receptors, including EGF receptor and insulin-like growth factor-1 receptor (IGF-1R). Activated PI3K induces activation of mTOR, which promotes cell division. Everolimus (RAD001) and ridaforolimus (AP23573) have been developed as mTOR inhibitors. IGF-1R has antiapoptotic and transforming activities and is involved in the development of metastasis, adhesion, migration, and invasion. Increased IGF-1R expression in prostate cancer cells can lead to androgen independence. Three monoclonal antibodies against IGF-1R, cixutumumab (IMC-A12), figitumumab (CP-751,871), and AMG-479, are being assessed in CRPC patients

15.6 PKA Pathway

The activated protein kinase A (PKA) is a serine/threonine kinase that reversibly phosphorylates numerous cytoplasmatic and nuclear proteins and it is strictly dependent on cAMP for its activity (Chin et al. 2002). PKA regulates the intracellular calcium intake, cell proliferation, inflammation and transcription (Francis and Corbin 1999) in addition, it regulates AR expression mainly by modulating its subcellular localization, and activate AR target genes in the absence of androgen or at low-androgen levels. This kinase is involved in the progression of prostate cancer through activation of protein kinase B (PKB) and expression of Bcl-2. It may induces also neuroendocrine differentiation of cancer cells. PKA then exerts a control on AR function.

The selective inhibition of PKA by H89 led to the cytoplasmic sequestration AR in the presence of a synthetic androgen in LNCaP cells. However, the possible role of PKA-driven nuclear entry in the progression of CRPC has still to be investigated.

PKA exerts its effects on several key transcriptional factors, as the cAMP response element (CRE)-binding protein 1 (CREB/CREB1). PKA phosphorylates CREB at serine 133 (Sands and Palmer 2008) Phosphorylated CREB (pCREB) modulates the expression of a large number of genes involved in cell growth and survival (Mayr and Montminy 2001; Sakamoto and Frank 2009), and its overexpression is a common event in several malignant tumors and in prostate cancer (Xiao et al. 2010). The activated CREB has been involved in bone prostate cancer metastasis, that display positive immunostaining for pCREB, unlike normal prostate tissue (Wu et al. 2007). Of even more interest, CREB exerts its pro-metastatic activity also by modulating the expression of multiple genes required for angiogenesis, as VEGF and HIF-1 (Wu et al. 2007). The expression of several PKA subunits have been examined in prostate cancer cell lines (Neary et al. 2004; Cho et al. 2002) and in prostate cancer specimens (Khor et al. 2008). Overexpression of PKARIα; subunit resulted correlated with the induction and up-regulation of genes required for proliferation and tumor progression in the aggressive PC3M cells (Cho et al. 2002). Moreover, it predicted outcome in prostate cancer patients treated with radiotherapy (RT) with or without short-term androgen deprivation therapy, and was also associated with metastatic disease.

It has still to be determined the specific role of the different PKA subunits such as AKAPs as novel biomarkers for predicting treatment response and outcome of CRPC. The transcription factor CREB1 appeared to be a potential therapeutic target (Chien et al. 2011). Preclinical and laboratory studies have shown that the combined targeting AR and PKA pathways inhibited growth of PCa cells. Further studies on larger series of cases are needed to confirm these preliminary data.

15.7 Immunotherapy

Although prostate cancer was not historically considered to be a particularly immune responsive cancer, recently clinical trials have demonstrated immunotherapy to be a good therapeutic strategy to improve overall survival (OS) in prostate cancer (Cha and Fong 2011). The most important studies include sipuleucel-T and PROSTVAC-VF, both determine a potentiation of immune system to target prostate proteins (Cha and Fong 2011). Before talking about these important results in the treatment of CRPC, let's explain the main immunogenic mechanisms shared by many tumors and also by prostate cancer are here briefly explained. Typical adaptive immune response sequence may be the following: first, the activation of antigen presenting cells (APCs) in presence of a target antigen presence; second, presentation of the antigen presentation to T cells; third, the targeting of antigen by activated T cells; and fourth, the down-regulation of T-cell response. The goal of the immunotherapy is to promote this effectors response against cancerous cells. Also co-stimulation by the B7 family of ligands on APCs interacting with the CD28 receptor on T cells delivers an important signal for activation. In the setting of malignancy there are several mechanisms of immune escape that lead tumor growth. For instance, Tumor cells can avoid maturation of DCs or prevent the expression of co- stimulatory molecules necessary for T-cell activation. MHC expression and peptide processing can be down-regulated to block the recognition by cytotoxic T cells. Sipuleucel T is one of the most important antigen targeted prostate cancer immunotherapy and it has been recently approved for the treatment of asymptomatic or minimally symptomatic HRPC.

Sipuleucel-T is an autologous cellular immunotherapy whose mechanism of action consists of stimulation of the patient's own immune system. Three randomized, double-blind, placebo-controlled phase III clinical trials of sipuleucel-T in patients with metastatic castration-resistant prostate cancer have shown improvement in overall survival vs control. The most important one that demonstrated a benefit is the IMPACT trial (Immunotherapy for Prostate Adenocarcinoma Treatment; D9902B). This was, a randomized, double-blind, placebo controlled phase III trial that enrolled 512 men. It showed a 4.1-month improvement in median OS (25.8 v 21.7 months) with no effect on TTP (14.6 v 14.4 weeks) (Lubaroff 2012). Survival benefit was showed despite a crossover design for placebo treated patients. Most frequent toxicities were infusion-related chills (54 %), nausea (28 %), fever (29 %), headache (16 %), and fatigue (39 %) within the first few days of treatment, although a trend toward increased but infrequent cerebrovascular events (2.4 v 1.8 %; P 1.0) was observed. ProstVac-VF is a PSA-directed vaccine approach, that was shown in early-phase trials demonstrated safety, induction of PSA-specific immune responses, and reduction in PSA velocity (Lubaroff 2012).

A phase II trial with PROSTVAC-VF in men with asymptomatic CRPC was conducted. This trial failed the primary end point of progression-free survival (PFS; 3.7 months in control arm v 3.8 months in treatment arm), but OS greatly favoured patients who received PROSTVAC-VF (25.1 v 16.6 months), with a 43 % reduction in death and 8.5-month improvement in median OS at 3 years post study. A phase III clinical trial in minimally symptomatic CRPC has been planned. Another tumor cell vaccines in prostate cancer is GVAX, which consists of two allogeneic prostate cancer cell lines (LNCaP and PC3) engineered to express GM-CSF. This approach could presumably deliver multiple antigens. Two phase III trials, one with GVAX alone and the other in combination with docetaxel, failed to show improvements in OS in patients treated with docetaxel plus prednisone (Lubaroff 2012). The reasons for failure are not clear, but probably for many variables that had not been addressed at the phase II level. Although there is preclinical evidence suggesting chemotherapy can induce immunomodulatory effects that may potentiate immunotherapy, how GVAX should be combined with docetaxel was not addressed before the phase III trial. Finally, the use of docetaxel in the control arms of both trials may not have been appropriate given the different kinetics of response seen with immunotherapies (Kantoff et al. 2010).

Another kind of immunologic approach is targeting immune checkpoint blockade. CTLA4 is a receptor on activated T cells that normally serves to inhibit further T-cell activation (Gerritsen and Sharma 2012). Ipilimumab (Yervoy; Bristol-Myers Squibb, New York, NY) is a humanized immunoglobulin G1 kappa monoclonal antibody that targets CTLA4. Two randomized phase III trials of Ipilimumab are underway in mCRPC; in the first trial (NCT00861614), patients who have progressed on docetaxel chemotherapy are randomized to low-dose palliative radiation therapy followed by Ipilimumab (10 mg/kg every 3 weeks for 4 doses) or to placebo.

A second trial, (NCT01057810), is enrolling patients with asymptomatic or minimally symptomatic chemotherapy-naive mCPRC, randomizing patients to either Ipilimumab or to placebo. The primary endpoint for both of these trials is overall survival, so mature data are not expected to be available for several years. Furthermore, some combination with Ipilimumab have been tested, such as with GM-CSF, demonstrating PSA responses as well as objective tumor responses in CRPC, with docetaxel, PROSTVAC-VF, and GVAX, with some clinical responses obtained. Interesting is Androgen ablation in combination with ipilimumab, and a phase II trial with this rationale has been designed (Gerritsen and Sharma 2012). Initial results show that the concurrent administration of ipilimumab with androgen ablation may have a synergistic action. In fact, patients in the combination arm had undetectable PSA by 3 months compared with the androgen ablation only group (55 versus 38 %). The timing of androgen ablation, in combination with other therapies, may prove to be crucial (Fig. 15.2).

15.8 Conclusion

Immunotherapy is an important part of the treatment for prostate cancer, but it has to be improved. Trial design continues to evolve in light of the biologic properties of immunologic agents. More agents become available to treat advanced prostate disease, the best clinical setting and how to administer immunotherapy will need to be investigated. For now, sipuleucel-T and PROSTVAC-VF have been primarily evaluated in patients with CRPC, yet immunotherapies should be more immunologically efficacious in earlier stages of disease (Gerritsen and Sharma 2012). Biochemical relapse after definitive therapy represents one setting in which these therapies should could be tested. A challenge will be to clarify the right combination and timing of the different therapeutic approaches described.

Application of guidelines to evaluate immune-mediated tumor responses radiographically and standardize cellular immune response assays may improve trial design and interpretation of results. A further goal will be to evaluate the possible efficacy of sipuleucel-T and PROSTVAC-VF also in earlier stages of disease, besides CRPC.

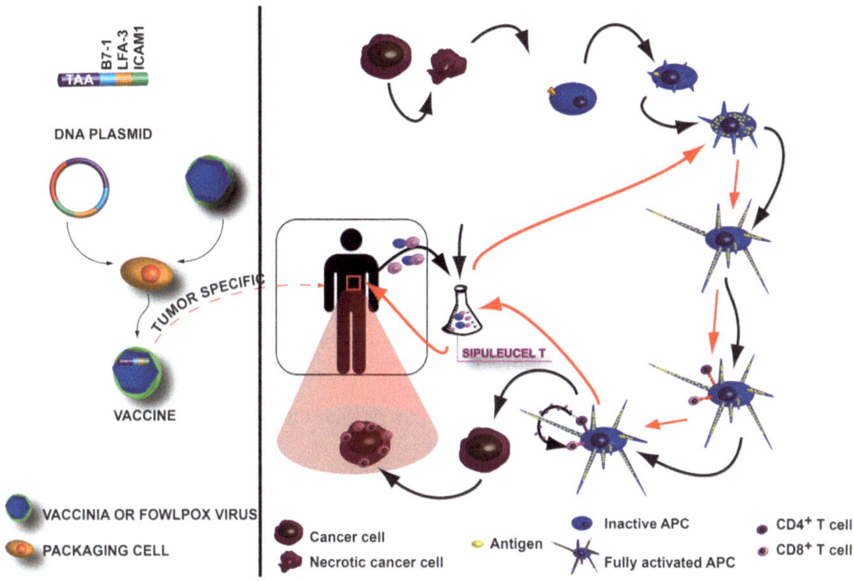

Fig. 15.2 Immunotherapy of prostate cancer. Tumors, and also prostate cancer, share immunogenic mechanisms. An immune response develops through a sequence of events, such as activation of antigen presenting cells (APCs) in the presence of a target antigen; presentation of the antigen to T cells; targeting of antigen by activated T cells; downregulation of T-cell response. The goal of immunotherapy is to promote this response against cancerous cells. Sipuleucel-T is a personalized cellular therapy that uses ex vivo antigen presentation to induce an antitumor immune response. PROSTVAC-VF is a viral vaccine that consists of a combination of recombinant vaccinia and fowlpox viruses that encode PSA and a triad of T-cell costimulatory molecules composed of lymphocyte function–associated antigen 3, intercellular adhesion molecule 1, and B7-1 (collectively labeled as TRICOM). This combination promotes tumor immunity

Anyhow, the rational current approach to prostate cancer therapy cannot overlook the biologic heterogeneity of CRPC and the biologic status of each individual tumor. Assessing gene expression and signaling activity will be soon fundamental to correctly design and direct a real individualized treatment. The better understanding of the molecular mechanisms underlying the progression of CRPC and treatment resistance is already full of promise.

References

Ang JE, Olmos D, de Bono JS (2009) CYP17 blockade by abiraterone: further evidence for frequent continued hormone-dependence in castration-resistant prostate cancer. Br J Cancer 100:671–675

Antonarakis ES, Armstrong AJ (2011) Emerging therapeutic approaches in the management of metastatic castration-resistant prostate cancer. Prostate Cancer Prostatic Dis 14(3):206–18, Epub 2011 May 17. Review

Araujo J, Logothetis C (2009) Targeting Src signaling in metastatic bone disease. Int J Cancer 124:1–6

Araujo J, Mathew P, Armstrong AJ et al (2009) Dasatinib and docetaxel combination treatment for patients with metastatic castration-resistant prostate cancer: analysis of study CA180–086 [abstract 7028]. Eur J Cancer 7:415S

Attar RM, Takimoto CH, Gottardis MM (2009) Castration-resistant prostate cancer: locking up the molecular escape routes. Clin Cancer Res 15:3251–3255

Attard G, Reid AH, Yap TA, Raynaud F, Dowsett M et al (2008) Phase I clinical trial of a selective inhibitor of CYP17, abiraterone acetate, confirms that castration-resistant prostate cancer commonly remains hormone driven. J Clin Oncol 26:4563–4571

Cairns P, Okami K, Halachmi S et al (1997) Frequent inactivation of PTEN/MMAC1 in primary prostate cancer. Cancer Res 57:4997–5000

Catz SD, Johnson JL (2003) BCL-2 in prostate cancer: a minireview. Apoptosis 8:29–37

Cha E, Fong L (2011) Immunotherapy for prostate cancer: biology and therapeutic approaches. J Clin Oncol 29(27):3677–85

Chang YM, Kung HJ, Evans CP (2007) Nonreceptor tyrosine kinases in prostate cancer. Neoplasia 9:90–100

Chang YM, Bai L, Liu S et al (2008) Src family kinase oncogenic potential and pathways in prostate cancer as revealed by AZD0530. Oncogene 27:6365–6375

Chen CD, Welsbie DS, Tran C, Baek SH, Chen R et al (2004) Molecular determinants of resistance to antiandrogen therapy. Nat Med 10:33–39

Chien Y et al (2011) Control of the senescence-associated secretory phenotype by NF-κB promotes senescence and enhances chemosensitivity. Genes Dev 25:2125–2136

Chin KV, Yang WL, Ravatn R, Kita T, Reitman E et al (2002) Reinventing the wheel of cyclic AMP: novel mechanisms of cAMP signaling. Ann N Y Acad Sci 968:49–64

Cho YS, Kim MK, Tan L, Srivastava R, Agrawal S et al (2002) Protein kinase A RIα antisense inhibition of PC3M prostate cancer cell growth: Bcl- 2 hyperphosphorylation, Bax up-regulation, and Bad-hypophosphorylation. Clin Cancer Res 8:607–614

Comstock CE, Knudsen KE (2007) The complex role of AR signaling after cytotoxic insult: implications for cell cycle based chemotherapeutics. Cell Cycle 6:1307–1313

Cunha GR (2008) Mesenchymal-epithelial interactions: past, present, and future. Differentiation 76:578–586

de Vries TJ, Mullender MG, van Duin MA et al (2009) The Src inhibitor AZD0530 reversibly inhibits the formation and activity of human osteoclasts. Mol Cancer Res 7:476–488

Edwards J, Bartlett JM (2005) The androgen receptor and signal-transduction pathways in hormone-refractory prostate cancer. Part 2: androgen-receptor cofactors and bypass pathways. BJU Int 95:1327–35

Evans CP, Bai L, Kung H et al (2011) Effect of the specific Src kinase inhibitor AZD0530 on osteolytic lesions in prostate cancer [abstract 170]. Presented at the 2008 American Society of Clinical Oncology Genitourinary Cancers symposium, San Francisco, 14–16 Feb 2008. Available at: http://www.asco.org/ASCOv2/Meetings/Abstracts. Accessed 8 Feb 2011

Fizazi K (2007) The role of Src in prostate cancer. Ann Oncol 18:1765–1773

Francis SH, Corbin JD (1999) Cyclic nucleotide-dependent protein kinases: intracellular receptors for cAMP and cGMP action. Crit Rev Clin Lab Sci 36:275–328

Gao J, Arnold JT, Isaacs JT (2001) Conversion from a paracrine to an autocrine mechanism of androgen-stimulated growth during malignant transformation of prostatic epithelial cells. Cancer Res 61:5038–5044

Gerritsen WR, Sharma P (2012) Current and emerging treatment options for castration-resistant prostate cancer: a focus on immunotherapy. J Clin Immunol 32(1):25–35

Grossmann ME, Huang H, Tindall DJ (2001) Androgen receptor signaling in androgen-refractory prostate cancer. J Natl Cancer Inst 93:1687–1697

Haag P, Bektic J, Bartsch G, Klocker H, Eder IE (2005) Androgen receptor down regulation by small interference RNA induces cell growth inhibition in androgen sensitive as well as in androgen independent prostate cancer cells. J Steroid Biochem Mol Biol 96:251–258

Holzbeierlein J, Lal P, LaTulippe E, Smith A, Satagopan J et al (2004) Gene expression analysis of human prostate carcinoma during hormonal therapy identifies androgen-responsive genes and mechanisms of therapy resistance. Am J Pathol 164:217–227

Jiao J, Wang S, Qiao R et al (2007) Murine cell lines derived from Pten null prostate cancer show the critical role of PTEN in hormone refractory prostate cancer development. Cancer Res 67:6083–6091

Kantoff PW, Schuetz TJ, Blumenstein BA, Glode LM, Bilhartz DL, Wyand M, Manson K, Panicali DL, Laus R, Schlom J, Dahut WL, Arlen PM, Gulley JL, Godfrey WR (2010) Overall survival analysis of a phase II randomized controlled trial of a Poxviral-based PSA-targeted immunotherapy in metastatic castration-resistant prostate cancer. J Clin Oncol 28(7):1099–105

Kawata H, Ishikura N, Watanabe M, Nishimoto A, Tsunenari T et al (2010) Prolonged treatment with bicalutamide induces androgen receptor overexpression and androgen hypersensitivity. Prostate 70:745–754

Khor LY, Bae K, Al-Saleem T, Hammond EH, Grignon DJ et al (2008) Protein kinase A RI-α predicts for prostate cancer outcome: analysis of radiation therapy oncology group trial 86-10. Int J Radiat Oncol Biol Phys 71:1309–1315

Kumari R (2012) Lighting up cancer studies. Drug Discov Devel 15(1):18

Kung HJ, Evans CP (2009) Oncogenic activation of androgen receptor. Urol Oncol 27:48–52

Li L, Ittmann MM, Ayala G, Tsai MJ, Amato RJ et al (2005) The emerging role of the PI3-K-Akt pathway in prostate cancer progression. Prostate Cancer Prostatic Dis 8:108–118

Linja MJ, Savinainen KJ, Saramäki OR, Tammela TL, Vessella RL et al (2001) Amplification and overexpression of androgen receptor gene in hormonerefractory prostate cancer. Cancer Res 61:3550–3555

Locke JA, Guns ES, Lubik AA, Adomat HH, Hendy SC et al (2008) Androgen levels increase by intratumoral *de novo* steroidogenesis during progression of castration-resistant prostate cancer. Cancer Res 68:6407–6415

LoPiccolo J, Blumenthal GM, Bernstein WB et al (2008) Targeting the PI3K/Akt/mTOR pathway: effective combinations and clinical considerations. Drug Resist Updat 11:32–50

Lubaroff DM (2012) Prostate cancer vaccines in clinical trials. Expert Rev Vaccines 11(7):857–868

Malinowska K, Neuwirt H, Cavarretta IT, Bektic J, Steiner H et al (2009) Interleukin-6 stimulation of growth of prostate cancer in vitro and in vivo through activation of the androgen receptor. Endocr Relat Cancer 16:155–169

Mayr B, Montminy M (2001) Transcriptional regulation by the phosphorylationdependent factor CREB. Nat Rev Mol Cell Biol 2:599–609

McMenamin ME, Soung P, Perera S et al (1999) Loss of PTEN expression in paraffin-embedded primary prostate cancer correlates with high Gleason score and advanced stage. Cancer Res 59:4291–4296

Montgomery RB, Mostaghel EA, Vessella R, Hess DL, Kalhorn TF et al (2008) Maintenance of intratumoral androgens in metastatic prostate cancer: a mechanism for castration-resistant tumor growth. Cancer Res 68:4447–4454

Morgan TM, Koreckij TD, Corey E (2009) Targeted therapy for advanced prostate cancer: inhibition of the PI3K/Akt/mTOR pathway. Curr Cancer Drug Targets 9:237–249

Nam S, Kim D, Cheng JQ et al (2005) Action of the Src family kinase inhibitor, dasatinib (BMS-354825), on human prostate cancer cells. Cancer Res 65:9185–9189

Neary CL, Nesterova M, Cho YS, Cheadle C, Becker KG et al (2004) Protein kinase A isozyme switching: eliciting differential cAMP signaling and tumor reversion. Oncogene 23:8847–8856

Nelson EC, Cambio AJ, Yang JC, Ok JH, Lara PN et al (2007) Clinical implications of neuroendocrine differentiation in prostate cancer. Prostate Cancer Prostatic Dis 10:6–14

Sakamoto KM, Frank DA (2009) CREB in the pathophysiology of cancer: implications for targeting transcription factors for cancer therapy. Clin Cancer Res 15:2583–2587

Sands WA, Palmer TM (2008) Regulating gene transcription in response to cyclic AMP elevation. Cell Signal 20:460–466

Sarwar M, Persson JL (2011) The protein kinase A (PKA) intracellular pathway and androgen receptor: a novel mechanism underlying the castration-resistant and metastatic prostate cancer. J Cancer Sci Ther S5:003

Scher HI, Sawyers CL (2005) Biology of progressive, castration-resistant prostate cancer: directed therapies targeting the androgen-receptor signaling axis. J Clin Oncol 23:8253–8261

Schmitz M, Grignard G, Margue C et al (2007) Complete loss of PTEN expression as a possible early prognostic marker for prostate cancer metastasis. Int J Cancer 120:1284–1292

Shen MM, Abate-Shen C (2007) Pten inactivation and the emergence of androgenindependent prostate cancer. Cancer Res 67:6535–6538

Sircar K, Yoshimoto M, Monzon FA et al (2009) PTEN genomic deletion is associated with p-Akt and AR signalling in poorer outcome, hormone refractory prostate cancer. J Pathol 218: 505–513

Sun S, Sprenger CC, Vessella RL, Haugk K, Soriano K et al (2010) Castration resistance in human prostate cancer is conferred by a frequently occurring androgen receptor splice variant. J Clin Invest 120:2715–2730

Taplin ME, Rajeshkumar B, Halabi S, Werner CP, Woda BA, Picus J, Stadler W, Hayes DF, Kantoff PW, Vogelzang NJ, Small EJ, Cancer and Leukemia Group B Study 9663 (2003) Androgen receptor mutations in androgen-independent prostate cancer: Cancer and Leukemia Group B Study 9663. J Clin Oncol 21(14):2673–2678

Urbanucci A, Waltering KK, Suikki HE, Helenius MA, Visakorpi T (2008) Androgen regulation of the androgen receptor coregulators. BMC Cancer 8:219

Vander Griend DJ, D'Antonio J, Gurel B, Antony L, Demarzo AM et al (2010) Cell-autonomous intracellular androgen receptor signaling drives the growth of human prostate cancer initiating cells. Prostate 70:90–99

Vandyke K, Dewar AL, Farrugia AN et al (2009) Therapeutic concentrations of dasatinib inhibit in vitro osteoclastogenesis. Leukemia 23:994–997

Waltering KK, Helenius MA, Sahu B, Manni V, Linja MJ et al (2009) Increased expression of androgen receptor sensitizes prostate cancer cells to low levels of androgens. Cancer Res 69:8141–8149

Wang Q, Li W, Zhang Y, Yuan X, Xu K et al (2009) Androgen receptor regulates a distinct transcription program in androgen-independent prostate cancer. Cell 138:245–256

Wu D, Zhau HE, Huang WC, Iqbal S, Habib FK et al (2007) cAMP-responsive element-binding protein regulates vascular endothelial growth factor expression: implication in human prostate cancer bone metastasis. Oncogene 26:5070–5077

Xiao X, Li BX, Mitton B, Ikeda A, Sakamoto KM (2010) Targeting CREB for cancer therapy: friend or foe. Curr Cancer Drug Targets 10:384–391

Yang JC, Ok JH, Busby JE, Borowsky AD, Kung HJ, Evans CP (2009) Aberrant activation of androgen receptor in a new neuropeptide-autocrine model of androgen-insensitive prostate cancer. Cancer Res 69(1):151–60

Yoshimoto M, Cunha IW, Coudry RA et al (2007) FISH analysis of 107 prostate cancers shows that PTEN genomic deletion is associated with poor clinical outcome. Br J Cancer 97:678–685

Yu EY, Wilding G, Posadas E et al (2009) Phase II study of dasatinib in patients with metastatic castration-resistant prostate cancer. Clin Cancer Res 15:7421–7428

Chapter 16
Therapeutic Targeting of the Bone Pre-metastatic Niche

Ester Simeone, Antonio Maria Grimaldi, Paolo Antonio Ascierto, Francesco Merolla, Gennaro Ilardi, Maria Siano, and Stefania Staibano

Abstract Most of relevant strategies designed for therapeutic targeting the highly lethal, bone-metastasizing and AR-resistant prostate cancers, have been reported and discussed elsewhere in this book. In this chapter, we aim to provide an overview of the rationale underlying the proposal of most of promising new therapeutic alternatives, most of which are still the early phase of evaluation.

Once more, we strongly wish to outline that the deeper understanding of the intricate crosstalks between the manifold molecular pathways responsible for the gain of invasive and metastasizing abilities of tumor cells, is giving rise to a previously unthinkable picture of the complex prostate cancer biology. A new, fascinating therapeutic era is opening up for the treatment of advanced prostate cancers.

16.1 Mechanisms of Bone Metastasis Onset

Ester Simeone (✉) • Antonio Maria Grimaldi • Paolo Antonio Ascierto

Bone metastases are the most frequent event in patients with advanced prostate cancer. They confer a high level of morbidity to patients, with a 5-year survival rate of 25 % and median survival of approximately 40 months. The molecular basis for the development of resistance to treatment is linked to some critical changes in the bone microenvironment that can confer on an advantage on cancer cell

E. Simeone (✉) • A.M. Grimaldi • P.A. Ascierto
Unit of Melanoma, Cancer Immunotherapy and Innovative Therapy, Department of Melanoma, National Institute of Tumors Fondazione "G. Pascale", Via Mariano Semmola 1, 80131 Naples, Italy
e-mail: ester.simeone@gmail.com; dott.a.m.grimaldi@gmail.com; paolo.ascierto@gmail.com

F. Merolla • G. Ilardi • M. Siano • S. Staibano
Department of Advanced Biomedical Sciences, Pathology Section, Faculty of Medicine and Surgery, University of Naples "Federico II", via S. Pansini, n.5, Naples, Italy
e-mail: merollafra@gmail.com; gennaro.ilardi@gmail.com; dysian@tin.it; staibano@unina.it

S. Staibano (ed.), *Prostate Cancer: Shifting from Morphology to Biology*,
DOI 10.1007/978-94-007-7149-9_16,

survival and proliferation (Bussard et al. 2008). These microenvironment-linked resistance mechanisms have just led us to consider new strategies to therapeutic bone- targeting.

The concept of "seed and soil", proposed by Stephen Paget in 1889, hypothesized the existence of an interplay between the metastatic properties of cancer cells (seed) and the favourable properties of the stromal/bone microenvironment (soil), conditioning the selective homing and growth of cancer cells. The "soil" must provide favourable conditions for cancers cells to successfully survive, clonally expand, and establish a nourishing vasculature (Bussard et al. 2008).

Further, the extension of this concept by David Lynch and colleagues, led to the formulation of the metastatic niche model. According to this hypothesis, the formation of a favourable microenvironment, (so-called "premetastatic niche"), before the tumor cells reach the metastatic destination, is critical for their engraftment. These niches facilitate then the formation of micrometastasis, and their subsequent transition into macrometastasis (Bussard et al. 2008).

Bone is an extremely metabolically active tissue. In order to maintain skeletal integrity, it undergoes continuous dynamic remodelling by sequential phases of bone resorption, mediated by osteoclasts, followed by osteoblast-mediated bone formation. The functions of osteoclasts and osteoblasts are tightly regulated in physiological conditions, to secure a perfect final balance between bone formation and degradation. To maintain skeletal homeostasis, multi-directional cross-talks among osteoblasts, osteoclasts, and hematopoietic cells are the rule. They take place through the mediation of systemic hormones and local bone-derived growth factors, including parathyroid hormone (PTH), 1,25-dihy-droxyvitaminD3, thyroxine, prostaglandins, bone morphogenic proteins (BMPs), TGFβ, IGF, and IL-1 and IL-6, in response to mechanical stresses and hormonal changes (Thudi et al. 2011; Kaplan et al. 2006).

Osteoblasts produce several local and systemic factors that are important for bone regulation, including receptors for PTH, prostaglandins, estrogen, vitamin D3, and PDGF, FGF and TG. Importantly, osteoblasts are intimately involved also in osteoclast differentiation, which involves the RANK/RANKL pathway.

RANKL production is influenced by osteotropic factors such as parathyroid hormone, 1,25-dihydroxyvitamin D, and prostaglandins. Crosstalks with the Wnt pathway allow a higher-layer control of osteoblast functions and bone formation. Current data indicate that the activation of Wnt/βbeta-catenin signaling is a major responsible for the increased bone mass. In particular, the overexpression of Wnt10b over-expression in animal models increases bone mass, while the over-expression of Wnt7B and beta-catenin in osteoblastic precursor cells induces their differentiation into mature osteoblasts (Thudi et al. 2011).

Vascular endothelial growth factor (VEGF) and its receptors (VEGFR-1, VEGFR-2 and VEGFR-3) provide for new vessel formation and their maintenance. VEGF also plays a crucial role within bone and bone marrow, with autocrine and paracrine mechanism. In the early metastatic bone modifications, hematopoietic progenitor cells (HPC) seem to have a main role. HPC and osteoprogenitor cells both express VEGFR-1, while endothelial progenitor cells (EPC) express

VEGFR-2. Pre-metastatic niches are particularly rich in tumor-derived VEGFA, placental growth factor (PlGF), and TGFβ, in response to which tumor-associated immune cells, such as HPC and macrophages cluster, prepare the "soil" for the imminent arrival adhesion and invasion of the tumor cell in the future metastatic sites (Schroten et al. 2012).

Accumulation of clusters of myeloid cells, fibronectin, growth factors and matrix remodelling proteins accelerate the micrometastatic process. The bone microenvironment is per se an important source of growth factors like TGFβ, IGF- 1, FGF, PDGF, BMPs, cytokines, chemokines, calcium ions, and cell adhesion molecules that contribute to make it a fertile soil conducive for the growth and proliferation of metastatic cancer cells. Moreover, the marrow stromal cells act also be in concert with the tumor cells in the homing, differentiation and proliferation processes, via the production of vascular cellular adhesion molecule-1 (VCAM-1), cadherin (**11**) and fibronectin. Thus, bones become a highly favourable microenvironments for prostate cancer cells, promoting their cell growth and proliferation (Schroten et al. 2012; Thudi et al. 2011).

Physical factors such as hypoxia, acidic pH, and high extracellular calcium concentrations also contribute to create this permissive environment for tumor growth. In order to continue this symbiotic relationship, cytokines and growth factors produced by cancer cells directly or indirectly impact osteoclastic bone resorption. This bidirectional interaction between the cancer cells and bone microenvironments results in the creation of a "vicious loop" that increases bone destruction to ultimately facilitate the establishment of cancer metastases in the bone.

The humanized RANKL monoclonal Denosumab, approved by FDA on November 2010 for the treatment of solid tumors with metastatic bone disease (MBD), or the tyrosine kinase SRC/BCR-ABL inhibitor dasatinib, already introduced in clinical trials, have been discussed elsewhere in this book.

16.2 Molecular Therapies for Metastasizing Disease

Francesco Merolla (✉) • Gennaro Ilardi • Maria Siano • Stefania Staibano

An impressive number of other new possible drugs and/or targets for alternative molecular therapies for metastasizing disease is actively in progress (Thudi et al. 2011).

Recently, the results of the ZEUS study indicated that we should reserve the use of potent osteoclast inhibition—with either zoledronic acid or denosumab—for men with bone metastatic prostate cancer (Smith et al. 2012).

Rather than preventing bone metastasis, zoledronic acid and denosumab have both been shown to significantly reduce the incidence of skeletal events, such as pathologic fractures and spinal cord compression, while the metastasis-free survival resulted only modestly prolonged by about 4 months in these patients.

A great deal of research has recently concerned the role of hypoxia-related factors HIF-1α and HIF-2α in prostate cancer progression (Mimeault and Batra 2013). Experimental evidences have revealed that they are key regulators of the adaptation of prostate bone-metastasis-initiating cells and corresponding differentiated progenies to oxygen and nutrient deprivation. HIFs are strongly induced by the niche-overexpressed (EGFR), insulin-like growth factor-1 receptor (IGF-1R), stem cell factor (SCF) receptor KIT, transforming growth factor-β receptors (TGF-βRs) and Notch, as well as by their downstream signalling elements such as phosphatidylinositol 3′-kinase (PI3K)/Akt/molecular target of rapamycin (mTOR).

Activated HIFs, in turn, induce and sustain the expression of induced pluripotency-associated transcription factors (Oct-3/4, Nanog and Sox-2), glycolysis- and epithelial-mesenchymal transition (EMT) programme-associated molecules, including CXC chemokine receptor 4 (CXCR4), snail and twist, microRNAs and, once more, VEGF.

This extraordinary melting of gene products create a powerful substrate which sustain self-renewal ability, survival, and treatment resistance of metastasizing prostate cancer cells.

On multivariate analysis, HIF1α recently emerged as an independent risk factor for progression to metastatic PC and development of CRPC in patients on androgen-deprivation therapy. No one prostate cancer not expressing HIF1α give rise to metastasis or developed CRPC (Ranasinghe et al. 2013).

Targeting of HIF signalling network represents then a very promising strategy to eradicate not only the bulk of prostate cancer, but also to directly hit bone-metastatic cancer cells, to prevent disease relapse and to increase the responsiveness of CRPCs to chemotherapy. Moreover, expression of HIF1α is a strong candidate for future new molecular screenings for the assessment of the risk to develop CRPC.

Among the possible tools for new screening tests aimed to the identification of the metastatic prostate cancer compartment, it has been proposed also the "old" member of the intermediate filament family of proteins, vimentin (Satelli and Li 2011).

Vimentin, is physiologically expressed in normal mesenchymal cells, where it maintain cellular integrity and provide resistance against stress. As well, this protein is overexpressed in prostate epithelial cancer, in which it correlates with high tumor growth, invasion, poor prognosis, and has been recognized also as a marker for epithelial-mesenchymal transition (EMT).

For these reasons, vimentin seem to be attractive for prostate cancer therapy, and this is particularly interesting, considering the recent discovery of a vimentin-binding mini-peptide of potential use for therapy. Further researches on this topic are in progress.

An atypical isoform of trypsin, PRSS3/mesotrypsin, represents another promising target for therapy of bone-metastasizing cancer cells. Its over-expression has been found associated with breast, lung, pancreatic cancers. In primary prostate tumors, it has shown prognostic significance, indicating systemic progression following prostatectomy. Mouse orthotopic model with bioluminescent imaging has confirmed that PRSS3/mesotrypsin is critical for prostate cancer metastasis.

To further support this idea, silencing of PRSS3 inhibits anchorage-independent growth of prostate cancer cells in soft agar assays, and suppresses invasiveness in Matrigel transwell assays and three-dimensional (3D) cell culture models. By converse, the treatment with recombinant mesotrypsin directly promotes an invasive cellular phenotype in prostate cancer cells.

This has fueled the search for new inhibitors of mesotrypsin activity to be used to suppress prostate cancer cell invasion (Hockla et al. 2012).

Maspin (mammary serine protease inhibitor) expression, has been correlated instead with a better prognosis in prostate, as well as in most of malignant solid tumors, as bladder, lung, gastric, colorectal, head and neck, thyroid and melanoma. In all these tumors, however, maspin is frequently down-regulated.

Maspin is a member of the serine protease superfamily, and a selectively increased adhesion by the presence of maspin may contribute to the inhibition of tumor metastasis. Possible therapeutic approaches could be to re-activate the system that inhibits the expression of maspin, identifying activating substances or possibly introducing maspin in cancer cell, up-regulating maspin to reduce the risk of metastasis.

However, the finding that maspin is usually over-expressed in pancreatic, gallbladder, colorectal, and thyroid cancers indicates that it may play different roles in human cancers, this deserving further studies to better define all the possible therapeutic implication of targeting maspin (Berardi et al. 2013).

Besides multiple reports indicating the association between the aberrant expression of EGF receptors with hormone-refractory and metastatic prostate cancer, to date the molecular mechanism linking EGF signaling to prostate cancer metastasis remains unclarified. Experimental models of PCa metastasis showed that EGF could induce epithelial-mesenchymal transition (EMT) and increase invasiveness, also through the extracellular signal-regulated kinase 1/2 (ERK1/2)-dependent phosphorylation, ubiquitination, and degradation of the epithelial protein lost in neoplasm (EPLIN), a putative suppressor of EMT and tumor metastasis. Pharmacological inhibition of the ERK1/2 pathway effectively antagonized EGF-induced EPLIN degradation. This indicates that blockade of EGF signaling could be useful to prevent and/or retard prostate cancer metastasis (Zhang et al. 2012).

As well, there was found a tendency for upregulation of the EGFR family members HER2, and EGFR and downregulation of HER3 in the prostate cancer lymph node metastases in comparison to the primary tumors. This indicate the existence of a rationale supporting possible combined strategies for EGFR- and HER2-targeted therapy of metastasizing prostate cancers, and further studies concerning this eveniences are in progress (Carlsson et al. 2013).

Of a particular interest, a role in predicting metastasis has been emerged also for non-cancerous prostate cancer, from several recent studies in vitro and on prostatectomy tumor tissue (Bijnsdorp et al. 2012).

The cell-communication protein connexin-26 (Cx26) has been suggested as a marker to predict the development of metastasis, when expressed in the adjacent noncancerous tissues (rather than cancer tissues) of prostatectomy sections. It appears then promising for select patients who may benefit from adjuvant therapy to decrease the risk of metastasis.

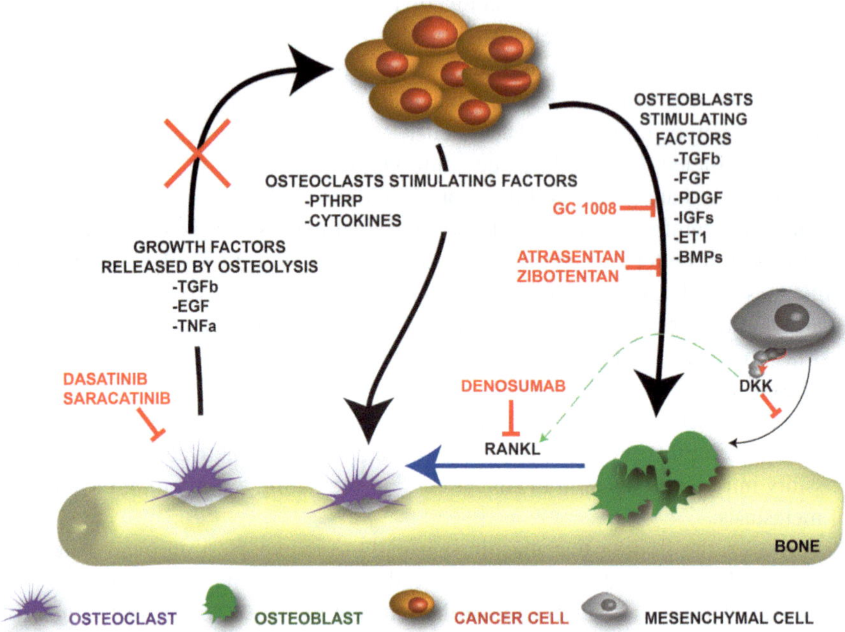

Fig. 16.1 Therapeutic targeting of the bone pre-metastatic niche. In the bone, the formation of a conducive microenvironment, called "premetastatic niche", before the tumor cells arrive at the metastatic destination, is critical for engraftment of the disseminated tumor cells and facilitate formation of micrometastasis, and subsequent transition into macrometastasis. To maintain skeletal homeostasis, cross-talk among osteoblasts and osteoclasts is necessary and it is possible through systemic hormones and local bone-derived growth factors, in response to mechanical stresses and hormonal changes. There is a bidirectional interaction between the cancer cells and bone microenvironment, mediated by production and release of growth factors like TGFβ, IGF-1, FGF, PDGF, BMPs, cytokines, chemokines, calcium ions, and cell adhesion molecules resulting in the creation of a "vicious cycle" that increases bone destruction, ultimately resulting in establishment of cancer metastases in the bone. Genetic ablation of Src results in osteopetrosis, with a decrease in osteoclast-mediated bone resorption and increased osteoblast differentiation and bone formation. TGF-β1 controls bone homeostasis coordinating the bone formation to sites where old bone degradation is occurring. Endhotelin-1 (ET-1) is a potent vasoconstrictor, that binds to its receptors A or B; it stimulates osteoblasts, causing the secretion of growth factors, with increase in tumor cells proliferation and ET-1 production, thus maintaining disease progression. Wnt signaling has a fundamental role in normal osteogenesis; the overexpression of Wnt or a deficiency in Wnt antagonist Dickkopf-related protein 1 (DKK1) result in an increase in bone formation. DKK-1 has been reported to be downregulated in prostate cancer patients in advanced stage, providing a strong basis for the further exploration of the Wnt signaling pathway as a future target in the treatment of bone metastasis in prostate cancer. In this figure, are briefly represented mechanism of action of some drugs, affecting the osteogenic pathway; in detail, denosumab is a RANKL monoclonal antibody; dasatinib and saracatinib inhibit SRC pathway; GC1008 is a monoclonal antibody directed against all three isoforms of TGF-β; Atrasentan and zibotentan are oral ET-A receptor antagonist

Moreover, circulating bone marrow-derived CD90, CD73, and CD105-expressing Mesenchymal Stem Cells (BM-MSCs) have been found to show an innate tropism for tumor tissue in response to the inflammatory microenvironment of prostate cancer tissue. MSCs represent 0.01–1.1 % of the total cells present in core biopsies from primary human prostatectomies. They not only may contribute to prostate carcinogenesis, but may also potentially be used to deliver cytotoxic or imaging agents for therapeutic and/or diagnostic purposes (Brennen et al. 2013).

The intriguing role of miRNAs in prostate cancer progression has been discussed elsewhere in this book. Nevertheless, we would further outline that they looks very promising as "multifunctional" tools for prostate cancer.

They act, in fact, either in suppressing or promoting prostate cancer growth, metastasis, and in maintaining the pluripotency of prostate cancer stem cells.

The low expression of miR-335 was significantly associated with high Gleason Score ($P = 0.04$), advanced clinical stage ($P = 0.04$), and positive metastasis ($P = 0.02$), but not with prognosis in PCa patients. By converse, overwhelming evidence establishes the role of microRNAs as essential actors in the metastasis generation of prostate cancers. Specific microRNAs then appear particularly attractive to be manipulated, either by mimicking or inhibition, to hit metastasizing prostate cancer cells. However, a lot of work is needed, to better understand their role in prostate physiology and cancer, before they may enter the clinics (Fang and Gao 2013; Xiong et al. 2013; Fenderico et al. 2013).

Overall, a great workload is still necessary to reach definitive data about new molecular therapeutic strategies toward metastatic prostate cancer, however, the works are actively on, and the endlessly emerging data are extremely exciting (Fig. 16.1).

References

Berardi R, Morgese F, Onofri A, Mazzanti P, Pistelli M, Ballatore Z, Savini A, De Lisa M, Caramanti M, Rinaldi S, Pagliaretta S, Santoni M, Pierantoni C, Cascinu S (2013) Role of maspin in cancer. Clin Transl Med 2(1):8. doi:10.1186/2001-1326-2-8

Bijnsdorp IV, Rozendaal L, van Moorselaar RJ, Geldof AA (2012) A predictive role for non-cancerous prostate cells: low connexin-26 expression in radical prostatectomy tissues predicts metastasis. Br J Cancer 107(12):1963–1968

Brennen WN, Chen S, Denmeade SR, Isaacs JT (2013) Quantification of Mesenchymal Stem Cells (MSCs) at sites of human prostate cancer. Oncotarget 4(1):106–117

Bussard KM, Gay CV, Mastro AM (2008) The bone microenvironment in metastasis; what is special about bone? Cancer Metastasis Rev 27(1):41–55, Review

Carlsson J, Shen L, Xiang J, Xu J, Wei Q (2013) Tendencies for higher co-expression of EGFR and HER2 and downregulation of HER3 in prostate cancer lymph node metastases compared with corresponding primary tumors. Oncol Lett 5(1):208–214

Fang YX, Gao WQ (2013) Roles of microRNAs during prostatic tumorigenesis and tumor progression. Oncogene. doi: 10.1038/onc.2013.54. Epub ahead of print

Fenderico N, Casamichele A, Profumo V, Zaffaroni N, Gandellini P (2013) MicroRNA-mediated control of prostate cancer metastasis: implications for the identification of novel biomarkers and therapeutic targets. Curr Med Chem 20(12):1566–1584

Hockla A, Miller E, Salameh MA, Copland JA, Radisky DC, Radisky ES (2012) PRSS3/mesotrypsin is a therapeutic target for metastatic prostate cancer. Mol Cancer Res 10(12):1555–1566

Kaplan RN, Psaila B, Lyden D (2006) Bone marrow cells in the 'pre-metastatic niche': within bone and beyond. Cancer Metastasis Rev 25(4):521–529, Review

Mimeault M, Batra SK (2013) Hypoxia-inducing factors as master regulators of stemness properties and altered metabolism of cancer- and metastasis-initiating cells. J Cell Mol Med 17(1):30–54, Epub 2013 Jan 10

Ranasinghe WK, Xiao L, Kovac S, Chang M, Michiels C, Bolton D, Shulkes A, Baldwin GS, Patel O (2013) The role of hypoxia-inducible factor 1α in determining the properties of castrate-resistant prostate cancers. PLoS One 8(1):e54251, Epub 2013 Jan 16

Satelli A, Li S (2011) Vimentin in cancer and its potential as a molecular target for cancer therapy. Cell Mol Life Sci 68(18):3033–3046, Epub 2011 Jun 3

Schroten C, Dits NF, Steyerberg EW, Kranse R, van Leenders AG, Bangma CH, Kraaij R (2012) The additional value of TGFβ1 and IL-7 to predict the course of prostate cancer progression. Cancer Immunol Immunother 61(6):905–910

Smith MR, Saad F, Coleman R, Shore N, Fizazi K, Tombal B, Miller K, Sieber P, Karsh L, Damião R, Tammela TL, Egerdie B, Van Poppel H, Chin J, Morote J, Gómez-Veiga F, Borkowski T, Ye Z, Kupic A, Dansey R, Goessl C (2012) Denosumab and bone-metastasis-free survival in men with castration-resistant prostate cancer: results of a phase 3, randomised, placebo-controlled trial. Lancet 379(9810):39–46

Thudi NK, Martin CK, Murahari S, Shu ST, Lanigan LG, Werbeck JL, Keller ET, McCauley LK, Pinzone JJ, Rosol TJ (2011) Dickkopf-1 (DKK-1) stimulated prostate cancer growth and metastasis and inhibited bone formation in osteoblastic bone metastases. Prostate 71(6): 615–625

Xiong SW, Lin TX, Xu KW, Dong W, Ling XH, Jiang FN, Chen G, Zhong WD, Huang J, Zhang S, Wang X, Iqbal S, Wang Y, Osunkoya AO, Chen Z, Chen Z, Shin DM, Yuan H (2013) MicroRNA-335 acts as a candidate tumor suppressor in prostate cancer. Pathol Oncol Res 19(3):529–537, Epub 2013 Mar 3

Zhang W, Haines BB, Efferson C, Zhu J, Ware C, Kunii K, Tammam J, Angagaw M, Hinton MC, Keilhack H, Paweletz CP, Zhang T, Winter C, Sathyanarayanan S, Cheng J, Zawel L, Fawell S, Gilliland G, Majumder PK (2012) Evidence of mTOR activation by an AKT-independent mechanism provides support for the combined treatment of PTEN-deficient prostate tumors with mTOR and AKT inhibitors. Transl Oncol 5(6):422–429

Chapter 17
Counteracting Hypoxia in Radio-Resistant Metastatic Lesions

Stefania Staibano

Abstract The identification of hypoxia-regulated genes and proteins, has provided the basis for the generation of new hypoxia-targeted drugs, conceived to re-oxygenate hypoxic tumor areas. In patients with advanced metastasizing prostate cancer (PC), these kinds of drugs are expected to optimize the effect of radiotherapy, reducing also its side effects. Immunohistochemistry, DNA, proteomic and, tissue array profiling, are increasingly providing us with exciting data, that could lead to the formulation of pre-treatment multimarker tests able to identify the individualized tumor response profiles to radiotherapy, basing on the specific cancer tissue hypoxia pattern and degree (Bussink et al., Radiother Oncol 67:3–15, 2003).

As an example, the recent discovery of the role of microRNA in PC tumor genesis points towards (Kulshreshtha et al., Cell Cycle 6(12):1426–1431, 2007) the, Inactivation of miRs affected by hypoxia as a promising synergistic therapeutic strategy for the radiotherapy-refractory subset of metastatic PC (Kulshreshtha et al., Cell Death Differ 15:667–671, 2008).

This chapter aims to give an outlook of the main hot-topics concerning the new trends of hypoxia-targeted molecular therapies for advanced metastasizing prostate cancers.

17.1 Background

Radiation therapy produces, in most cases, a durable disease control of prostate cancer. According to a recent study on 3,546 patients, the 10-year disease-free survival rates of patients treated with radiation is 75 %, similar to that registered for radical prostatectomy, without all the inconveniences and complications linked

S. Staibano (✉)
Department of Advanced Biomedical Sciences, Pathology Section, Faculty of Medicine and Surgery, University of Naples "Federico II", via S. Pansini, n.5, 80131 Naples, Italy
e-mail: staibano@unina.it

S. Staibano (ed.), *Prostate Cancer: Shifting from Morphology to Biology*,
DOI 10.1007/978-94-007-7149-9_17, © Springer Science+Business Media Dordrecht 2013

to surgical intervention. However, a subset of prostate cancers can recur, in a great extent (>75 %) in the first 5 years after treatment, and in about 5 % during or even after the 10-year follow-up (Critz et al. 2013).

These recurrent cancers are aggressive, metastasizing and radio-resistant. Ionizing radiations exert their therapeutic effects on tumors mainly via the production of cytotoxic reactive oxygen species (ROS), leading to irreversible DNA damage (Morales et al. 1998), and cell death (Wilson and Hay 2011).

This accounts, for a large extent, for the resistance to radiotherapy showed by metastasizing solid tumors, characterized by the frequent presence of multiple hypoxic areas (Moeller and Dewhirst 2006).

In hypoxic tumors, ROS down-regulate the permeability of the mitochondrial outer membrane, shutting-down the mitochondria-driven apoptosis (Moll et al. 2006; Kroemer 2006).

In addition, ROS activate PI3K (stabilizing HIF-1a and favoring the up-regulation of glycolysis with anaerobic ATP production) (Muzandu et al. 2005; Kaelin 2005).

Tumor hypoxia is due to the presence of dysfunctional, abnormal blood vessels and artero-venous shunts responsible for the overall low and heterogeneous blood-flow. Besides low oxygen tension, cells of malignant tumors have to face also high extracellular hydrostatic pressure and low pH (Helmlinger et al. 1997; Jain 1999). A growing body of data indicates that the correction of these tumor microenvironment alterations may mitigate, or even reverse, the malignant phenotype of cancers (Kenny and Bissell 2003). Successful approaches have been developed to counteract tumor hypoxia, as (Bussink et al. 2003) the use of radio- or chemotherapy combined with hyperbaric oxygen or hypoxic cell sensitizers (Henk 1986; Overgaard et al. 1998; Watson et al. 1978).

Several phase III trials are currently investigating new strategies. However, most of these treatments showed an increase in side-effects. To date, treatments targeting tumor hypoxia, widely accepted in clinical practice, are unavailable. This is particularly challenging for advanced, metastatic prostate cancer (PC) which, as previously outlined, frequently shows a poor response to radiotherapy, with an overall worse outcome for most patients. There is the urgent need to reliably predict the risk of tumor recurrence after radiotherapy, to enable the selection of high-risk PC patients that would be candidates for treatment with novel investigational agents. Recently, it has been shown that PC, in spite of a median blood flow three times higher than in normal prostate (NP), exhibits mean pO2 values fewer than one fourth than in NP, with an extremely heterogeneous distribution of intratumor oxygen, independent from clinical or pathological features (Vaupel and Kelleher 2013).

In solid cancers, when the tumor's bulk exceeds 1 mm^3 in volume, neoplastic cell growth progressively overcomes neo-angiogenesis (Shannon et al. 2003).

Tumor blood vessels, surrounded by a rapidly remodeling connective tissue, become leakier than the vessels of the corresponding healthy tissue, and allow serum proteins to infiltrate extracellular matrix, contributing to the elevation of interstitial pressure (Dvorak 1986). Thus, a hypoxic tumor environment takes place

(Huang et al. 1998b); the intrinsic structural and functional abnormalities lead to repeated shut-down of the tumor microvessels, with a parallel decrease of oxygen gradients and nutrients, and, even, to the local reversion of blood flow (acute hypoxia) (Vaupel et al. 1989, 2001; Dewhirst et al. 1999). At the same time, the increase in diffusion distances results in chronic hypoxia, leaving cells chronically deprived of oxygen and nutrients (Vaupel et al. 2001).

Acute, chronic, and, also, regions of intermediate hypoxia, are common findings in advanced metastasizing cancers. Tumor cells react to the hypoxic environment through the paroxysmal activation of physiological responses: up-regulating the expression of a network of gene products that advantage to tumor growth in this adverse conditions and favour resistance to radiotherapy (Anastasiadis et al. 2003).

The frequent resistance to radiation therapy, of hypoxic cancer cells, is largely thought to be caused by the lack of oxygen as a source of radiation induced radicals. As a rule, in normal cells, the highly reactive free radicals produced by radiation induce cell death through the generation of DNA double-strand breaks. The presence of oxygen stabilise the free radicals, further increasing DNA damage and impairing the DNA repair (Höckel and Vaupel 2001; Horsman and Overgaard 2002; Isa et al. 2006).

Intratumoral oxygen levels are, then, directly correlated with the cytotoxic effects of radiation. By converse, hypoxia induces rapid changes in cancer cells gene expression, altering proliferation kinetics by inhibiting DNA repair and apoptosis. Moreover, it increases anaerobic glycolysis that ensures the fast proliferation of hypoxic tumor cells (Harrison and Blackwell 2004; Wouters et al. 2004).

This has been recently demonstrated in human soft tissue sarcomas, in which the more hypoxic tumors showed the fastest proliferating tumor cells (Nordsmark et al. 1996; Bussink et al. 1999; Kennedy et al. 1997; Ljungkvist et al. 2002; Schmaltz et al. 1998; Webster et al. 1995).

All these changes facilitate the onset of radiation resistance and/or cytotoxic drugs treatment, up to a level three times greater than non-hypoxic cancer cells. Tumor hypoxia, then, represents a major cause of treatment failure and poor outcome of human malignancies and, thus, is to be considered for prognostic evaluation of tumors and therapeutic options for cancer patients (Lundgren et al. 2007; Le and Courter 2008).

The modification of tumor hypoxia significantly improved radiotherapy outcome in several tumor types, as head and neck carcinomas (Overgaard and Horsman 1996).

Based on immunohistochemistry and direct oxygen-electrode measurements, hypoxia may be detected in 30–90 % of prostate cancers. The presence of hypoxic regions has been associated with radio-resistance and poor clinical outcome, of prostate cancer patients, representing a very troublesome concern, for the treatment of this tumor (Thomlinson and Gray 1955; Höckel et al. 1991).

Radiation therapy (RT) is commonly used as a primary treatment for prostate cancer, sometimes combined with neoadjuvant and adjuvant hormone therapy (Srigley et al. 2012).

Patients with locally advanced PC, in fact, frequently develop relapse after RT. In addition, recurrence, morbidity, and toxicity, often, complicate radiotherapy, used either as the primary radical treatment of prostate cancer, or as an adjunctive therapy after radical prostatectomy, or palliative therapy.

Recently, it has been confirmed that hypoxia correlates with tumor stage and long-term biochemical outcome, in prostate cancer patients treated with brachytherapy. Several molecules have been proposed as specific biomarkers of the hypoxia-induced response of prostate cancer cells. Among these, Hypoxia inducible factor-1 (HIF-1), VEGF, and osteopontin are among the most frequently reported as overexpressed in aggressive prostate cancers (Huang et al. 2012)

The pathways activated by ionising radiation and hypoxia, include also several pro-caspases, anti-apoptotic proteins and the transcription factor NFKB (Rugo and Schiestl 2004; Gilbert and Knox 1997; Chresta et al. 1996; Romashkova and Makarov 1999).

However, HIF-1 is the only DNA regulatory element truly regulated by oxygen.

17.2 HIF-1

HIF-1 consists of alpha and beta subunits. This transcription factor becomes activated during alpha/beta dimerisation, bounding to p300 to form a complex. This active complex rapidly targets the so-called "hypoxia- response element (HRE)" of more than 60 hypoxia-inducible genes such as erythropoietin (Epo), VEGF, glucose transporter-1 (GLUT-1) (Semenza et al. 1991; Shweiki et al. 1992; Levy et al. 1996; Bashan et al. 1992), and multidrug resistance (MDR) gene, (Comerford et al. 2002).

In prostate cells, HIF-1 expression has been shown to be induced under normoxic conditions (Park et al. 2007). In normal prostatic tissue, HIF-1 probably acts as an intrinsic defender of prostate cells against a zinc-rich environment. Normal prostate tissue and prostatic fluid are, in fact, extremely rich of zinc (Costello et al. 2005), (1,000–3,000 and 9,000 mol/kg, respectively), and this may explain the ability of prostate cells to stabilize HIFs (Ku et al. 2010).

HIF-1 beta is constitutively expressed in all normal cell types, whereas HIF-1 alpha is rapidly post-transcriptionally hydroxylated in the presence of oxygen (Yasuda 2008), and targeted for ubiquitation, a process directly mediated by the von Hippel-Lindau (VHL) tumor suppressor ubiquitin-ligase protein (Maxwell et al. 1999).

HIF-1α is overexpressed in 70 % of human cancers and their metastases provided the first clinical evidence supporting the role played by HIF-1 in human cancer progression (Zhong et al. 1999).

Expression of HIF-1 alpha, assessed by immunohistochemistry, has been recently shown to predict tumor aggressiveness of several primary malignant tumors, (Koukourakis et al. 2002; Aebersold et al. 2001) including prostate cancer, and has been found overexpressed also in their corresponding metastases (Zhong et al. 1999; Lekas et al. 2006).

Several other reports confirmed the association between the immunohistochemical overexpression of HIF-1alpha, increased prostate cancer patient mortality and radiotherapy failure (Aebersold et al. 2001; Quintero et al. 2004).

Evidences on rat and human prostate cell lines have further supported this. Very intriguingly, knocking down HIF- 1α expression with small interfering RNA (siRNA), sensitizes to radiation the androgen independent, highly metastatic PC3 cell line, enhancing their apoptotic rate. HIF-1α siRNA transfection resulted in a significant decrease in the G0/G1 phase, with an accompanied cell cycle arrest at the proliferative phase. It is common knowledge that cells at the proliferative phase are more sensitive to therapy, including irradiation, than in the resting phase. The increase in both interphase death and reproductive death after irradiation, apoptotic potential, and cell cycle arrest (at the proliferative phase) contribute to its radiosensitizing effect (Huang et al. 2012).

HIF-1α inhibition looks, then, promising as an effective molecular therapy to sensitize PC to RT. Hormone-independent prostate cancers have mutations in a critical regulatory domain of the HIF-1α protein, oxygen-dependent degradation domain, which may have a great relevance for the development of therapeutic androgen blockade (Anastasiadis et al. 2002) resistance. HIF-1α inhibition might have more anti-apoptotic effect in hormonal-independent prostate cancers. This however deserves further confirmatory studies.

17.3 miRNA

The hypoxia-inducible factor-1 seems to be involved also in determining a particular kind of hypoxic signature of prostate cancer cells, constituted by a specific mark on microRNA profiles (hypoxia-regulated microRNAs, HRMs). In eukaryotic cells, microRNAs regulate the expression of most genes, participating in cell differentiation, proliferation, metabolism and death (Bartel 2004; Calin et al. 2004, 2005) through translational repression and/or mRNA degradation (Cheng et al. 2005; Croce and Calin 2005).

Sensible microRNA changes have been described in human cancers, sometimes correlated with the clinico-pathological features of tumors (Iorio et al. 2005; Yanaihara et al. 2006).

To date, the pathogenetic events underlying this phenomenon are largely unknown. However, there are evidence that miRNA take a part in the hypoxia-mediated gene repression, contributing to cell survival in low-oxygen conditions. Specific microRNA patterns are a signature of normal and/or neoplastic hypoxic cells. They include miR-23, -24, -26, -27, -103, -107, -181, -210, and -213; miR-26, -107, and -210 are also overexpressed in a variety of human hypoxic tumors, in which they are thought to have a role in tumor-genesis, via the decrease of proapoptotic signaling (Volinia et al. 2006).

The great number of HRMs that are overexpressed in hypoxic tumors suggests that hypoxia represents a driving force leading to microRNA alterations in cancer.

Besides HIF-1, additional transcription factors responsive to hypoxia/anoxia, such as p53 and NF-kB, may induce the expression of several specific microRNAs (Yanaihara et al. 2006; Zhao et al. 2005).

As well, VEGF is considered a potential target for a series of hypoxia-responsive candidate regulatory microRNAs, as miR-16, miR-20, let-7b, miR-17-5p, miR-27, miR-106, miR-107, miR-193, miR-210, miR-320 and miR-361. The patterns of microRNA alterations reported in cancer versus normal tissues, is likely the consequence of the alteration of complex interacting molecular pathways induced, at least in part, by hypoxia. These different patterns, however, are relatively scarce, if compared with the plethora of genes and proteins commonly altered in tumors. To date, a definitive explanation of this phenomenon does not exist. In the meantime, this could be facilitating the possible future applications of miRNA for cancer therapy. The recent availability of microRNA derivatives with increased half life and binding efficiency, such as AMOs (anti microRNA oligonucleotides), LNAs (locked nucleic acids) and antagomirs represents potentially important developments for such purpose (Kulshreshtha et al. 2007; Weiler et al. 2006; Orom et al. 2006; Krutzfeldt et al. 2007).

It has to be pointed out that, in any case, multiple microRNAs are involved in the hypoxic response, this implying that the various attempts of therapies miRNA-specific should be performed through the simultaneous combination of several selected microRNAs (Bartel 2004).

This strategy could improve the outcome of conventional therapies, as early studies have recently reported. In prostate cancers, as in a large number of other tumors with different histogenesis (breast, lung, colon, stomach), specific alterations of microRNA expression have been identified. Very interesting, in prostate cancer, miR-210 seems to be an interesting marker of chronic hypoxia, irrespective of the androgen dependency and should, therefore, be tested as a prognostic marker in high risk prostate cancer patients (Volinia et al. 2006).

Considering that, recently, miR-210 has been detected in serum of lymphoma patients as well as in sera of patients with solid tumors, the hypothesis of the future development of non-invasive cancer new diagnostic tests utilizing miRNAs, could be considered as feasible (Crosby et al. 2009).

17.4 VEGF

The Vascular Endothelial Growth Factor (VEGF) is a master growth factor driving angiogenesis and tumor cell growth, promoting the increase of blood vessel permeability, endothelial cell growth, proliferation, migration, and differentiation (Senger et al. 1983; Ferrara 1995; Hicklin and Ellis 2005). It is regulated by a plethora of cytokine growth factors (EGF, PDGF, bFGF, TGFalpha). Tumor hypoxia directly up-regulates VEGF transcription through the increase of HIF-1 alpha levels (Pugh and Ratcliffe 2003; Harris 2002).

Recent studies demonstrated that the increasing levels of hypoxia correlated with the highest tumor expression of VEGF (Cvetkovic et al. 2001) and predicted biochemical failure after radiotherapy, showing an independent prognostic value (Movsas et al. 2002; Moeller and Dewhirst 2006).

This implies that the VEGF expression directly reflects tumor hypoxia and reduced radiosensitivity of prostate tumor cells (Gray et al. 1953).

This observation could justify the hypothesis that the assessment of tumor VEGF expression on pretreatment diagnostic biopsies, may identify with reasonable probability patients non-responder to radiotherapy, that need to be treated with more aggressive radiation treatments and/or anti-angiogenic or hypoxia-targeted drugs. Further studies on prostate cancer treated with radiotherapy or radical prostatectomy showed that biochemical failure might be better predicted by the increased expression of both HIF-1 alpha and VEGF, independently of T stage, Gleason score, and PSA levels. HIF-1alpha and VEGF, then, are both to be considered very promising as new therapeutic targets for aggressive prostate cancers (Kimbro and Simons 2006; Semenza 2003; Rohwer and Cramer 2011).

The protein kinases inhibitors (TKIs) are the ideal candidates for this purpose. They block the tyrosine kinase-dependent pathways in a mono-specific manner (one TKI directed against a single type of TK) or can be directed toward several tyrosine kinase receptors, thus being able to inhibit multiple signaling pathways. The use of multikinase inhibitors, like Sunitinib and sorafenib, that interfere with several HIF-1 related signaling pathways (i.e. VEGFR/PDGFR) (Merino et al. 2011; Nilsson et al. 2010), is showing encouraging results, even combined with small molecules targeting HIF-1 (Nordgren and Tavassoli 2011).

As for traditional therapies, also in this case, the combination of molecule targeted toward different targets seems to produce a positive, synergistic, effect in counteracting aggressive cancer cells. Even in localised prostate cancers, the extent of tumor hypoxia (Parker et al. 2004) seems to be correlated with long-term poor outcome of prostate cancer patients (Denhardt and Guo 1993; Shweiki et al. 1992), when the immunohistochemical over-expression of HIF-1 alpha and VEGF is found associated with the hypoxia-induced secreted phosphoglycoprotein osteopontin (Zhong et al. 1999; Strohmeyer et al. 2004; Forootan et al. 2006).

Further investigations are needed to evaluate the better way to therapeutically regulate the multi-layer cross-talks between HIF-1alpha and VEGF pathways in hypoxic prostate cancers. Several aspects of these interactions are still matter of active investigation. Among these, the role of the reciprocal interactions between HIF-1, VEGF and the androgen/androgen receptor axis is of particular relevance. Androgens influence tumor vasculature through several mechanisms, enclosing a paracrine signalling mediated through androgen receptors expressed by endothelial cells (Godoy et al. 2008, 2011).

Hypoxia induces androgen hypersensitivity. The transition from androgen-dependence to androgen-independence is a key event in prostate cancer. Patients with clinically localized prostate cancer showed a reduction in the hypoxic fraction following androgen withdrawal (Milosevic et al. 2007) that appear to correlate with downregulation of VEGF expression (Aslan et al. 2005), Tumor hypoxia progressively decreases due to the "normalization" of prostate cancer vasculature allowed by the down-regulation of VEGF (Shweiki et al. 1992; Overgaard et al. 2005).

Quantitative dynamic contrast-enhanced magnetic resonance imaging (DCE MRI) detected an increased and highly functional vascular network in experimental prostate tumors after ADT, which was confirmed by quantitative analysis of fluorescent immunohistochemistry qIHC. However, long-term androgen deprivation induces transient acute hypoxia, and this may be involved in the transition to androgen independence (Rothermund et al. 2005; Park et al. 2006).

Anti-androgens bind to AR in the tumor, but have affinity for pituitary and hypothalamus receptors, stimulating over-secretion of androgens; moreover, in recurrent prostate cancers, an increased expression, stability and translocation of the AR takes place, making tumor cells hypersensible to the growth-promoting effect of dehydrotestoterone (DHT) that, in turn, stabilizes HIF-1a, contributing to the hypoxic response (Mabjeesh et al. 2003; Bakin et al. 2003).

The chronic activation of the androgen receptor (AR) has also shown to upregulate HIF-1 alpha and VEGF in prostate cells through the autocrine receptor tyrosine kinase receptor/PIP3K/AKT-1/mTor signaling (Culig and Bartsch 2006). A prolonged androgen withdrawal leads, as side-effect, to the over-activation of Akt signalling, increasing the ultimate apoptosis-resistance of prostate cancer cells. This partially explains why hormone-resistant prostate cancers are also resistant to most other forms of therapy, comprising the inhibition of PI3K-mediated response (Pfeil et al. 2004).

Although apoptosis is not the main biological effect of ionising radiation, apoptosis resistance has been correlated with radiation therapy failure and proposed as an effective marker for the radioresponse of prostate tumors (Szostak and Kyprianou 2000; Zhivotovsky et al. 1999; Wang et al. 2004).

In prostate cancer cells as in radical prostatectomy specimens (as detected by immunochemistry), Bcl-2 overexpression was positively associated with a high risk of biochemical failure in clinically localized prostate cancer, and poor therapeutic response to radiation therapy (Xie et al. 2006; Revelos et al. 2005; Huang et al. 1998a; Scherr et al. 1999).

These findings indicate that AR-induced HIFs-VEGF-overactivation may represent a potential source of pitfall for experimental trials utilizing VEGF sequestrants, as Bevacizumab, in patients with hormone- resistant prostate cancer.

All these aspects have to be considered to achieve a beneficial combination therapy based upon RT, anti-angiogenic/vascular disrupting therapy, and ADT, in advanced PC. Indeed, strategies aimed at restoring apoptosis pathways in prostate tumor cells seem a pivotal feature of new radiotherapy treatment protocols. The ultimate question that still arises after about two decades of intensive research, concerns the nature, hierarchy and timing of the prevalent acquired mechanisms of radio-resistance in metastasizing, hypoxic prostate cancers. Pathway redundancy, molecular crossing-over and the progressive selection of hypoxic-resistant tumor cells are among the most promising pathogenetic factors for this phenomenon.

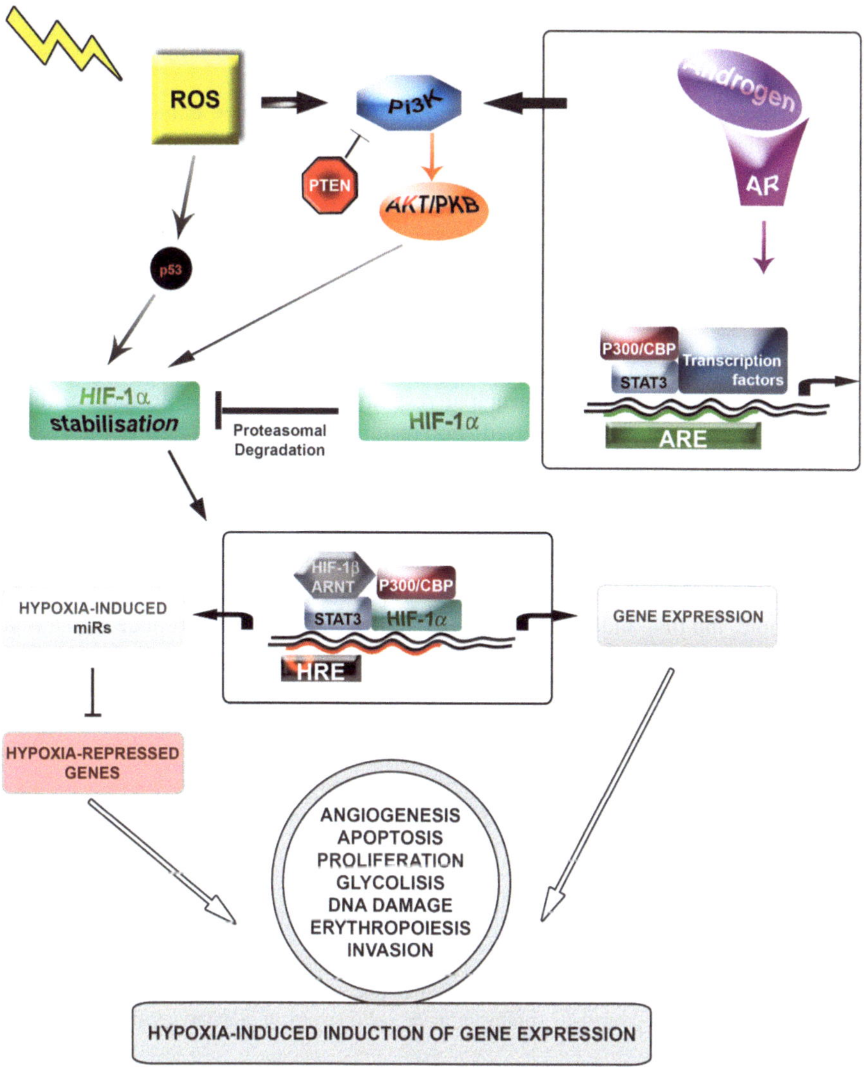

Fig. 17.1 Hypoxia regulated genes and proteins. The main mechanism of action of ionizing radiations in the therapy of tumors is the production of cytotoxic reactive oxygen species (ROS), leading to irreversible DNA damage and cell death. ROS activate PI3K that drives the stabilization of HIF-1a and the up-regulation of glycolysis. Hypoxia modifies gene expression of cancer cells, causing inhibition of DNA repair and apoptosis and increases anaerobic glycolysis which favors the proliferation of hypoxic tumor cells. Moreover, ionising radiation and hypoxia regulate the expression of several pro-caspases and anti-apoptotic proteins. HIF-1 is the most reliable marker of hypoxia. It targets hypoxia-inducible genes and is involved in determining a specific mark on microRNA profiles (hypoxia-regulated microRNAs, HRMs). miRNA take a part in the hypoxia-mediated gene repression, contributing to cell survival in low-oxygen conditions

It seems likely that the plethora of hypoxia-induced transcription factors, operating in hypoxic prostate cancers, may lead to radio-resistance also through the generation or expansion of cancer stem cells.

Moreover, it has recently been shown that, under hypoxic conditions, prostate cancer cell lines interact with p16ink4a via HIF1-alpha, preventing cells entering the senescent state and thereby increasing tumor radioresistance. The existing data are still far to be conclusive. Nevertheless, Hypoxia is now recognized as a major factor driving malignant progression and resistance to treatments of a considerable amount of prostate cancer and the impressive amount of reports in the recent literature support the hypothesis that, counteracting hypoxia through its molecular mediators, is a valid way to fight aggressive prostate cancers. The incomplete comprehension of the molecular events responsible for the cellular reaction to hypoxia, can be reasonable considered as the principal responsible of the variable failure of the majority of the treatment proposed. For instance, still unexploited is the role of a homologous member of the HIF family, : HIF-2. HIF-1 and HIF-2, differ in their transactivation domains, this suggesting that may regulate distinct target genes (Hu et al. 2003; Koukourakis et al. 2006).

As well, the inter-relations between the different members of hypoxia-related pathways should be interpreted further, to optimize the therapeutic approach. As an example, the first attempts of antiangiogenic therapies have produced, as a final effect, an elevated tumor hypoxia with HIF-alpha up-regulation and further gain-of-aggressiveness and radio-resistance of tumors. Furthermore, the use of molecules interacting with most of physiological processes, frequently leads to a relevant toxicity as medium (or long) term side effect. However, the time in which targeting hypoxia will be routinely addressed in the management of aggressive prostate cancer, is rapidly coming in (Fig. 17.1).

References

Aebersold DM, Burri P, Beer KT et al (2001) Expression of hypoxiainducible factor-1alpha: a novel predictive and prognostic parameter in the radiotherapy of oropharyngeal cancer. Cancer Res 61:2911–2916

Anastasiadis AG, Ghafar MA, Salomon L et al (2002) Human hormone refractory prostate cancers can harbor mutations in the O(2)- dependent degradation domain of hypoxia inducible factor-1alpha (HIF-1alpha). J Cancer Res Clin Oncol 128:358–362

Anastasiadis AG, Bemis DL, Stisser BC, Salomon L, Ghafar MA, Buttyan R (2003) Tumor cell hypoxia and the hypoxia-response signaling system as a target for prostate cancer therapy. Curr Drug Targets 4(3):191–196

Aslan G, Cimen S, Yorukoglu K, Tuna B, Sonmez D, Mungan U et al (2005) Vascular endothelial growth factor expression in untreated and androgen-deprived patients with prostate cancer. Pathol Res Pract 201:593–598

Bakin RE, Gioeli D, Sikes RA, Bissonette EA, Weber MJ (2003) Constitutive activation of the Ras/mitogen-activated protein kinase signaling pathway promotes androgen hypersensitivity in LNCaP prostate cancer cells. Cancer Res 63:1981–1989

Bartel DP (2004) MicroRNAs: genomics, biogenesis, mechanism, and function. Cell 116(2): 281–297. Review

Bashan N, Burdett E, Hundal HS, Klip A (1992) Regulation of glucose transport and GLUT1 glucose transporter expression by O2 in muscle cells in culture. Am J Physiol 262(3 Pt 1):C682–C690

Bussink J, Kaanders JHAM, Rijken PFJW et al (1999) Vascular architecture and microenvironmental parameters in human squamous cell carcinoma xenografts: effects of carbogen and nicotinamide. Radiother Oncol 50:173–184

Bussink J, Kaanders JHAM, van der Kogel AJ (2003) Tumor hypoxia at the micro-regional level: clinical relevance and predictive value of exogenous and endogenous hypoxic cell markers. Radiother Oncol 67:3–15

Calin GA, Liu CG, Sevignani C, Ferracin M, Felli N, Dumitru CD, Shimizu M, Cimmino A, Zupo S, Dono M, Dell'Aquila ML, Alder H, Rassenti L, Kipps TJ, Bullrich F, Negrini M, Croce CM (2004) MicroRNA profiling reveals distinct signatures in B cell chronic lymphocytic leukemias. Proc Natl Acad Sci USA 101:11755–11760

Calin GA, Ferracin M, Cimmino A, Di Leva G, Shimizu M, Wojcik SE, Iorio MV, Visone R, Sever NI, Fabbri M, Iuliano R, Palumbo T, Pichiorri F, Roldo C, Garzon R, Sevignani C, Rassenti L, Alder H, Volinia S, Liu CG, Kipps TJ, Negrini M, Croce CM (2005) A microRNA signature associated with prognosis and progression in chronic lymphocytic leukemia. N Engl J Med 353:1793–1801

Cheng AM, Byrom MW, Shelton J, Ford LP (2005) Antisense inhibition of human miRNAs and indications for an involvement of miRNA in cell growth and apoptosis. Nucleic Acids Res 33:1290–1297

Chresta CM, Masters JR, Hickman JA (1996) Hypersensitivity of human testicular tumors to etoposide-induced apoptosis is associated with functional p53 and a high Bax: Bcl-2 ratio. Cancer Res 56:1834–1841

Comerford KM, Wallace TJ, Karhausen J, Louis NA, Montalto MC, Colgan SP (2002) Hypoxia-inducible factor-1-dependent regulation of the multidrug resistance (MDR1) gene. Cancer Res 62:3387–3394

Costello LC, Franklin RB, Feng P et al (2005) Zinc and prostate cancer: a critical scientific, medical, and public interest issue (United States). Cancer Causes Control 16:901–915

Critz FA, Benton JB, Shrake P, Merlin ML (2013) 25-year disease-free survival rate after irradiation for prostate cancer calculated with the prostate specific antigen definition of recurrence used for radical prostatectomy. J Urol 189(3):878–883

Croce CM, Calin GA (2005) miRNAs, cancer, and stem cell division. Cell 122:6–7

Crosby ME, Devlin CM, Glazer PM, Calin GA, Ivan M (2009) Emerging roles of microRNAs in the molecular responses to hypoxia. Curr Pharm Des 15(33):3861–3866

Culig Z, Bartsch G (2006) Androgen axis in prostate cancer. J Cell Biochem 99:373–381

Cvetkovic D, Movsas B, Dicker AP et al (2001) Increased hypoxia correlates with increased expression of the angiogenesis marker vascular endothelial growth factor in human prostate cancer. Urology 57:821–825

Denhardt DT, Guo X (1993) Osteopontin: a protein with diverse functions. FASEB J 7:1475–1482

Dewhirst MW, Ong ET, Braun RD et al (1999) Quantification of longitudinal tissue pO2 gradients in window chamber tumours: impact on tumour hypoxia. Br J Cancer 79:1717–1722

Dvorak HF (1986) Tumors: wounds that do not heal. Similarities between tumor stroma generation and wound healing. N Engl J Med 25:1650–1659

Ferrara N (1995) The role of vascular endothelial growth factor in pathological angiogenesis. Breast Cancer Res Treat 36:127–137

Forootan SS, Foster CS, Aachi VR et al (2006) Prognostic significance of osteopontin expression in human prostate cancer. Int J Cancer 118:2255–2261

Gilbert M, Knox S (1997) Influence of Bcl-2 overexpression on Na+/K(+)-ATPase pump activity: correlation with radiationinduced programmed cell death. J Cell Physiol 171:299–304

Godoy A, Watts A, Sotomayor P, Montecinos VP, Huss WJ, Onate SA, Smith GJ (2008) Androgen receptor is causally involved in the homeostasis of the human prostate endothelial cell. Endocrinology 149:2959–2969

Godoy A, Montecinos VP, Gray DR, Sotomayor P, Yau JM, Vethanayagam RR, Singh S, Mohler JL, Smith GJ (2011) Androgen deprivation induces rapid involution and recovery of human prostate vasculature. Am J Physiol Endocrinol Metab 300:E263–E275

Gray LH, Conger AD, Ebert M, Hornsey S, Scott OC (1953) The concentration of oxygen dissolved in tissues at the time of irradiation as a factor in radiotherapy. Br J Radiol 26:638–648

Harris AL (2002) Hypoxia—a key regulatory factor in tumour growth. Nat Rev Cancer 2:38–47

Harrison L, Blackwell K (2004) Hypoxia and anemia: factors in decreased sensitivity to radiation therapy and chemotherapy. Oncologist 9(suppl 5):31–40

Helmlinger G, Yuan F, Dellian M, Jain RK (1997) Interstitial pH and pO2 gradients in solid tumors *in vivo*: high-resolution measurements reveal a lack of correlation. Nat Med 3:177–182

Henk JM (1986) Late results of a trial of hyperbaric oxygen and radiotherapy in head and neck cancer: a rationale for hypoxic cell sensitizers? Int J Radiat Oncol Biol Phys 12:1339–1341

Hicklin DJ, Ellis LM (2005) Role of the vascular endothelial growth factor pathway in tumor growth and angiogenesis. J Clin Oncol 23:1011–1027

Höckel M, Vaupel P (2001) Tumor hypoxia: definitions and current clinical, biological and molecular aspects. J Natl Cancer Inst 93:266–276

Höckel M, Schlenger K, Knoop C, Vaupel P (1991) Oxygenation of carcinomas of the uterine cervix: evaluation by computerized O2 tension measurements. Cancer Res 51:6098–6102

Horsman MR, Overgaard J (2002) The oxygen effect and tumour microenvironment. In: Steel GG (ed) Basic clinical radiobiology. Arnold, London, pp 158–168

Hu CJ, Wang LY, Chodosh LA, Keith B, Simon MC (2003) Differential roles of hypoxia-inducible factor 1α (HIF-1α) and HIF-2α in hypoxic gene regulation. Mol Cell Biol 23:9361–9374

Huang A, Gandour-Edwards R, Rosenthal SA, Siders DB, Deitch RW, White RW (1998a) p53 and bcl-2 immunohistochemical alterations in prostate cancer treated with radiation therapy. Urology 51:346–351

Huang LE, Jie GU, Schau M, Bunn HF (1998b) Regulation of hypoxia-inducible factor-1a is mediated by an O2-dependent degradation domain via the ubiquitinproteasome pathway. Proc Natl Acad Sci USA 95:7989–7992

Huang Y, Yu J, Yan C, Hou J, Pu J, Zhang G, Fu Z, Wang X (2012) Effect of small interfering RNA targeting hypoxia-inducible factor-1α on radiosensitivity of PC3 cell line. Urology 79(3):744.e17–744.e24

Iorio MV, Ferracin M, Liu CG, Veronese A, Spizzo R, Sabbioni S, Magri E, Pedriali M, Fabbri M, Campiglio M, Menard S, Palazzo JP, Rosenberg A, Musiani P, Volinia S, Nenci I, Calin GA, Querzoli P, Negrini M, Croce CM (2005) MicroRNA gene expression deregulation in human breast cancer. Cancer Res 65:7065–7070

Isa AY, Ward TH, West CML, Slevin NJ, Homer JJ (2006) Hypoxia in head and neck cancer. Br J Radiol 79:791–798

Jain RK (1999) Transport of molecules, particles, and cells in solid tumors. Annu Rev Biomed Eng 1:241–263

Kaelin WG Jr (2005) ROS: really involved in oxygen sensing. Cell Metab 1:357–358

Kennedy AS, Raleigh JA, Perez GM et al (1997) Proliferation and hypoxia in human squamous cell carcinoma of the cervix: first report of combined immunohistochemical assays. Int J Radiat Oncol Biol Phys 37:897–905

Kenny PA, Bissell MJ (2003) Tumor reversion: correction of malignant behavior by microenvironmental cues. Int J Cancer 107:688–695

Kimbro KS, Simons JW (2006) Hypoxia-inducible factor-1 in human breast and prostate cancer. Endocr Relat Cancer 13:739–749

Koukourakis MI, Giatromanolaki A, Sivridis E et al (2002) Hypoxiainducible factor (HIF1A and HIF2A), angiogenesis, and chemoradiotherapy outcome of squamous cell head-and-neck cancer. Int J Radiat Oncol Biol Phys 53:1192–1202

Koukourakis MI, Bentzen SM, Giatromanolaki A et al (2006) Endogenous markers of two separate hypoxia response pathways (hypoxia inducible factor 2 alpha and carbonic anhydrase 9) are associated with radiotherapy failure in head and neck cancer patients recruited in the CHART randomized trial. J Clin Oncol 24:727–735

Kroemer G (2006) Mitochondria in cancer. Oncogene 25:4630–4632
Krutzfeldt J, Kuwajima S, Braich R, Rajeev KG, Pena J, Tuschl T, Manoharan M, Stoffel M (2007) Specificity, duplex degradation and subcellular localization of antagomirs. Nucleic Acids Res 35:2885–2892
Ku JH, Seo SY, Kwak C et al (2010) The role of survivin and Bcl-2 in zinc-induced apoptosis in prostate cancer cells. Urol Oncol 30:562–568
Kulshreshtha R, Ferracin M, Negrini M, Calin GA, Davuluri RV, Ivan M (2007) Regulation of microRNA expression the hypoxic component. Cell Cycle 6(12):1426–1431
Kulshreshtha R, Davuluri RV, Calin GA, Ivan MA (2008) microRNA component of the hypoxic response. Cell Death Differ 15:667–671
Le QT, Courter D (2008) Clinical biomarkers for hypoxia targeting. Cancer Metastasis Rev 27(3):351–362
Lekas A, Lazaris AC, Deliveliotis C, Chrisofos M, Zoubouli C, Lapas D et al (2006) The expression of hypoxia-inducible factor- 1alpha (HIF-1alpha) and angiogenesis markers in hyperplastic and malignant prostate tissue. Anticancer Res 26:2989–2993
Levy AP, Levy NS, Goldberg MA (1996) Hypoxia-inducible protein binding to vascular endothelial growth factor mRNA and its modulation by the von Hippel-Lindau protein. J Biol Chem 271:25492–25497
Ljungkvist ASE, Bussink J, Rijken PFJW, Kaanders JHAM, van der Kogel AJ, Denekamp J (2002) Vascular architecture, hypoxia, and proliferation in the first passage of xenografts of human head and neck squamous cell carcinomas. Int J Radiat Oncol Biol Phys 54:215–228
Lundgren K, Holm C, Landberg G (2007) Hypoxia and breast cancer: prognostic and therapeutic implications. Cell Mol Life Sci 64(24):3233–3247
Mabjeesh NJ, Willard MT, Frederickson CE, Zhong H, Simons JW (2003) Androgens stimulate hypoxia-inducible factor 1 activation via autocrine loop of tyrosine kinase receptor/phosphatidylinositol 30-kinase/protein kinase B in prostate cancer cells. Clin Cancer Res 9:2416–2425
Maxwell PH, Wiesener MS, Chang GW et al (1999) The tumour suppressor protein VHL targets hypoxia-inducible factors for oxygen-dependent proteolysis. Nature 399:271–275
Merino M, Pinto A, González R et al (2011) Antiangiogenic agents and endothelin antagonists in advanced castration resistant prostate cancer. Eur J Cancer 47:1846–1851
Milosevic M, Chung P, Parker C, Bristow R, Toi A, Panzarella T et al (2007) Androgen withdrawal in patients reduces prostate cancer hypoxia: implications for disease progression and radiation response. Cancer Res 67:6022–6025
Moeller BJ, Dewhirst MW (2006) HIF-1 and tumour radiosensitivity. Br J Cancer 95:1–5. (Review)
Moll UM, Marchenko N, Zhang XK (2006) p53 and Nur77/TR3 – transcription factors that directly target mitochondria for cell death induction. Oncogene 25:4725–4743
Morales A, Miranda M, Sánchez-Reyes A, Biete A, Fernández-Checa JC (1998) Oxidative damage of mitochondrial and nuclear DNA induced by ionizing radiation in human hepatoblastoma cells. Int J Radiat Oncol Biol Phys 42:191–203
Movsas B, Chapman JD, Hanlon AL et al (2002) Hypoxic prostate/muscle pO2 ratio predicts for biochemical failure in patients with prostate cancer: preliminary findings. Urology 60:634–639
Muzandu K, Shaban Z, Ishizuka M, Kazusaka A, Fujita S (2005) Nitric oxide enhances catechol estrogen-induced oxidative stress in LNCaP cells. Free Radic Res 39:389–398
Nilsson MB, Zage PE, Zeng L et al (2010) Multiple receptor tyrosine kinases regulate HIF-1alpha and HIF-2alpha in normoxia and hypoxia in neuroblastoma: implications for anti-angiogenic mechanisms of multikinase inhibitors. Oncogene 29:2938–2949
Nordgren IK, Tavassoli A (2011) Targeting tumour angiogenesis with small molecule inhibitors of hypoxia inducible factor. Chem Soc Rev 40:4307–4317
Nordsmark M, Hoyer M, Keller J, Nielsen OS, Jensen OM, Overgaard J (1996) The relationship between tumor oxygenation and cell proliferation in human soft tissue sarcomas. Int J Radiat Oncol Biol Phys 35:701–708
Orom UA, Kauppinen S, Lund AH (2006) LNA-modified oligonucleotides mediate specific inhibition of microRNA function. Gene 372:137–141

Overgaard J, Horsman MR (1996) Modification of hypoxia-induced radioresistance in tumors by the use of oxygen and sensitizers. Semin Radiat Oncol 6:10–21
Overgaard J, Hansen HS, Overgaard M et al (1998) A randomized double blind phase III study of nimorazole as a hypoxic radiosensitizer of primary radiotherapy in supraglottic larynx and pharynx carcinoma. Results of the Danish Head and Neck Cancer Study (DAHANCA) Protocol 5–85. Radiother Oncol 46:135–146
Overgaard J, Eriksen JG, Nordsmark M, Alsner J, Horsman MR (2005) Plasma osteopontin, hypoxia, and response to the hypoxia sensitiser nimorazole in radiotherapy of head and neck cancer: results from the DAHANCA 5 randomised doubleblind placebo-controlled trial. Lancet Oncol 6:757–764
Park SY, Kim YJ, Gao AC, Mohler JL, Onate SA, Hidalgo AA et al (2006) Hypoxia increases androgen receptor activity in prostate cancer cells. Cancer Res 66:5121–5129
Park SE, Park JW, Cho YS et al (2007) HIF-1&;_ promotes survival of prostate cells at a high zinc environment. Prostate 67:1514–1523
Parker C, Milosevic M, Toi A et al (2004) Polarographic electrode study of tumour oxygenation in clinically localised prostate cancer. Int J Radiat Oncol Biol Phys 58:750–757
Pfeil K, Eder IE, Putz T, Ramoner R, Culig Z, Ueberall F et al (2004) Long-term androgen-ablation causes increased resistance to PI3K/Akt pathway inhibition in prostate cancer cells. Prostate 58:259–268
Pugh CW, Ratcliffe PJ (2003) Regulation of angiogenesis by hypoxia: role of the HIF system. Nat Med 9:677–684
Quintero M, Mackenzie N, Brennan PA (2004) Hypoxia-inducible factor 1 (HIF-1) in cancer. Eur J Surg Oncol 30:465–468
Revelos K, Petraki C, Gregorakis A, Scorilas A, Papanastasiou M, Koutsilieris M (2005) Immunohistochemical expression of Bcl-2 is an independent predictor of time-to-biochemical failure in patients with clinically localized prostate cancer following radical prostatectomy. Anticancer Res 25:3123–3133
Rohwer N, Cramer T (2011) Hypoxia-mediated drug resistance: novel insights on the functional interaction of HIFs and cell death pathways. Drug Resist Updat 14:191–201
Romashkova JA, Makarov SS (1999) NF-kappaB is a target of AKT in anti-apoptotic PDGF signalling. Nature 401:86–90
Rothermund CA, Gopalakrishnan VK, Eudy JD, Vishwanatha JK (2005) Casodex treatment induces hypoxia-related gene expression in the LNCaP prostate cancer progression model. BMC Urol 5:5
Rugo RE, Schiestl RH (2004) Increases in oxidative stress in the progeny of X-irradiated cells. Radiat Res 162:416–425
Scherr DS, Vaughan ED Jr, Wei J, Chung M, Felsen D, Allbright R et al (1999) BCL-2 and p53 expression in clinically localized prostate cancer predicts response to external beam radiotherapy. J Urol 162:12–16
Schmaltz C, Hardenbergh PH, Wells A, Fisher DE (1998) Regulation of proliferation-survival decisions during tumor cell hypoxia. Mol Cell Biol 18:2845–2854
Semenza GL (2003) Targeting HIF-1 for cancer therapy. Nat Rev Cancer 3:721–732
Semenza GL, Nejfelt MK, Chi SM, Antonarakis SE (1991) Hypoxiainducible nuclear factors bind to an enhancer element located 30 to the human erythropoietin gene. Proc Natl Acad Sci USA 88:5680–5684
Senger DR, Galli SJ, Dvorak AM et al (1983) Tumor cells secrete a vascular permeability factor that promotes accumulation of ascites fluid. Science 219:983–985
Shannon AM, Bouchier-Hayes DJ, Condron CM, Toomey D (2003) Tumour hypoxia, chemotherapeutic resistance and hypoxia-related therapies. Cancer Treat Rev 29:297–307
Shweiki D, Itin A, Soffer D, Keshet E (1992) Vascular endothelial growth factor induced by hypoxia may mediate hypoxia-initiated angiogenesis. Nature 359:843–845
Srigley JR, Delahunt B, Evans AJ (2012) Therapy-associated effects in the prostate gland. Histopathology 60:153–165

Strohmeyer D, Strauss F, Rossing C et al (2004) Expression of bFGF, VEGF and c-met and their correlation with microvessel density and progression in prostate carcinoma. Anticancer Res 24:1797–1804

Szostak MJ, Kyprianou N (2000) Radiation-induced apoptosis: predictive and therapeutic significance in radiotherapy of prostate cancer (review). Oncol Rep 7:699–706

Thomlinson RH, Gray LH (1955) The histological structure of some human lung cancers and the possible implications for radiotherapy. Br J Cancer 9:539–549

Vaupel P, Kelleher DK (2013) Blood flow and oxygenation status of prostate cancers. Adv Exp Med Biol 765:299–305

Vaupel P, Kallinowski F, Okunieff P (1989) Blood flow, oxygen and nutrient supply, and metabolic microenvironment of human tumors: a review. Cancer Res 49:6449–6465

Vaupel P, Thews O, Hoeckel M (2001) Treatment resistance of solid tumors. Role of hypoxia and anemia. Med Oncol 18:243–259

Volinia S, Calin GA, Liu CG, Ambs S, Cimmino A, Petrocca F et al (2006) A microRNA expression signature of human solid tumors defines cancer gene targets. Proc Natl Acad Sci USA 103:2257–2261

Wang G, Reed E, Li QQ (2004) Apoptosis in prostate cancer: progressive and therapeutic implications (Review). Int J Mol Med 14:23–34

Watson ER, Halnan KE, Dische S et al (1978) Hyperbaric oxygen and radiotherapy: a Medical Research Council trial in carcinoma of the cervix. Br J Radiol 51:879–887

Webster L, Hodgkiss RJ, Wilson GD (1995) Simultaneous triple staining for hypoxia, proliferation, and DNA content in murine tumours. Cytometry 21:344–351

Weiler J, Hunziker J, Hall J (2006) Anti-miRNA oligonucleotides (AMOs): ammunition to target miRNAs implicated in human disease? Gene Ther 13:496–502

Wilson WR, Hay MP (2011) Targeting hypoxia in cancer therapy. Nat Rev Cancer 11:393–410

Wouters BG, van den Beucken T, Magagnin MG, Lambin P, Koumenis C (2004) Targeting hypoxia tolerance in cancer. Drug Resist Updat 7:25–40

Xie Y, Xu K, Dai B, Guo Z, Jiang T, Chen H et al (2006) The 44 kDa Pim-1 kinase directly interacts with tyrosine kinase Etk/BMX and protects human prostate cancer cells from apoptosis induced by chemotherapeutic drugs. Oncogene 25:70–78

Yanaihara N, Caplen N, Bowman E, Seike M, Kumamoto K, Yi M, Stephens RM, Okamoto A, Yokota J, Tanaka T, Calin GA, Liu CG, Croce CM, Harris CC (2006) Unique microRNA molecular profiles in lung cancer diagnosis and prognosis. Cancer Cell 9:189–198

Yasuda H (2008) Solid tumor physiology and hypoxia-induced chemo/radio-resistance: novel strategy for cancer therapy: nitric oxide donor as a therapeutic enhancer. Nitric Oxide 19(2):205–216

Zhao Y, Samal E, Srivastava D (2005) Serum response factor regulates a muscle-specific microRNA that targets Hand2 during cardiogenesis. Nature 436:214–220

Zhivotovsky B, Joseph B, Orrenius S (1999) Tumor radiosensitivity and apoptosis. Exp Cell Res 248:10–17

Zhong H, De Marzo AM, Laughner E, Lim M, Hilton DA, Zagzag D et al (1999) Overexpression of hypoxia-inducible factor 1alpha in common human cancers and their metastases. Cancer Res 59:5830–5835

Chapter 18
"Synthetic Lethality": Molecular Co-targeting to Restore the DNA Repair Mechanisms in Prostate Cancer Cells

Gennaro Ilardi and Stefania Staibano

Abstract Resistance to anticancer radiation treatment has a strong negative impact upon morbidity and mortality related to prostate cancer (Liu et al., Radiother Oncol 88(2):258–268, 2008).

This justifies the great interest in the advancing efforts toward the design of new molecularly-targeted agents which could improve the therapeutic ratio for aggressive prostate cancers via tumor radio-sensitization (Fan et al., Cancer Res 64(23):8526–8533, 2004).

Tumor progression of prostate cancer is associated, as in most of human malignancies, with the sequential loss of function of genes that normally protect against DNA damage.

Malignant prostate cells respond to both endogenous and exogenous DNA damage through complex signaling responses. Due to a specific genetic background, or in an acquired manner during tumor progression, PC cell clones show defect in either DNA single-strand break (SSB) and/or double-strand break (DSB) repair, and/or base damage repair (Stewart et al., Biochem Pharmacol 81(2):203–210, 2011), DSBs are the principal responsible for cell killing due to ionizing radiation (Ward 1988).

A defective DNA double-strand break repair increases genetic instability of PC cells, could be considered as part of their "mutator" phenotype (Tyson et al., Prostate 67:1601–1613, 2007).

During the last decades, it has emerged the concept of "synthetic lethality" (Chalmers et al., Semin Radiat Oncol 20(4):274–281, 2010).

This concept derives from the observation that the use of a single inhibitor of a DNA repair enzyme leads to the selective killing of tumor cells, bearing a

G. Ilardi (✉) • S. Staibano
Department of Advanced Biomedical Sciences, Pathology Section, Faculty of Medicine and Surgery, University of Naples "Federico II", via S. Pansini, n.5, Naples, Italy
e-mail: gennaro.ilardi@gmail.com; staibano@unina.it

S. Staibano (ed.), *Prostate Cancer: Shifting from Morphology to Biology*,
DOI 10.1007/978-94-007-7149-9_18, © Springer Science+Business Media Dordrecht 2013

second DNA repair defect (Bryant et al., Nature 434(7035):913–917, 2005; Jones and Plummer, Br J Radiol 81(Spec No 1):S2–S5, 2008; Fong et al., N Engl J Med 361:123–134, 2009).

To this end, PARP inhibitors are the well-known class of drugs that have recently been proposed to reach synthetic lethality in DNA repair-defective, radio-resistant prostate tumors.

This chapter aims to provide a framework for understanding the recent therapeutic trends designed to overcome radioresistance in prostate cancer via synthetic lethality, we review what it is actually known about the structures and functions of the members of the PARP family of enzymes, outlining a series of open questions that should be addressed in the short time to better guide the development (and the safe clinical use) of PARP inhibitors as new anticancer agents for prostate cancer (Cybulski et al., Cancer Res 64:1215–1219, 2004; Stewart et al., Biochem Pharmacol 81(2):203–210, 2011).

Radiotherapy, either in the form of external beam radiotherapy or brachytherapy, still represents a key therapeutic option for localized or locally advanced prostate cancers.

The effectiveness of radiotherapy is strongly influenced by the occurrence of adverse effects on surrounding normal tissues, ranging from radiation-induced cystitis and/or proctitis, up to erectile dysfunction (Stewart et al. 2011; Barreto-Andrade et al. 2011).

This acute and chronic by-stander toxicity has been greatly reduced with the introduction of the intensity-modulated radiation therapy (IMRT), which allows a more specific targeting of the tumor area.

Despite the advances in radiotherapy techniques (Esgueva et al. 2012; Meng et al. 2005) up to 30 % of radio-treated (Pollack et al. 2003) (intermediate and high risk Zelefsky et al. 2006; Bill-Axelson et al. 2005) prostate cancer patients experience a very aggressive, metastatic disease (Esgueva et al. 2012).

Radioresistance of PC cells is thought to be due to complex inter-relationships between intrinsic genetic and micro-environmental factors (Bristow and Hill 1998; Bristow et al. 2007).

This scenario is further complicated by the significant variability in normal tissue reactions to the radiation-induced DNA damage among prostate cancer patients.

Ionizing radiation kills eukaryotic cells mainly through the induction of DNA-double-strand breaks (DNA-DSBs) (Ward 1988) and, with a lesser extent, (Bristow et al. 2007) *via* DNA single-strand breaks (DNA-SSBs), alteration/loss of DNA bases or DNA-DNA/DNA-protein cross-links (Chalmers et al. 2010).

The ratio of SSBs and DSBs generated by therapeutic ionizing radiation is about 25:1, but DNA double-strand breaks (DSBs) are by far the most potent inducers of cancer cell death (Chalmers et al. 2010).

DNA damage detection and repair require several well-characterized epigenetic events, represented, in first instance, by the relaxation of chromatin and phosphorylation of histone H2AX on the chromatin area lining the DNA lesions, followed either by methylation/acetylation, depending on the specific damaged residue (Escargueil et al. 2008).

DSBs result from the collision of base damage or SSBs with the advancing replication fork, and represent the most cytotoxic lesions (Curtin 2012).

They are usually repaired through two interacting pathways: the homologous recombination (HR) and the non-homologous end joining (NHEJ)-one.

HR utilizes the undamaged sister chromatid (or chromosomal homologue) as a template (Sonoda et al. 2006). This means that HR can take place only in S and G2 phases (Bertrand and Saintigny 2004) operating then during DNA replication (Bernstein et al. 2002; Hansen and Kelly 2000; Hoeijmakers 2001).

NHEJ rapidly binds directly to broken DNA ends during all phases of the cell cycle (Weterings and van Gent 2004; Collis et al. 2005; Riballo et al. 2004; Fan et al. 2004; Rothkamm et al. 2003; Willers et al. 2004) but it lacks the ability to restore any DNA that is lost during the breakage event or subsequent processing, thus resulting in error prone (Sonoda et al. 2006; Chalmers et al. 2010).

These two DNA repair pathways are hyper-activated in normal cells in response to radiation-induced DNA damage (Bromfield et al. 2003).

The non-repair or mis-repair of radiotherapy-induced DNA-DSBs, due to the inhibition of HR or NHEJ, leads to chromosomal deletions, translocations and rearrangements (Bertrand et al. 2004; Bindra and Glazer 2005; Guirouilh-Barbat et al. 2004; Richardson et al. 2004), favouring the onset of genetic instability (Collis et al. 2005; Weterings and van Gent 2004) DNA-DSBs repair has been found to be defective in prostate cancer cell lines (Yuan et al. 1999; Collis et al. 2002; Trzeciak et al. 2004; Fan et al. 2004).

Furthermore, models of prostate carcinogenesis have shown the association with increased levels of chromosomal aberrations and instability can drive the progression from high-grade PIN to PC (Elliott and Jasin 2002; Pihan et al. 2001; Vukovic et al. 2003). Accumulating evidences indicate that the defective DNA double-strand break repair could be considered as part of the "mutator" phenotype of PC cells (Loeb et al. 2003; Bristow et al. 2007).

This has particular relevance, if we consider that the fractionated prostate radiotherapy protocols lead to the generation of a huge number of DNA-DSBs.

During the last few years, in order to overcome PC aggressiveness and radioresistance (Overgaard 2007; Wouters et al. 2002), in fact a positive trend toward multiple promising kinds of "combined" therapeutic approaches has been registered.

Intriguing therapeutic approaches to radiosensitize hypoxic, metastasizing and highly lethal PC cells are focusing on the concept of "synthetic lethality". This definition refers to a situation where the simultaneous presence of two genes mutation results in cell death, whereas each mutation *per se* does not impair cell viability (Curtin 2012).

This phenomenon has inspired new fascinating chances for cancer treatment.

The most promising clinical translations of synthetic lethality concern cancers with specific defects in the HR-mediated repair of double-strand breaks (Antonarakis and Armstrong 2011), as the tumor suppressors BRCA1 and BRCA2 mutant, hereditary breast or ovarian cancers (Venkitaraman 2002).

These tumors represent the first successful examples of treatments based on the use of a single inhibitor of a DNA repair enzyme to selectively kill tumor cells with a second complementary DNA repair pathway defect (Fong et al. 2009; Bryant et al. 2005).

The absolute requirement of HR for DSB repair results in an extreme dependency of BRCA-mutated tumors on PARP-1 action and BER to maintain genomic integrity (Chalmers et al. 2010; Saleh-Gohari et al. 2005).

PARP-1 is the prototypical member of the "poly(ADP-ribose) polymerases (PARPs) superfamily", highly active in protecting cells from endogenous and/or therapeutically induced DNA damage (Curtin 2012).

This large family of enzymes is characterized by the "PARP signature" (GenBank XP_037275 residues 796–1014 de Murcia and Ménissier de Murcia 1994): a 50-amino acid sequence within the enzymatic domain, which catalyzes the cleavage of NAD + into nicotinamide and ADP-ribose. This latter is used to synthesize long, branching, negatively charged polymers, which are then covalently attached to a variety of partner nuclear proteins, as core histones, linker histone H1 (Giner et al. 1992; Grube et al. 1991), HMG proteins, topoisomerases I and II, DNA helicases, single-strandbreak repair (SSBR) and base-excision repair (BER) factors, various transcription factors and PARP-1 itself, involved in DNA damage signalling and repair (Pleschke et al. 2000; Ruf et al. 1996; Oliver et al. 2004), proximally to the DNA breaks.

This poly(ADP-ribosyl)ation leads to the loosening of chromatin structure (Schreiber et al. 2006) allowing the spatial organization of DNA repair through the exposure to the cellular DNA repair machinery (Grube et al. 1991).

PARP-1 is *a 113-kDa* nuclear protein that accounts for at least 80 % of human cellular PARP activity. It is a highly conserved, multifunctional enzyme (Schreiber et al. 2006), with a modular structure (Pfieffer et al. 1999). Under normal conditions, PARP-1 is found associated with histones, DNA and other chromatin associated factors.

In response to DNA damage, it acts as a molecular sensor for DNA-breaks through two zinc-finger motifs, referred to as zf-PARP (Tulin et al. 2002; Menissier et al. 1997), undergoing conformational change and becoming activated.

The binding to DNA breaks, either single-strand break (SSB) and double-strand break (DSB), rapidly stimulated its catalytic activity more than 500-fold. PARP1, as well as his isoenzyme PARP2, acts in SSBs repair mostly by activating base-excision repair (BER).

PARP-1–deficient (or inhibited) cells show, in fact, reduced BER activity (Dantzer et al. 2000) and hypersensitivity to SSB-inducing agents (Horton and Wilson 2007).

If PARP-1 fails to promote SSBs repair, replication forks collapse, converting the DNA damage into replication-associated DSBs, which PARP-1 and PARP-2 attempt to repair either via HR and NHEJ (Chalmers et al. 2010).

The success of PARPs action is strictly dependent upon the extent of the DNA damage. This is due to the transient action of PARP-1 and 2, caused by the rapid degradation of poly-ADP chains due to the poly(ADP-ribose) glycohydrolase (D'Amours et al. 1999).

The half-life of polyADPribose ranges from seconds to minutes, and the hyperactivation of PARP1 consumes the cell pool of NAD + to generate pADPr, lowering cellular energy. So, low-to-moderate DNA damage triggers polyADPribose-dependent DNA repair (Rouleau et al. 2010).

As a complementary effect, pADPr diminishes the affinity for DNA of PARP1, which is then removed from DNA, favouring the post-repair chromatin compaction (Timinszky et al. 2009; Ogata et al. 1980).

In case of excessive DNA damage, PARP1 hyperactivation leads to the excessive NAD + consumption (Juarez-Salinas et al. 1979; Berger et al. 1986; Carson et al. 1988), inducing the catastrophic events that trigger cell death through mechanisms ranging from parthanatos (David et al. 2009; Andrabi et al. 2006), which is directly driven by the longest pADPr chains, to necrosis (Berger et al. 1986; Carson et al. 1988; Zong et al. 2004) or to the establishment of an autophagic state (Huang et al. 2009; Huang and Shen 2009; Munoz-Gamez et al. 2009) of damaged cells.

Due to its fundamental role in DNA-repair, PARP-1 has been identified as the ideal therapeutic target to either specifically kill cancer cells lacking HRR function, and increase the efficacy of radio/chemotherapy in terms of selective tumor cytotoxicity (Farmer et al. 2005; Bryant et al. 2005).

Consistently with these postulates, PARP inhibition in (Bryant et al. 2005) BRCA1 and/or BRCA2-mutated cancer cell (Antonarakis and Armstrong 2011) leads to accumulation of single-strand DNA breaks and (Chalmers et al. 2010) impairs the efficient resolution of collapsed replication forks, impeding the release of PARP molecules from damaged sites, leading to double-strand DNA breaks at replication forks (Antonarakis and Armstrong 2011).

The result is chromosomal instability, cell cycle arrest and subsequent apoptosis caused by the persistence of DNA lesions (Bryant et al. 2005). In other words, the synthetic lethality has been reached.

This synthetic lethality approach has been validated in a multitude of preclinical models, *in vitro* and *in vivo* (Bryant et al. 2005; Farmer et al. 2005), and several PARP inhibitors (olaparib-AZD2281, 3-AB, ISQ, NU1025, KU0058684 or AG14361) have shown promising results, when used as single-agents against *BRCA1*- or *BRCA2*-mutant tumors in clinical testing (Carnell et al. 2006), and promises, as radio-sensitizers in these tumors (Bristow et al. 2007). However, BRCA2 and BRCA1 germ-line mutation carriers have a higher risk to develop PC respect to the normal population. Respectively, prostate cancer relative risk ranges from 2.5 to 7.5 in BRCA2-mutated and <2.0 in BRCA1-carriers; a data particularly significant in tumors diagnosed in younger patients (between ages 40 and 45) (Dong 2006; Levy-Lahad and Friedman 2007).

However, these subsets of PC are relatively poorly differentiated, with poor prognosis (Horsburgh et al. 2005). Additionally, a new BRCA2-interacting protein, PALB2, has been found to be associated with an increased risk of prostate cancer (Erkko et al. 2007).

Olaparib was the first PARP inhibitor to reach human clinical testing in patients with *BRCA1/2*-mutated tumors. In a phase I study, oral olaparib allowed a >50 % PSA drop with resolution of bone metastases in a man with *BRCA2*-related CRPC

(Antonarakis and Armstrong 2011). Nevertheless, BRCA1 and BRCA2 mutations are not considered a major cause of familial or sporadic prostate cancer. A number of other mutations that decrease HR and NHEJ DNA repair responses, that can also sensitize PC cells for synthetic lethality induced by PARP inhibitors, are in fact being increasingly detected (Barreto-Andrade et al. 2011; Plummer et al. 2008; Miknyoczki et al. 2003; Calabrese et al. 2003). They include the phosphatase and tensin homolog gene (PTEN) that is located on chromosome 10, frequently deleted in human cancers and commonly inactivated in prostate cancer (Delaney et al. 2000).

PTEN is a tumor suppressor which, besides inactivating the P13-K/AKT pathway, controls chromosomal integrity and regulates the expression of the repair protein Rad51, reducing the incidence of spontaneous double strand breaks (Shen et al. 2007).

PTEN-deficient tumors exhibit genomic instability due to the down-regulation of Rad51 and the impaired homologous recombination, and result extremely sensitive to PARP inhibitors (Antonarakis and Armstrong 2011).

This sounds of even particular interest, if we consider that, during PC progression the impairment of DNA repair processes mediated by tumor hypoxia greatly contributes to the increase of the genetic instability of prostate cancer (Fan et al. 2004).

Hypoxia, in fact, occurs in 30–90 % of prostate tumors (Chan et al. 2007; Stewart et al. 2010).

Chronic hypoxia, down-regulates the expression and function of many of the DNA-dsb-associated genes, as RAD51 decreasing homologous recombination and DNA double-strand break repair (Vaupel and Mayer 2007).

Thus contributing to the overall genetic instability and aggressiveness of prostate cancer cells.

Tumor hypoxia is indeed progressively emerging as a common feature of prostate tumors associated with poor prognosis.

In-line with these findings, hypoxia-induced metastatic lesions are characterized by gene amplification, point mutation, hyper-mutagenesis and a large amount of DNA strand breaks (Tannock et al. 2005).

Thus, it is becoming increasingly clear that multiple approaches may be hypothesized to overcome radio-resistance of PC cells.

Targeting the hypoxic response has been shown to sensitize PC cells to ionizing radiation *in vitro* (Russell et al. 2003; Slupianek et al. 2001) and may be effective as a complement to radiotherapy of prostate cancer patients. The first attempts of RAD51 expression inhibition by imatinib mesylate (Gleevec) have provided encouraging results (Bristow et al. 2007).

As well, treatment with ABT-888 (Veliparib) has shown some efficacy in PTEN defective PC-3 prostate cancer cells (Barreto-Andrade et al. 2011; Mendes-Pereira et al. 2009).

In addition, ABT-888 enhanced the antitumor activity of TMZ in orthotopic human breast and prostate xenografts.

Radiosensitization *in vivo* has also been demonstrated using several PARPi: AG14361, GPI 15427, ABT-88 and E7016; all showed good radio-sensitization against colon, head and neck, lung and prostate cancer xenografts (Donawho et al. 2007; Calabrese et al. 2004a, b; Palma et al. 2009; Barreto-Andrade et al. 2011).

It could be then hypothesized that PARPi could be successfully used as monotherapy to achieve tumor control, and/or as radio-sensitizers of PTEN-deficient prostate tumors (Barreto-Andrade et al. 2011; Curtin 2012).

A further, extremely interesting, finding derives from a recent report indicating that PARPi can increase the vascular perfusion of tumors through direct vasoactive effects, thus increasing their oxygenation and radio-sensitivity (Liu et al. 2008), leading us to conclude that it may be possible that some PARP inhibitors could induce short-term, vasodilatory effects by virtue of their structural similarities to nicotinamide.

Nicotinamide, a weak PARPi, inhibits contraction of vascular smooth muscle, and its utility in combination with carbogen is being tested in radiotherapy clinical trials (Horsman 1995).

Recently, the AG14361 PARPi was shown to improve intra-tumoral perfusion and possibly reduce tumor hypoxia in mouse xenografts (Calabrese et al. 2004b). Additionally, pharmacological inhibition of PARP has been recently demonstrated to impair HIF-1a induction and angiogenesis (Martin-Oliva et al. 2006; Rajesh et al. 2006a, b).

ABT-888 has been shown to inhibit endothelial tubule formation as well as decreased tumor vascular density (Albert et al. 2007).

This could enhance tumor growth delay after radiotherapy by increasing tumor blood flow, enhancing drug penetration, and increasing oxygen concentrations to offset hypoxic cell radio-resistance. Vasoactive properties and/or anti-endothelial effects have also been documented for AG14361 and ABT888 (Albert et al. 2007; Calabrese et al. 2004a, b; Ali et al. 2009).

Accumulating clinical evidence indicates, in addition, that the short-term use of PARP inhibitors would be extremely well tolerated, even in patients who have undergone multiple previous cytotoxic therapies (Curtin 2012).

Several small-molecule PARP1 and PARP2 inhibitors are currently in preclinical and clinical trials, alone or in combination with DNA-damaging agents (Rodon et al. 2009; Rouleau et al. 2010).

The use of radio-sensitizers to target recognition and repair of DNA damage is becoming an emerging strategy to improve the efficacy of radiotherapy at lower IR doses (Ljungman 2009).

PARP is activated by ionizing radiation (IR) and chemotherapy agents, and this has provided the rationale to examine the combined effects of PARP inhibitors and genotoxic therapy in tumor models and in clinical trials (Donawho et al. 2007; Plummer et al. 2008; Powell et al. 2010).

Several aspects concerning the use of PARPi as radiosensitizers for PC are still to be better elucidated. Here we will briefly examine the most debated ones.

1. PARP inhibition has been shown to radiosensitize mainly replicating cells through the increase of unrepaired DSB (Noel et al. 2006; Dungey et al. 2008).

   This favors the increase of the therapeutic index of radiation therapy in highly replicating tumors (Kastan and Bartek 2004) as recently demonstrated *in vivo,* in colon, head and neck, lung and prostate cancer xenografts using several PARPi, as ABT-88, AG14361, E7016, GPI 15427 (Donawho et al. 2007; Calabrese et al. 2004a, b; Palma et al. 2009; Barreto-Andrade et al. 2011).

   However, the long-term success of radiation therapy of PC depends upon the eradication of mostly non-replicating tumor stem cells, that constitute up to 1 % of total tumor cells (Wong and Hill 1998).

   As it is well-known, cancer stem cells typically reside in hypoxic niches.

   Some PARPi, such as ABT-888, have shown the ability to radio-sensitize *in vivo* also hypoxic tumor clonogens.

   It seems, then, that at least some PARPi could improve the therapeutic ratio of clinical radiotherapy by overcoming both oxic and hypoxic radioresistance.

   These findings are still to be confirmed on large case–control studies. Nevertheless, they look very promising. Moreover, these PARPi would be used also as a complement for new biological imaging-guided hypoxic tumor regions-targeted, high doses-radiotherapy ("dose painting") (Liu et al. 2008) and, some evidences, show that radiotherapy may induce prostate cancer cell death also through a terminal growth arrest (Schwarze et al. 2001). As indicated by the overexpression of markers of senescence, such as p21WAF1/Cip1 and p16INK4a (Stein et al. 1999), therapy-induced senescence is increasingly being reported as an alternative mode of cell death.

   It can result from several inducers, including accumulation of unrepaired DNA damage and is proposed to contribute to tumor control following treatment with cytotoxic agents (Roninson 2003; Lleonart et al. 2009). Some results in PC-3 cells and tumors have suggested that accelerated senescence may be a factor in the therapeutic response of some human tumors to IR combined with PARP inhibition (Efimova et al. 2010).

   A terminal growth arrest should probably be considered an adjunctive end-point and novel therapeutic approach for radiotherapy of prostate carcinoma. PARP inhibitors are among the favorite candidates for inducing this purpose.

   However, this point still deserves consideration in clinical trials and, mainly due to the lack of reliable senescence-inducing agents, this area constitutes an open field for further research (Barreto-Andrade et al. 2011).
2. First generation PARP inhibitors have produced defects in lymphocytes and muscle cells differentiation in several cases. This side-effect may be due to the need of inhibit >90 % of PARP activity to produce a therapeutic impair of DNA repair (Satoh et al. 1994; Farzaneh et al. 1982; Johnstone and Williams 1982).

   However, the third generation of highly potent and specific PARP inhibitors has not produced these adverse effects, suggesting that they might have been only a result of an off-target effect specific for the first type of inhibitors.

However, given the high potency of the new generation of PARPi, the systemic effects of near-complete PARP inhibition should be tested with additional studies on animal model (105. Konishi et al. 1986; Takahashi et al. 1984).

As an example, PARP1 is required either for the protection of the cardiovascular system and for the development of memory (Pacher and Szabo 2007; Goldberg et al. 2009).

As well, long-term PARP1 inhibition could also lead to secondary malignancies, particularly when inhibitors are administered with DNA damaging agents. This hypothesis is supported by several reports. Recently, a high incidence of cancers in mice knocked-out for Parp1 has been reported (Morrison et al. 1997; Tong et al. 2002, 2003).

Reasonably, in each case the risk of occurrence of secondary tumors should be challenged against the chance of improving the therapeutic ratio of currently lethal cancers (Bristow et al. 2007).

The better understanding of both acute and late effects of therapeutic DNA repair inhibition may allow oncologists to focus on the possible way to prevent second malignancies in PARPi-treated patients, by chemopreventive strategies or alternative pathway activation.

Studies on large collections of tumor specimens will be essential to evaluate, in situ, new potential targets for complementary therapies (Antonarakis and Armstrong 2011).

Lastly, it will be interesting to see how durable will be the response rates of patients treated with PARPi.

It appears worrisome, in fact, that resistance to PARP inhibitors has been described in *BRCA1*- or *BRCA2*-deficient cancer cells, following to the reactivation of these genes by secondary mutations (Ashworth 2008; Sakai et al. 2008; Edwards et al. 2008).

3. Little is actually known about the effects of inhibiting PARPs other than PARP-1 and 2 (Rouleau et al. 2010).

   Among the 17 members of the 'PARP superfamily' identified to date, only PARP3, V-PARP and Tankyrase-1 and -2 (TNKS-1 and -2) have the ADP-ribose polymerizing activity (Hakame et al. 2008) PARP-3 co-operate with PARP-1 in the response to DNA double strand breaks (Boehler et al. 2011).

   Tankyrases (TNKS) 1 and 2 are involved in telomere maintenance (Midorikawa et al. 2006) and V-PARP is associated with large ribonuclear protein structures (cytoplasmic vaults), which are amplified in some drug resistance models (Cohen-Armon et al. 2004; Kickhoefer et al. 1999; Fang et al. 2006).

   A potentially specific tankyrase inhibitor, XAV-939, has been identified, raising the possibility that *BRCA1*- or *BRCA2*-mutant tumors might be successfully targeted without inhibiting PARP1 (Ju et al. 2004).

   To date, however, it is unclear to what extent the inhibition of other PARPs contributes to the cellular effects of PARP inhibitors. Only a few studies exist. Moreover, specific inhibitors of all the different PARP-family members are still incompletely available (Curtin 2005).

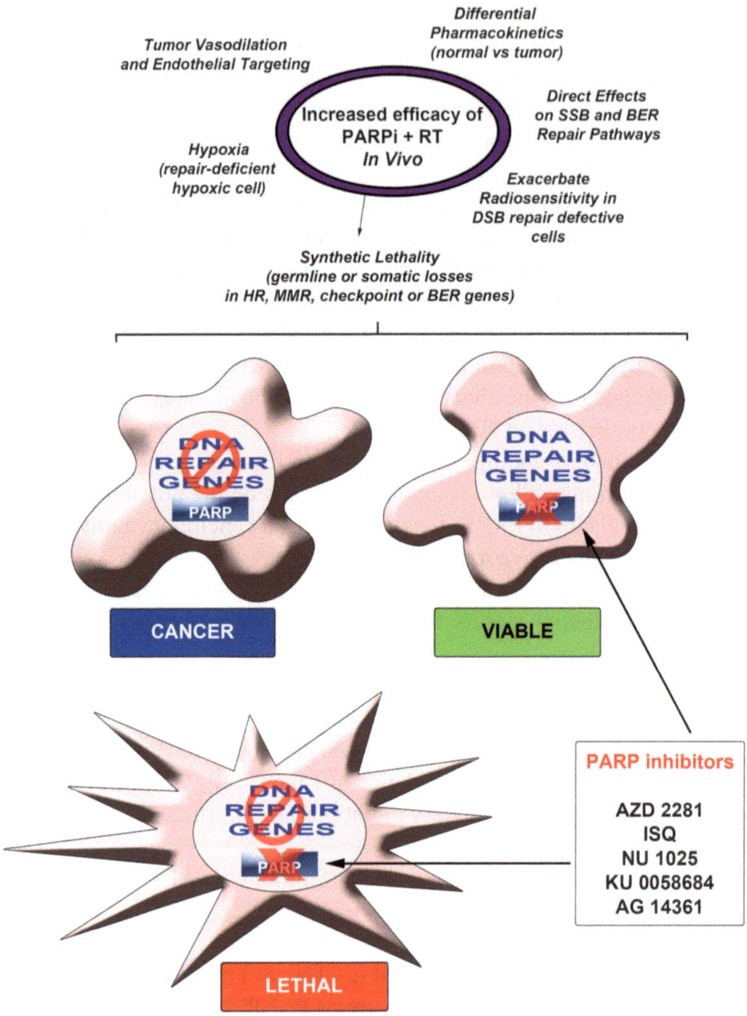

Fig. 18.1 "Synthetic lethality" in the therapy of PCa. Ionizing radiation kills eukaryotic cells through the induction of DNA- double-strand breaks (DNA-dsbs) that represent the most cytotoxic lesions. They are usually repaired through the homologous recombination (HR) and the non-homologous end joining (NHEJ) pathways that, in fact, result hyper-activated in normal cells in response to radiation-induced DNA damage. The non-repair of radiotherapy-induced DNA-dsbs causes genomic instability. The term "synthetic lethality" refers to a condition of simultaneous presence of mutation of two genes resulting in cell death. The most promising clinical translations of synthetic lethality concern cancers with specific defects in the HR-mediated repair of double-strand breaks that results in an extreme dependency of tumors on PARP1 action and BER to maintain genomic integrity. Basing on its fundamental role in DNA-repair, PARP-1 represents the ideal therapeutic target to kill cancer cells with loss of HRR function and to increase the efficacy of radio/chemotherapy. PARP inhibition leads to chromosomal instability, cell cycle arrest and apoptosis caused by the persistence of DNA lesions

This topic, in any case, will certainly be a matter of intense investigation in the near future.

At present, the prediction of radio-responsiveness of prostate cancer is based upon the pre-treatment PSA level/doubling time, Gleason score and T-stage (Nichol et al. 2005).

Novel therapeutic approaches, differentially targeting HR and/or NHEJ DNA-dsb repair, could necessitate of new identifiers of DNA repair (i.e., single nucleotide polymorphisms (SNPs), protein expression, functional assays for DNA damage sensing and repair) related to normal and tumor radio-sensitivity, to drive for individual prostate cancer therapy (Bristow et al. 2007).

These tests may result useful as biomarkers of the genetic instability, malignant progression and aggressiveness of tumor (Choudhury et al. 2006).

This approach may protect normal tissues, allowing the delivery of high doses of radiation and DNA repair inhibitors exclusively on tumor areas targeted by hypoxic signals from MRI, CT or PET-based imaging.

This is an exciting time for oncologists, radio-therapists and pathologist now able to surf over the mounting data concerning the molecular interactions responsible for DNA repair, to discover and apply new therapies based upon a direct collaboration between basic science, industry, academia, and regulatory agencies.

The chances to achieve a new integrative and interdisciplinary approach to prostate cancer patient care, based upon translational oncology, are indeed rapidly becoming reality.

We are now almost ready to take on the challenge to apply next-generation discovered biomarkers able to drive a successful control of previously untreatable, radio-resistant prostate cancers (Fig. 18.1).

References

Albert JM, Cao C, Kim KW et al (2007) Inhibition of poly(ADPRibose) polymerase enhances cell death and improves tumor growth delay in irradiated lung cancer models. Clin Cancer Res 13:3033–3042

Ali M, Telfer BA, McCrudden C et al (2009) Vasoactivity of AG014699, a clinically active small molecule inhibitor of poly(ADP-ribose) polymerase: a contributory factor to chemopotentiation *in vivo*? Clin Cancer Res 15:6106–6112

Andrabi SA et al (2006) Poly(ADP-ribose) (PAR) polymer is a death signal. Proc Natl Acad Sci USA 103:18308–18313

Antonarakis ES, Armstrong AJ (2011) Emerging therapeutic approaches in the management of metastatic castration-resistant prostate cancer. Prostate Cancer Prostatic Dis 14(3):206–218, Review

Ashworth A (2008) Drug resistance caused by reversion mutation. Cancer Res 68:10021–10023

Barreto-Andrade JC, Efimova EV, Mauceri HJ, Beckett MA, Sutton HG, Darga TE, Vokes EE, Posner MC, Kron SJ, Weichselbaum RR (2011) Response of human prostate cancer cells and tumors to combining PARP inhibition with ionizing radiation. Mol Cancer Ther 10(7):1185–1193

Berger SJ, Sudar DC, Berger NA (1986) Metabolic consequences of DNA damage: DNA damage induces alterations in glucose metabolism by activation of poly (ADP-ribose) polymerase. Biochem Biophys Res Commun 134:227–232

Bernstein C, Bernstein H, Payne CM, Garewal H (2002) DNA repair/pro-apoptotic dual-role proteins in five major DNA repair pathways: fail-safe protection against carcinogenesis. Mutat Res 511(2):145–178

Bertrand P, Saintigny Y et al (2004) p53's double life: transactivation-independent repression of homologous recombination. Trends Genet 20:235–243

Bill-Axelson A, Holmberg L, Ruutu M, Haggman M, Andersson SO, Bratell S et al (2005) Radical prostatectomy versus watchful waiting in early prostate cancer. N Engl J Med 352:1977–1984

Bindra RS, Glazer PM (2005) Genetic instability and the tumor microenvironment: towards the concept of microenvironment-induced mutagenesis. Mutat Res 569:75–85

Boehler C et al (2011) Poly(ADP-ribose) polymerase 3 (PARP3), a newcomer in cellular response to DNA damage and mitotic progression. Proc Natl Acad Sci U S A 108:2783–2788

Bristow RG, Hill P (1998) Molecular and cellular basis of radiotherapy. In: Tannock IF, Hill RP (eds) The basic science of oncology. McGraw-Hill, Toronto, pp 295–321

Bristow RG, Ozcelik H, Jalali F, Chan N, Vesprini D (2007) Homologous recombination and prostate cancer: a model for novel DNA repair targets and therapies. Radiother Oncol 83:220–230

Bromfield GP, Meng A, Warde P, Bristow RG (2003) Cell death in irradiated prostate epithelial cells: role of apoptotic and clonogenic cell kill. Prostate Cancer Prostatic Dis 6:73–85

Bryant HE, Schultz N, Thomas HD, Parker KM, Flower D, Lopez E, Kyle S, Meuth M, Curtin NJ, Helleday T (2005) Specific killing of BRCA2-deficient tumours with inhibitors of poly(ADP-ribose) polymerase. Nature 434(7035):913–917

Calabrese CR et al (2003) Identification of potent non-toxic poly(ADPribose) polymerase-1 (PARP-1) inhibitors: chemopotentiation and pharmacological studies. Clin Cancer Res 9:2711–2718

Calabrese CR, Almassy R, Barton S et al (2004a) Anticancer chemosensitization and radiosensitization by the novel poly(ADPribose) polymerase-1 inhibitor AG14361. J Natl Cancer Inst 96:56–67

Calabrese CR et al (2004b) Preclinical evaluation of a novel poly(ADPribose) polymerase-1 (PARP-1) inhibitor, AG14361, with significant anticancer chemo- and radio-sensitization activity. J Natl Cancer Inst 96:56–67

Carnell DM, Smith RE, Daley FM, Saunders MI, Bentzen SM, Hoskin PJ (2006) An immunohistochemical assessment of hypoxia in prostate carcinoma using pimonidazole: implications for radioresistance. Int J Radiat Oncol Biol Phys 65:91–99

Carson DA, Carrera CJ, Wasson DB, Yamanaka H (1988) Programmed cell death and adenine deoxynucleotide metabolism in human lymphocytes. Adv Enzyme Regul 27:395–404

Chalmers AJ, Lakshman M, Chan N, Bristow RG (2010) Poly(ADP-ribose) polymerase inhibition as a model for synthetic lethality in developing radiation oncology targets. Semin Radiat Oncol 20(4):274–281

Chan N, Milosevic M, Bristow RG (2007) Tumor hypoxia, DNA repair and prostate cancer progression: new targets and new therapies. Future Oncol 3:329–341

Choudhury A, Cuddihy A, Bristow RG (2006) Radiation and new molecular agents part I: targeting ATM-ATR checkpoints, DNA repair, and the proteasome. Semin Radiat Oncol 16:51–58

Cohen-Armon M et al (2004) Long-term memory requires polyADP-ribosylation. Science 304:1820–1822

Collis SJ, Sangar VK, Tighe A et al (2002) Development of a novel rapid assay to assess the fidelity of DNA double-strand-break repair in human tumour cells. Nucleic Acids Res 30:E1

Collis SJ, DeWeese TL et al (2005) The life and death of DNA-PK. Oncogene 24:949–961

Curtin NJ (2005) PARP inhibitors for cancer therapy. Expert Rev Mol Med 7:1–20. Together with reference 15, excellent reviews describing the therapeutic promise of PARP Identification of a PAR-binding motif that mediates selective interaction between PAR and protein partners. inhibitors in cancer treatment or in inflammatory diseases

Curtin NJ (2012) Poly(ADP-ribose) polymerase (PARP) and PARP inhibitors. Drug Discov Today Dis Mod Target DNA Repair 9(2):e51–e58

Cybulski C, Gorski B, Debniak T et al (2004) NBS1 is a prostate cancer susceptibility gene. Cancer Res 64:1215–1219

D'Amours D, Desnoyers S, D'Silva I et al (1999) Poly(ADP-ribosyl)ation reactions in the regulation of nuclear functions. Biochem J 342:249–268

Dantzer F, de la Rubia G, Menissier-De Murcia J et al (2000) Base excision repair is impaired in mammalian cells lacking poly(ADP-ribose) polymerase-1. Biochemistry 39:7559–7569

David KK, Andrabi SA, Dawson TM, Dawson VL (2009) Parthanatos, a messenger of death. Front Biosci 14:1116–1128

de Murcia G, Ménissier de Murcia J (1994) Poly(ADP-ribose) polymerase: a molecular nick-sensor. Trends Biochem Sci 19:172–176

Delaney CA et al (2000) Potentiation of temozolomide and topotecan growth inhibition and cytotoxicity by novel poly (adenosine diphosphoribose) polymerase inhibitors in a panel of human tumor cell lines. Clin Cancer Res 6:2860–2867

Donawho CK, Luo Y, Luo Y, Penning TD, Bauch JL, Bouska JJ et al (2007) ABT-888, an orally active poly (ADP-ribose) polymerase inhibitor that potentiates DNA-damaging agents in preclinical tumor models. Clin Cancer Res 13:2728–2737

Dong JT (2006) Prevalent mutations in prostate cancer. J Cell Biochem 97:433–447

Dungey FA, Loser DA, Chalmers AJ (2008) Replication-dependent radiosensitization of human glioma cells by inhibition of poly(ADP-ribose) polymerase: mechanisms and Therapeutic potential. Int J Radiat Oncol Biol Phys 72:1188–1197

Edwards SL et al (2008) Resistance to therapy caused by intragenic deletion in BRCA2. Nature 451:1111–1115

Efimova EV, Mauceri HJ, Golden DW, Labay E, Bindokas VP, Darga TE et al (2010) Poly(ADP-ribose) polymerase inhibitor induces accelerated senescence in irradiated breast cancer cells and tumors. Cancer Res 70:6277–6282

Elliott B, Jasin M (2002) Double-strand breaks and translocations in cancer. Cell Mol Life Sci 59:373–385

Erkko H, Xia B, Nikkila J et al (2007) A recurrent mutation in PALB2 in Finnish cancer families. Nature 446:316–319

Escargueil AE, Soares DG, Salvador M, Larsen AK, Henriques JA (2008) What histone code for DNA repair? Mutat Res 658(3):259–270

Esgueva R, Park K, Kim R, Kitabayashi N, Barbieri CE, Dorsey PJ Jr, Abraham C, Banerjee S, Leung RA, Tewari AK, Terry S, Shevchuk MM, Rickman DS, Rubin MA, Weill Cornell Medical College (2012) Next-generation prostate cancer biobanking: toward a processing protocol amenable for the International Cancer Genome Consortium. Diagn Mol Pathol 21(2):61–68

Fan R, Kumaravel TS, Jalali F, Marrano P, Squire JA, Bristow RG (2004) Defective DNA strand break repair after DNA damage in prostate cancer cells: implications for genetic instability and prostate cancer progression. Cancer Res 64(23):8526–8533

Fang Y et al (2006) BubR1 is involved in regulation of DNA damage responses. Oncogene 25:3598–3605. doi:10.1038/sj.onc.1209392

Farmer H, McCabe N, Lord CJ, Tutt AN, Johnson DA, Richardson TB et al (2005) Targeting the DNA repair defect in BRCA mutant cells as a therapeutic strategy. Nature 434:917–921

Farzaneh F, Zalin R, Brill D, Shall S (1982) DNA strand breaks and ADP-ribosyl transferase activation during cell differentiation. Nature 300:362–366

Fong PC, Boss DS, Yap TA et al (2009) Inhibition of poly(ADP-ribose) polymerase in tumors from BRCA mutation carriers. N Engl J Med 361:123–134

Giner H et al (1992) Overproduction and large-scale purification of the human poly(ADP-ribose) polymerase using a baculovirus expression system. Gene 114:279–283

Goldberg S, Visochek L, Giladi E, Gozes I, Cohen-Armon M (2009) PolyADP-ribosylation is required for long-term memory formation in mammals. J Neurochem 111:72–79

Grube K et al (1991) Direct stimulation of poly(ADP-ribose) polymerase in permeabilised cells by double-stranded DNA oligomers. Anal Biochem 193:236–239
Guirouilh-Barbat J, Huck S et al (2004) Impact of the KU80 pathway on NHEJ-induced genome rearrangements in mammalian cells. Mol Cell 14:611–623
Hakame A et al (2008) The expanding field of poly(ADP-ribosyl)ation reactions. EMBO Rep 9:1094–1100
Hansen K, Kelly M (2000) Review of mammalian DNA repair and translational implications. J Pharmacol Exp Ther 295(1):1–9
Hoeijmakers JH (2001) Genome maintenance mechanisms for preventing cancer. Nature 411:360–374
Horsburgh S, Matthew A, Bristow RG, Trachtenberg J (2005) Male BRCA1 and BRCA2 mutation carriers: a pilot study investigating medical characteristics of patients participating in a prostate cancer prevention clinic. Prostate 65:124–129
Horsman MR (1995) Nicotinamide and other benzamide analogs as agents for overcoming hypoxic cell radiation resistance in tumours. A review. Acta Oncol 34:571–587
Horton JK, Wilson SH (2007) Hypersensitivity phenotypes associated with genetic and synthetic inhibitor-induced base excision repair deficiency. DNA Repair (Amst) 6:530–543
Huang Q, Shen HM (2009) To die or to live: the dual role of poly(ADP-ribose) polymerase-1 in autophagy and necrosis under oxidative stress and DNA damage. Autophagy 5:273–276
Huang Q, Wu YT, Tan HL, Ong CN, Shen HM (2009) A novel function of poly(ADP-ribose) polymerase-1 in modulation of autophagy and necrosis under oxidative stress. Cell Death Differ 16:264–277
Johnstone AP, Williams GT (1982) Role of DNA breaks and ADP-ribosyl transferase activity in eukaryotic differentiation demonstrated in human lymphocytes. Nature 300:368–370
Jones C, Plummer ER (2008) PARP inhibitors and cancer therapy - early results and potential applications. Br J Radiol 81(Spec No 1):S2–S5
Ju BG et al (2004) Activating the PARP-1 sensor component of the groucho–TLE1 corepressor complex mediates a CaMKinase IIδ-dependent neurogenic gene activation pathway. Cell 119:815–829
Juarez-Salinas H, Sims JL, Jacobson MK (1979) Poly(ADP-ribose) levels in carcinogen-treated cells. Nature 282:740–741
Kastan MB, Bartek J (2004) Cell-cycle checkpoints and cancer. Nature 432:316–323
Kickhoefer VA, Siva AC, Kedersha NL, Inman EM, Ruland C, Streuli M, Rome LH (1999) The 193-kD vault protein, VPARP, is a novel poly(ADP-ribose) polymerase. J Cell Biol 146(5):917–928
Konishi Y et al (1986) Possible model of liver carcinogenesis using inhibitors of NAD + ADP ribosyl transferase in rats. Toxicol Pathol 14:483–488
Levy-Lahad E, Friedman E (2007) Cancer risks among BRCA1 and BRCA2 mutation carriers. Br J Cancer 96:11–15
Liu SK, Coackley C, Krause M, Jalali F, Chan N, Bristow RG (2008) A novel poly(ADP-ribose) polymerase inhibitor, ABT-888, radiosensitizes malignant human cell lines under hypoxia. Radiother Oncol 88(2):258–268
Ljungman M (2009) Targeting the DNA damage response in cancer. Chem Rev 109:2929–2950
Lleonart ME, Artero-Castro A, Kondoh H (2009) Senescence induction; a possible cancer therapy. Mol Cancer 8:3
Loeb LA, Loeb KR, Anderson JP (2003) Multiple mutations and cancer. Proc Natl Acad Sci USA 100:776–781
Martin-Oliva D, Aguilar-Quesada R, O'Valle F et al (2006) Inhibition of poly(ADP-ribose) polymerase modulates tumor-related gene expression, including hypoxia-inducible factor-1 activation, during skin carcinogenesis. Cancer Res 66:5744–5756
Mendes-Pereira AM, Martin SA, Brough R, McCarthy A, Taylor JR, Kim JS et al (2009) Synthetic lethal targeting of PTEN mutant cells with PARP inhibitors. EMBO Mol Med 1:315–322

Meng AX, Jalalia F, Cuddihya A, Chan N, Bindrab RS, Glazerb PM, Robert G (2005) Bristow Hypoxia down-regulates DNA double strand break repair gene expression in prostate cancer cells. Radiother Oncol 76:168–176
Menissier de Murcia J et al (1997) Requirement of poly(ADP-ribose) polymerase in recovery from DNA damage in mice and in cells. Proc Natl Acad Sci USA 94:7303–7307
Midorikawa R, Takei Y, Hirokawa N (2006) KIF4 motor regulates activity-dependent neuronal survival by suppressing PARP-1 enzymatic activity. Cell 125:371–383
Miknyoczki SJ et al (2003) Chemopotentiation of temozolomide, irinotecan, and cisplatin activity by CEP-6800, a poly(ADP-ribose) polymerase inhibitor. Mol Cancer Ther 2:371–382
Morrison C et al (1997) Genetic interaction between PARP and DNA-PK in V(D)J. Recombination and tumorigenesis. Nat Genet 17:479–482
Munoz-Gamez JA et al (2009) PARP-1 is involved in autophagy induced by DNA damage. Autophagy 5:61–74
Nichol AM, Warde P, Bristow RG (2005) Optimal treatment of intermediate-risk prostate carcinoma with radiotherapy: clinical and translational issues. Cancer 104:891–905
Noel G, Godon C, Fernet M et al (2006) Radiosensitization by the poly(ADPribose) polymerase inhibitor 4-amino-1,8-naphthalimide is specific of the S phase of the cell cycle and involves arrest of DNA synthesis. Mol Cancer Ther 5:564–574
Ogata N, Ueda K, Kagamiyama H, Hayaishi O (1980) ADP-ribosylation of histone H1. Identification of glutamic acid residues 2, 14, and the COOH-terminal lysine residue as modification sites. J Biol Chem 255:7616–7620
Oliver AW et al (2004) Crystal structure of the catalytic fragment of murine poly(ADP-ribose) polymerase-2. Nucleic Acids Res 32:456–464
Overgaard J (2007) Hypoxic radiosensitization: adored and ignored. J Clin Oncol 25:4066–4074
Pacher P, Szabo C (2007) Role of poly(ADP-ribose) polymerase 1 (PARP-1) in cardiovascular diseases: the therapeutic potential of PARP inhibitors. Cardiovasc Drug Rev 25:235–260
Palma JP et al (2009) ABT-888 confers broad *in vivo* activity in combination with temozolomide in diverse tumours. Clin Cancer Res 15:7277–7290
Pfieffer R et al (1999) Quantitative nonisotopic immuno-dot-blot method for the assessment cellular poly(ADP-ribosyl)ation capacity. Anal Biochem 275:118–122
Pihan GA, Purohit A, Wallace J, Malhotra R, Liotta L, Doxsey SJ (2001) Centrosome defects can account for cellular and genetic changes that characterize prostate cancer progression. Cancer Res 61:2212–2219
Pleschke JM, Kleczkowska HE, Strohm M, Althaus FR (2000) Poly(ADP-ribose) binds to specific domains in DNA damage checkpoint proteins. J Biol Chem 275:40974–40980
Plummer R, Jones C, Middleton M, Wilson R, Evans J, Olsen A et al (2008) Phase I study of the poly(ADP-ribose) polymerase inhibitor, AG014699, in combination with temozolomide in patients with advanced solid tumors. Clin Cancer Res 14:7917–7923
Pollack A, Hanlon A et al (2003) Radiation therapy dose escalation for prostate cancer: a rationale for IMRT. World J Urol 21:200–208
Powell C, Mikropoulos C, Kaye SB, Nutting CM, Bhide SA, Newbold K et al (2010) Pre-clinical and clinical evaluation of PARP inhibitors as tumor-specific radiosensitizers. Cancer Treat Rev 36:566–575
Rajesh M, Mukhopadhyay P, Batkai S et al (2006a) Pharmacological inhibition of poly(ADP-ribose) polymerase inhibits angiogenesis. Biochem Biophys Res Commun 350:352–357
Rajesh M, Mukhopadhyay P, Godlewski G et al (2006b) Poly(ADPribose) polymerase inhibition decreases angiogenesis. Biochem Biophys Res Commun 350:1056–1062
Riballo E, Kuhne M et al (2004) A pathway of double-strand break rejoining dependent upon ATM, Artemis, and proteins locating to gamma-H2AX foci. Mol Cell 16:715–724
Richardson C, Stark JM et al (2004) Rad51 overexpression promotes alternative double-strand break repair pathways and genome instability. Oncogene 23:546–553
Rodon J, Iniesta MD, Papadopoulos K (2009) Development of PARP inhibitors in oncology. Expert Opin Investig Drugs 18:31–43
Roninson IB (2003) Tumor cell senescence in cancer treatment. Cancer Res 63:2705–2715

Rothkamm K, Kruger I et al (2003) Pathways of DNA double-strand break repair during the mammalian cell cycle. Mol Cell Biol 23:5706–5715
Rouleau M, Patel A, Hendzel MJ, Kaufmann SH, Poirier GG (2010) PARP inhibition: PARP1 and beyond. Nat Rev Cancer 10(4):293–301, Review
Ruf A, Mennissier de Murcia J, de Murcia G, Schulz GE (1996) Structure of the catalytic fragment of poly(AD-ribose) polymerase from chicken. Proc Natl Acad Sci USA 93:7481–7485
Russell JS, Brady K, Burgan WE et al (2003) Gleevec-mediated inhibition of Rad51 expression and enhancement of tumor cell radiosensitivity. Cancer Res 63:7377–7383
Sakai W et al (2008) Secondary mutations as a mechanism of cisplatin resistance in BRCA2-mutated cancers. Nature 451:1116–1120
Saleh-Gohari N et al (2005) Spontaneous homologous recombination is induced by collapsed replication forks that are caused by endogenous DNA single-strand breaks. Mol Cell Biol 25:7158–7169
Satoh MS, Poirier GG, Lindahl T (1994) Dual function for poly(ADP-ribose) synthesis in response to DNA strand breakage. Biochemistry 33:7099–7106
Schreiber V, Dantzer F, Ame JC, de Murcia G (2006) Poly(ADP-ribose): novel functions for an old molecule. Nat Rev Mol Cell Biol 7(7):517–528, Review
Schwarze SR et al (2001) Role of cyclin-dependent kinase inhibitors in the growth arrest at senescence in human prostate epithelial and uroepithelial cells. Oncogene 20:8184–8192
Shen WH, Balajee AS, Wang J, Wu H, Eng C, Pandolfi PP et al (2007) Essential role for nuclear PTEN in maintaining chromosomal integrity. Cell 128:157–170
Slupianek A, Schmutte C, Tombline G et al (2001) BCR/ABL regulates mammalian RecA homologs, resulting in drug resistance. Mol Cell 8:795–806
Sonoda E, Hochegger H, Saberi A, Taniguchi Y, Takeda S (2006) Differential usage of non-homologous end-joining and homologous recombination in double strand break repair. DNA Repair (Amst) 5:1021–1029
Stein GH, Drullinger LF, Soulard A, Dulic V (1999) Differential roles for cyclin-dependent kinase inhibitors p21 and p16 in the mechanisms of senescence and differentiation in human fibroblasts. Mol Cell Biol 19:2109–2117
Stewart GD, Ross JA, McLaren DB, Parker CC, Habib FK, Riddick AC (2010) The relevance of a hypoxic tumour microenvironment in prostate cancer. BJU Int 105:8–13
Stewart GD, Nanda J, Katz E, Bowman KJ, Christie JG, Brown DJ, McLaren DB, Riddick AC, Ross JA, Jones GD, Habib FK (2011) DNA strand breaks and hypoxia response inhibition mediate the radiosensitisation effect of nitric oxide donors on prostate cancer under varying oxygen conditions. Biochem Pharmacol 81(2):203–210
Takahashi S et al (1984) Enhancement of DEN initiation of liver carcinogenesis by inhibitors of NAD + ADP ribosyl transferase in rats. Carcinogenesis 5:901–906
Tannock IF, Hill RP, Bristow RG, Harrington L (2005) The basic science of oncology. McGraw-Hill Professional, New York
Timinszky G et al (2009) A macrodomain-containing histone rearranges chromatin upon sensing PARP1 activation. Nat Struct Mol Biol 16:923–929
Tong WM et al (2002) Synergistic role of Ku80 and poly(ADP-ribose) polymerase in suppressing chromosomal aberrations and liver cancer formation. Cancer Res 62:6990–6996
Tong WM et al (2003) Null mutation of DNA strand break-binding molecule poly(ADP-ribose) polymerase causes medulloblastomas in p53−/− mice. Am J Pathol 162:343–352
Trzeciak AR, Nyaga SG, Jaruga P, Lohani A, Dizdaroglu M, Evans MK (2004) Cellular repair of oxidatively induced DNA base lesions is defective in prostate cancer cell lines, PC-3 and DU-145. Carcinogenesis 25:1359–1370
Tulin A, Stewart D, Spradling AC (2002) The *Drosophila* heterochromatic gene encoding poly(ADP-ribose) polymerase (PARP) is required to modulate chromatin structure during development. Genes Dev 16:2108–2119
Tyson DR, Inokuchi J, Tsunoda T, Lau A, Ornstein DK (2007) Culture requirements of prostatic epithelial cell lines for acinar morphogenesis and lumen formation *in vitro*: role of extracellular calcium. Prostate 67:1601–1613

Vaupel P, Mayer A (2007) Hypoxia in cancer: significance and impact on clinical outcome. Cancer Metastasis Rev 26:225–239
Venkitaraman AR (2002) Cancer susceptibility and the functions of BRCA1 and BRCA2. Cell 108:171–182
Vukovic B, Park PC, Al-Maghrabi J et al (2003) Evidence of multifocality of telomere erosion in high-grade prostatic intraepithelial neoplasia (HPIN) and concurrent carcinoma. Oncogene 22:1978–1987
Ward JF (1988) DNA damage produced by ionizing radiation in mammalian cells: identities, mechanisms of formation, and reparability. Prog Nucleic Acid Res Mol Biol 35:95–125
Weterings E, van Gent DC (2004) The mechanism of non-homologous end-joining: a synopsis of synapsis. DNA Repair (Amst) 3:1425–1435
Willers H, Dahm-Daphi J et al (2004) Repair of radiation damage to DNA. Br J Cancer 90: 1297–1301
Wong CS, Hill RP (1998) Experimental radiotherapy. In: Tannock IF, Hill RP (eds) The basic science of oncology, 3rd edn. McGraw-Hill, Toronto, pp 322–349
Wouters BG, Weppler SA, Koritzinsky M, Landuyt W, Nuyts S, Theys J, Chiu RK, Lambin P (2002) Hypoxia as a target for combined modality treatments. Eur J Cancer 38:240–257
Yuan R, Fan S, Wang JA et al (1999) Coordinate alterations in the expression of BRCA1, BRCA2, p300, and Rad51 in response to genotoxic and other stresses in human prostate cancer cells. Prostate 40:37–49
Zelefsky MJ, Chan H, Hunt M, Yamada Y, Shippy AM, Amols H (2006) Longterm outcome of high dose intensity modulated radiation therapy for patients with clinically localized prostate cancer. J Urol 176:1415–1419
Zong WX, Ditsworth D, Bauer DE, Wang ZQ, Thompson CB (2004) Alkylating DNA damage stimulates a regulated form of necrotic cell death. Genes Dev 18:1272–1282

Concluding Remarks

The recent explosion of "translational" molecular-based technologies has dramatically enhanced our ability to subclassify cancers based on their genetic context. This represents the apotheosis for current biology, which is based upon the "molecular reductionist" approach linking clinical research with molecular diagnostics and histopathology, or next generation "omics" technologies with bioinformatics and drug discovery, in order to maximize the benefits for the cancer patient.

Pathology is rapidly changing, embracing the new biological knowledge and tools. Its new, adjunctive role, is to validate new disease stratification and targeted therapeutic interventions derived from the endless flow of information provided by basic research. However, our first challenge is in our evolving concept of cancer. Hopefully, we now have to think to prostate cancer considering that it represents the results of a global disturbance of the dynamic cellular network of molecularly-driven, multidirectional flow of information connecting prostate epithelial cells each other and with their external microenvironment. We cannot continue to look separately to genetic alterations, epimutations, or stromal alterations occurring in prostate cancer tissue. Gene profiling is now possible on impressively small amounts of tissue or on single cells, allowing heterogeneity to be assessed at the regional and cellular level. New agents with therapeutic potential continue to arise.

The translational value of these discoveries is currently being tested in controlled preclinical studies and clinical trials.

Personalized genetic and epigenetic therapy for prostate cancer is being taken to a new level.

Nevertheless, prostate cancer may represents the end-stage of multiple chronic stressful events and that stressful life can affect cancer growth and metastasis by modulating nervous, endocrine, and immune systems, as pointed out by a recent report in animal models The key challenge, then, will be to step back and consider prostate cancer by an holistic point-of-view, to cure the patient with his cancer, not simply a cancer.

S. Staibano (ed.), *Prostate Cancer: Shifting from Morphology to Biology*,
DOI 10.1007/978-94-007-7149-9,

Index

DOI 10.1007/978-94-007-7149-9, © Springer Science+Business Media Dordrecht 2013

MIX
Papier aus verantwortungsvollen Quellen
Paper from responsible sources
FSC® C105338
FSC
www.fsc.org

If you have any concerns about our products,
you can contact us on
ProductSafety@springernature.com

In case Publisher is established outside the EU,
the EU authorized representative is:
Springer Nature Customer Service Center GmbH
Europaplatz 3, 69115 Heidelberg, Germany

Printed by Libri Plureos GmbH
in Hamburg, Germany